ENVIRONMENTAL HAZARDS

ENVIRONMENTAL HAZARDS

Communicating Risks as a Social Process

SHELDON KRIMSKY
ALONZO PLOUGH

Tufts University

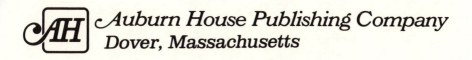
Auburn House Publishing Company
Dover, Massachusetts

Library of Congress Cataloging in Publication Data

Krimsky, Sheldon.
  Environmental risks.

  Bibliography: p.
  Includes index.
  1. Risk communication.   I. Plough, Alonzo L.
II. Title.
T10.68.K75   1988   363.1   88-14467
ISBN 0-86569-184-3

Printed in the United States of America

To my children, Alyssa and Eliot;
my wife, Carolyn Boriss-Krimsky;
and her mother, Eve Boriss (SK).

To my son, Louis;
my wife, Jeanette Valentine;
and my mother, Rosemary Busch Harrington (AP).

CONTENTS

FOREWORD

The burgeoning field of risk analysis has been divided between those who view risk as a technical concept and those who emphasize its cultural and social dimensions. Some people talk of risk in terms of cost-effective solutions, of efficiency; others use the language of "rights," emphasizing moral issues and questions of social responsibility, justice, and obligation. Some measure risks and evaluate them in statistical terms; others talk of "victims," or "real people." Some define risk as a problem that requires expert opinion; others seek more participatory control.

These differences are most evident in the discussions of risk communication. The persistence of public disputes over science and technology and the continued disagreement over the character of risk have baffled those involved in technological projects. It has become clear that there is far more to the communication of risk than simply the disclosure of technical information, and far more to the response to risk information than simply technical understanding. Sheldon Krimsky and Alonzo Plough have sought to illuminate this complexity through a series of five case studies of risk communication. These studies, carefully structured to illustrate the social and cultural dimensions of risk, suggest the diversity in the modes of risk communication and document the difficulties of predicting the public response. In each case, the authors show how this response is influenced by social, cultural, and political considerations.

An important contribution of this book, as the reader will discover, lies in its clarification of the differences between technical and cultural rationality in the perception of risk. Social perceptions of risk are difficult to deal with, and they therefore are often ignored in favor of more straightforward technical issues. But, as the case studies demonstrate, such perceptions are of central importance if we are to understand the dynamics of risk disputes.

It has become increasingly evident that communication about

xiii

the risks as well as the benefits of technology has strong bearing on institutional credibility and the public's trust. Accidents—even minor incidents—have contributed to an erosion of belief in the value of technology as an instrument of progress and to a decline of public confidence in the legitimacy of decision-making institutions. Neither the problems of technological risk nor the concerns about risk communication are new issues, but they have assumed new dimensions. Problems of risk have been magnified by the increasing dangers brought about by new technologies and the scale of new developments. Understanding the forces that shape the perception and communication of risk is thus increasingly important to the efforts to define and implement acceptable policies for technological change. The cases and analytical framework presented in *Environmental Hazards* bring an important new perspective to our understanding of the complex factors that shape risk communication.

DOROTHY NELKIN
New York University

ACKNOWLEDGMENTS

This book is in large part the result of a grant from the U.S. Environmental Protection Agency to the Center for Environmental Management at Tufts University under assistance agreement CR813481-01-1. The Center funded our proposal on "Improving Risk Communication," and this book emerged from that project. We are particularly appreciative of the supporting role both the EPA and the Center have played in our work. Through the grant format we had considerable latitude in approaching the subject matter as covered in this work in as fresh and creative a way as our intellects would permit. We alone bear the responsibility for any shortcomings in the way the problems of environmental risk communication are framed or for any errors in the analysis.

From the Tufts Center for Environmental Management, we give special thanks to Anthony D. Cortese, Ann Rappaport, Gene T. Blake, and Dennis A. Crivello.

At EPA many people supported this work, but a few deserve special mention. Our project monitor, Daphne Kamely, provided insights, inspiration, and encouragement. John H. Skinner, Director of the Office of Environmental Engineering and Technology, Darwin Wright, project officer for the EPA-Tufts cooperative agreement, and Frederick W. Allen of the agency's Office of Policy, Planning and Evaluation showed a special interest in the approach we took. Others at EPA who reviewed drafts of the cases or who provided much needed information are cited below.

We wish to thank: John D. Powell, Professor of Political Science at Tufts, for being part of the risk communication project and contributing Chapter 6; Jennifer Helmick, a graduate student in the Department of Urban and Environmental Policy, for her role in the radon case and her important contributions to Chapter 3; Eileen Schell for her project management, data collection, and extensive interviewing; Sharon Becker and Naomi Friedman for their research assistance; Roger Geller for his invaluable assistance in production, editing, bibliographic work, and preparation of the manuscript; Ann Gerroir for word processing; Hope Kingma and Arville Grady for their help in organizing the archival materials; and Patricia Flaherty for her brilliant editing of the first draft.

The following members of the Project Advisory Committee provided important guidance that helped us in selecting the cases: June Fesenden-Raden, Cornell University; Steven Greene, Digital Corporation; Roger Kasperson, Clark University; John O'Conner, National Campaign Against Toxic Hazards; Maria Pavlova, EPA, Region II; Peter Sandman, Rutgers University.

We are grateful to the following individuals who read portions of the manuscript in its early stages and provided valuable comments: Mary Anderson, EPA, Region 1; Robert Ajax, EPA, North Carolina; Patricia Donahue, Department of Environmental Quality Engineering, Massachusetts; David Gute and Steven Havas, Department of Public Health, Massachusetts; Allan Mazur, Syracuse University; Clark Guilding, EPA, Region 10; David Patrick, Clements Associates; Alex Smith, EPA, Region 7; Trevor V. Suslow, Advanced Genetic Sciences. From EPA headquarters: Frederick S. Betz, Terry Dinan, Ann Fisher, Elizabeth Milewski, Roy Sjoblad, Gregory A. Thies. Also, a very special thanks is given to Glen Nathan of EPA's Television Section who provided videotapes under short notice.

Finally, we offer thanks to the many people who agreed to lengthy interviews and shared important documents with us. These individuals are listed at the end of the respective chapters.

THE AUTHORS

Chapter 1

INTRODUCTION: THE MEANINGS OF RISK COMMUNICATION

In the past quarter century the evaluation and management of risks have taken on an increasingly important role in the industrialized nations. New professions, legal structures, scientific disciplines, bureaucracies, and extensive regulations have been created in response to public demand for a safer environment. In the United States, risk assessments are conducted widely for many consumer products, commercial processes, new technologies, industrial emissions, and the siting of potentially hazardous facilities. The results of these risk assessments are eventually translated and communicated to the public in many forms. These risk messages contain information transmitted from technical to nontechnical sectors of society. There are also circumstances when the identification of risk originates from lay citizens or the media, and from these sources eventually reach the professional risk analysts and regulators. In either case, the creation, transfer, and response to risk information are not new.

Throughout history individuals and groups, in their struggles to survive and promote health and well-being, have had to contend with a variety of risks. An ever-changing environment, incurable diseases, famine, and man-made threats such as war have always presented risks to both individual and collective survival. The ori-

---

Portions of this chapter were adapted from A. Plough and S. Krimsky, "The Emergence of Risk Communication Studies: Social and Political Context," *Science, Technology & Human Values* 12 (3&4): 4–10 (Summer/Fall 1987).

gins of structured risk analysis have been traced to the Babylonians in 3200 B.C.[1] Methods used by the ancients to predict risks and to communicate knowledge about avoiding hazards were based on myths, metaphors, and ritual. Risk communication was embedded in folk discourse.

While there has always been a role for risk assessment and risk communication in human societies, the *study* of risk communication as a distinct social phenomenon is new. Prior to 1986, there were only a few essays in the scholarly and policy literature with "risk communication" in the title. Since that year, however, scores of titles with the term "risk communication" have appeared in the published literature, along with conferences, special sessions in scientific meetings, agency-sponsored workshops, and grants.[2]

The concept of "risk communication" evolved from two distinct disciplines—communication studies and risk studies. It is not unusual for disciplinary marriages of this nature to occur. New problems often capture the attention of researchers in separate fields and, for a period of time, become the centerpiece of a new hybrid of academic and policy research.

But risk communication is not just another fashionable rubric for the activity of a specialized group of researchers. If that were the case, then its emergence as an area of study would be of interest primarily to historians and sociologists of science. The interest in risk communication as an area of study was prompted by dilemmas arising in public policy, particularly conflicts arising between and among scientific experts and lay citizens centered on the issue of risk. These tensions are played out in disputes between different research traditions over fundamental questions regarding the perception of risk and the essential nature of human rationality. They are also played out in the political arena when social discord develops over the proper management of technology.

Federal regulatory agencies with responsibilities in the health and environmental areas draw on a diverse and at times conflicting body of research on risk. Selectively, they choose analytical frameworks that are most compatible with their primary agendas. Citizen groups, less concerned about formal theories, have become increasingly aware that getting a message across to government in disputes over health and environmental hazards is essentially a political activity. Experts in risk analysis, on the other hand, have come to appreciate the gap between their analytically derived conclusions and the conceptions of risk expressed in the popular culture. Thus, risk communication is more than a research agenda. It has become closely associated with a set of techniques that are strongly marketed by special interest groups and used opportunistically to achieve particular ends.

At the federal level, the Environmental Protection Agency (EPA) has done the most to advance the use of risk communication techniques in the regulatory sector. Two-time EPA administrator William Ruckelshaus was a principal conceptualizer and promoter of risk communication for the agency. Ruckelshaus foresaw the importance of risk communication for resolving the conflicts that arise between the technosphere (those legally responsible for making risk decisions) and the demosphere (those on whose behalf risk decisions are made) in the context of new laws that give citizens greater rights of participation in the decision-making process. He stated:

> *My point here is not to say whether a sharing of the power to make risk management decisions is right or wrong; it is simply to state that it is a fact of life in the United States. We have decided, in an unprecedented way, that the decision-making responsibility involving risk issues must be shared with the American people, and we are very unlikely to back away from that decision. So the question before us is not whether there is going to be a sharing, whether we will have participatory democracy with regard to the management of risk, but how.[3]*

Ruckelshaus's mandate on risk communication was supported by his successor, Lee Thomas, who stated: "On the national level we will build risk communication into regulatory policy whenever possible."[4] The EPA has elevated risk communication to a strategic level of importance in both its regulatory activities and its research agenda. Industries that are regulated by EPA also regard risk communication as a key policy and management issue.

In the public health field the terminology of risk communication is slightly different (federal health officials speak of educating the public about risks), but the concept is equally important. Current policy on preventing health problems such as AIDS, teen pregnancy, and substance abuse focuses on communicating risk to a target population.[5]

This emphasis has gone beyond the health education strategies of an earlier era in public health. The present attempt to change "unhealthy" personal behaviors incorporates sophisticated national media campaigns developed by public relations or advertising firms in conjunction with science-based strategies.

Risk communication has always been an important component in medical practice, particularly in doctor-patient relationships. More recently, the risk of chronic diseases has emerged as a major health problem. The uncertainties associated with the risk factors for these diseases and the course of treated illnesses have given risk communication a more central and visible role in medicine.[6]

Contributions to the study of risk communication fall into several categories. First, there are studies exploring how risk is communicated by the media to the general public. This involves an analysis of the media for accuracy, emotional and informational content, and the ability to influence readers and viewers. A second group of risk communication studies evolved from the risk perception literature. These studies examine how the public's perception of risk is affected by different forms of communication. An underlying assumption of this research program is that a reasonably clear distinction can be drawn between objective and subjective risk. Here, good risk communication is defined as a message that brings public perception of risk closer to the scientific determination.

A third approach to studying risk communication, fostered by the needs of the regulatory and industrial sectors, is prescriptive and pragmatic. It attempts to establish canons for "effective risk communication." The canons are based on accepted institutional goals and are designed for use by risk communicators. The fields of social and cognitive psychology, communication theory, and public relations contribute to this approach.

A fourth group of studies addresses the conceptual foundations for the problems of risk communication. Within a period of a few years a dominant paradigm has emerged that likens risk communication to the transfer of electronic signals through a mechanical system. The paradigm draws heavily from the physical sciences—particularly, electrical engineering, information theory, and cybernetics. A four-part scheme is introduced consisting of (1) an information source (transmitter); a channel of communications (transducer); an audience (receiver); and finally the message itself that "flows" from the source through a channel to an audience. The problems of risk communication are analyzed according to this mechanistic model.[7]

This book provides an alternative approach to risk communication to those frameworks described above. It draws on the sociological and anthropological traditions exemplified by the recent works of Mary Douglas[8] and James F. Short.[9] It treats risk events as comprised partly of physical processes and partly of socially constructed phenomena. The assessment, interpretation, communication, and response to such events are examined in their social context to obtain a full accounting of the factors involved. The scientific aspects of risk are embedded in a complex sociopolitical tapestry in which there are not only different voices but different perceptions of the problem. We have chosen a case study approach to illuminate the cultural aspects of risk communication—aspects that are not satisfactorily addressed in the existing literature.

In developing the cases, we are aware that our approach is out of phase with the current trends in research. We are also cognizant that case studies are often criticized for being parochial and failing to provide generalizable findings. However, the risk communication literature lacks detailed analytic cases that describe the genesis and development of a risk event and the associated messages, meanings, and perceptions that become risk communication. The foundations of this nascent field have thus far been dominated by a narrow and sometimes mechanistic paradigm. Our goal in this book is to expand the boundaries of analysis. We begin with a few key definitions that will serve as guides to the case studies.

The term "risk communication" has both a conventional and symbolic meaning. In its conventional meaning, which reflects the use of the term in risk management, risk communication is the transmission of technical or scientific information from elites to the general public. In its symbolic meaning, which derives from the role of risk in political discourse, risk communication can refer to any public or private communication that informs individuals about the existence, nature, form, severity, or acceptability of risks.

The conventional meaning of risk communication emphasizes the intentionality and the quality of the message. According to Covello et al., risk communication is "any purposeful exchange of scientific information between interested parties regarding health or environmental risks."[10]

To frame the concept of risk communication exclusively in conventional terms restricts its meaning to surface behavior or what natural scientists like to call "the phenomenon of the event." The conventional account neglects cultural themes, motivations, and symbolic meanings, which may be of equal or greater importance to the technical understanding of how and why a risk message gets transmitted.

In the definition we propose, risk communication may be directive and purposeful or nondirective and fortuitous. It may describe the controlled release of information by official sources toward certain well-defined ends or it may represent the unintended consequences of informal messages about risks. This broad interpretation has been adopted by Kasperson and Palmlund. They state that risk communication "enters our lives in a multitude of forms, sometimes part of the imagery of advertising, sometimes a local corporation's formal statement, or its failure to say anything, sometimes a multi-volumed and impenetrable technical risk assessment."[11] Al-

Table 1–1   Definition Latitude of Risk Communication

	Broad	*Narrow*
Intentionality	Risk communication goal unnecessary	Intentional and directed; outcome expectations about the risk message
Content	Any form of individual or social risk	Health and environmental risks
Audience directed	Targeted audience not necessary	Targeted audience
Source of information	Any source	Scientists and technical experts
Flow of message	From any source to any recipient through any channel	From experts to nonexperts through designated channels

SOURCE: A. Plough and S. Krimsky, "The Emergence of Risk Communication Studies: Social and Political Context," *Science, Technology & Human Values* 12(3&4):4–10 (Summer/Fall, 1987).

most any communication from any source that speaks to the issue of risk satisfies our definition.

Risk communication has five components to its meaning: intention of the communicator; content of the message; nature of the audience; source of the message; and direction of the message. Different definitions of the term are narrow or broad, depending on the latitude of interpretation of these components (see Table 1–1). In its most restricted meaning, risk communication is a plan executed by a regulatory body targeted to a particular lay audience and embodying specific outcome goals for behavior or attitudinal change. Alternatively, a broad meaning of risk communication includes messages from any source, such as folk wisdom or experiential reports, to any audience, whether intended or not.

To understand the symbolic meaning of risk communication, we have to study the risk event *in situ* with all of its cultural expressions. A scientist speaking to a community about the health effects of a hazardous waste site is part of a political ritual that aims to evoke confidence and respect. The technical information in the message is secondary to the real goal of the communicator: "Have faith; we are in charge." Local residents citing a litany of symptomatology that they attribute to contaminated drinking water use risk communication to channel their anger and anxieties about environmental overload. "Popular" or "barefoot" epidemiologists are lay people who spot disease clusters and then use risk communication to arouse community sentiment and negotiate toxic waste cleanup.

For a company, risk communication is not a message about risks but an affirmation about safety and confidence. For community organizers, risk communication is a strategy for building a grassroots movement. Kasperson notes that in local controversies "risk communication becomes a vehicle of protest by which community groups create resources with which to bargain with government in the risk management process."[12] The literal meaning of risk communication is found in the various risk messages. But the symbolic meaning is embedded in the political discourse of a controversy.

In this book, five environmental risk events were chosen for an in-depth analysis of risk communication. The purpose of the case studies is to illuminate the processes through which environmental risks are communicated to citizens, the response of the public to the risk messages, and the factors that impede or effectuate successful risk communication. The five risk events were selected because they are illustrative of different forms of environmental risk and because they exemplify the diverse dimensions of risk communication activities, including public activism, agency response, high media visibility, and the uncertainties of risk assessment. The events are studied in their historical and regulatory context, making it possible to view the development of risk communication over time and to understand it both as a technical and cultural phenomenon.

The cases illustrate different modalities of risk and varying risk communication problems such as:

1. *Pesticide residues in food.* The uncertainties are about the risks of consuming, for a period of time, minute quantities of toxic chemicals. The chemicals are introduced into the food chain by grain distributors and farmers. The affected population is distributed throughout the nation and includes people in countries where these products are exported.

2. *Potential hazards of unknown consequence with a new technology.* The uncertainties pertain to speculative and hypothetical risk scenarios of releasing genetically modified organisms into the environment. The affected population consists of communities in proximity to the field experiments, in addition to regional and national groups concerned about broad ecological impacts of such environmental releases.

3. A *natural environmental hazard that is distributed erratically and ubiquitously.* The uncertainties are over the potential health effects of lifetime exposure to the radioactive gas radon that accumulates in homes. The exposed population is undefined, but it has been estimated that more than 8 million homes may have unsafe levels of the gas.

4. *Point source airborne pollution of a highly toxic and carcino-genic agent*. A copper-smelting plant released toxic emissions that posed a health risk to people living in its vicinity. Risk communication problems centered on defining an acceptable level of stack emissions and assessing the health effects of arsenic in the air, water, and soil.
5. *Hazards associated with a toxic waste site*. Chlorinated hydro-carbons and toxic metals accumulated in the ground, saturated soils, and percolated into the groundwater after years of care-less disposal practices. The polluting parties have disap-peared. The primary risks are from soil contact and drinking water. Central concerns are over cleaning up the site and as-sessing the residual health effects of prior chemical exposure.

The first case involves the pesticide and soil fumigant ethylene dibromide (EDB). At issue are the permissible residues of EDB in grain products. Differences arose among state and federal regula-tors on the safe levels of EDB in food. Since EDB is used by the major grain distributors, the risks of consuming contaminated prod-ucts has a nationwide impact. The case reflects the general category of risks associated with low-dose and long-term exposure to pesti-cide residues in produce and prepared food. The study draws a comparison between federal and state risk communication activi-ties and contrasts the responses of Hawaii, an agricultural state, and Massachusetts, an industrial state.

The second case discusses a risk event associated with an emerg-ing technology. Scientists at the University of California at Berke-ley and at Advanced Genetic Sciences Inc. of Oakland, California, sought regulatory permits to field test a genetically modified plant bacterium. The bacterium is called ice minus to signify the removal of a gene associated with ice nucleation activity. The proposed test involved several communities in northern California, and repre-sented, symbolically, if not actually, the first environmental release of a product emerging from the new field of biotechnology. As a generic class of risk events, there is little empirical information about the potential adverse consequences of releasing genetically modified organisms into the biosphere. The case examines how the hypothetical risks associated with the proposed field test were de-fined and communicated by scientists, industry, regulatory agen-cies, the media, and other parties to the controversy.

A third case examines risk communication for a natural and con-tinuous geological process distributed widely across the United States. Radon gas forms naturally in the earth's bedrock and finds its way into homes through tiny fissures in basement floors. While no human cause is responsible for its production, the build-up of radon in homes may be exacerbated by new home construction technologies and greater emphasis on energy efficiency.

Unlike the other cases where legal authority exists to prevent or reduce the source of the hazard, geological radon represents a regulatory enigma. First, the formation of radon cannot be prevented. Second, no regulatory body has authority over environmental hazards in the home. Third, with millions of homes across every state potentially affected, the radon risk is not physically circumscribed in the manner of a hazardous waste site, a polluting source, or a system of food distribution. This case discusses EPA's role in providing public information about the radon risks, compares Massachusetts' and Pennsylvania's approaches to risk communication, analyzes the response by the media, and discusses public and community reactions to radon risk communication.

The fourth case discusses the ASARCO copper smelter in the Tacoma, Washington, area. ASARCO was both a primary source of employment for the Tacoma area and a source of airborne arsenic emissions. The smelter was the focus of a long and protracted regulatory process, beginning in the late 1960s and intensifying with the passage of the Clean Air and Occupational Safety and Health Acts of 1970. While the risks of high exposure to arsenic are well documented, there is considerable uncertainty about the health effects of long-term exposure to low doses of the chemical. Under existing laws and regulations, EPA was obligated to set arsenic standards at ASARCO. If the agency set standards that lowered exposure to near zero levels on the grounds that arsenic is not safe at any dose, plant closure was threatened. Higher emission standards, based upon best available control technology, however, brought intense community opposition. As part of the standard-setting process, officials at EPA headquarters developed an intensive public information strategy and communicated directly with the local citizenry in what has been called the "Tacoma process." The case study examines how that process functioned as a risk communication program.

The fifth and final case concerns the Nyanza hazardous waste site in Ashland, Massachusetts, which was placed on EPA's Superfund priority list in 1981. Unlike some toxic waste sites that received national and international public attention (Love Canal, Times Beach, Woburn), the risk communication around Nyanza was predominantly directed to the adjacent community. As a local Superfund risk communication event, Nyanza is representative of the scores of potentially hazardous chemical contamination sites scattered throughout the United States. Typically, these sites have a long and complex history of chemical dumping and regulatory response.

The risks are primarily the result of long-term exposure to low doses of toxic and carcinogenic chemicals. The health effects of exposure to low doses of the chemicals are uncertain but apply to a reasonably well defined population either in direct contact with the contaminated soil or who consume the groundwater. Public perceptions and technical risk assessments invariably are at considerable odds for community chemical waste sites. In Nyanza, a local community comes to terms with the legacy of its industrial past. The case examines how the community dealt with the existing site, the forms of risk communication used, the public response to fears of adverse health effects, and the role of the media in shaping and reflecting public perceptions.

Each case was developed from a common framework consisting of (1) a detailed social history of the risk event; (2) an overview of the risk assessment as constructed by experts; (3) the regulatory background; (4) risk communications, including media analysis; and (5) conclusions, with a discussion of the lessons learned from the case.

To construct the cases, primary data were obtained from on-site and phone interviews with key participants involved in risk assessment, risk communication, news coverage, and citizen response. For each study, an archive was developed consisting of audiotape interviews, videotapes (of media coverage, news conferences, public meetings), reports, news articles, and analytical studies.

The historical sections describe the conditions under which the risk events became fully realized as public problems. The events around risk communication are analyzed from their nascent to their mature stages. The roles of key actors, organizations, state and federal agencies, and media are examined and placed in their causal and temporal sequence. A detailed chronology follows each case.

The section on risk assessment reviews the assumptions and analytical frameworks that underlie the evaluation of human health effects or environmental consequences. Areas of agreement and disagreement among experts over risk estimates are discussed. The cases also describe citizen and media response to risk estimates and proposed "acceptable" standards.

Each case contains a review of the regulatory background, which includes the relevant statutes, responsible agencies, and the evolution of state and federal regulatory responses.

The main segment of each case is devoted to how the factors described above influenced risk communications. Taking our broad definition of "risk communication," the analysis includes regulatory communications to the public (goals, form, effectiveness); private sector communications; media analysis of risk events (print and television); institutional video productions (as a public education instrument); the socioeconomic context of risk communication; methods of communicating uncertainty (symbols, metaphors, analogies, and comparative risk estimates); and risk messages emanating from citizen groups and public interest organizations.

The concluding remarks for each case were guided by a series of queries: What were the primary modes and channels of risk communication? Were the goals of risk communication, relative to some group or agency, realized? What factors inhibited or effectuated good risk communication? (The criteria for "good risk communication" are contextual.) What cultural and socioeconomic elements are important in understanding the social responses to the risk events?

The cases present the communication activity of all important participants in the process. None of the cases is intended as an evaluation of EPA's or other institutions' risk communication approaches but instead describe how regulatory bodies became part of a diverse risk communication context involving many voices. Our emphasis is on the evolution of a risk communication event and the factors that influence the overall pattern of risk communication.

Endnotes

1. Vincent Covello and Jerry Mumpower, "Risk Analysis and Risk Management: An Historical Perspective," *Risk Analysis* 5(2):103–120 (June 1985).
2. Vincent Covello, "Introductory Remarks," Workshop on the Role of Government in Health Risk Communication and Public Education, Washington, D.C., January 21–23, 1987.
3. William D. Ruckelshaus, "Overview of the Problem: Communicating about Risk," in J. Clarence Davies, Vincent T. Covello, and Frederick W. Allen (eds.), *Risk Communication, Proceedings of the National Conference on Risk Communication*, Washington, D.C., January 29–31, 1986 (Washington, D.C.: The Conservation Foundation, 1987).
4. Lee Thomas, "Risk Communication: Why We Must Talk About Risk," *Environment* 28(2):40 (March 1986).
5. See, for example, U.S. Public Health Service, *Promoting Health/Preventing Disease: Objectives for the Nation* (Washington, D.C.: U.S. Public Health Service, 1980).
6. Alonzo Plough, *Borrowed Time: Artificial Organs and the Politics of Extending Lives* (Philadelphia: Temple University Press, 1986).
7. V. Covello, D. von Winterfeldt, and P. Slovic, "Communicating Scientific

Information About Health and Environmental Risks: Problems and Opportunities from a Social and Behavioral Perspective," in V. Covello, L. Lave, A. Moghissi, and V. Uppuluri (eds.), *Uncertainty in Risk Assessment and Risk Management and Decisionmaking* (New York: Plenum Press, 1987).

8. Mary Douglas, *Risk Acceptability According to the Social Sciences* (New York: Russell Sage, 1981).

9. James F. Short, "The Social Fabric at Risk: Toward the Social Transformation of Risk Analysis," *American Sociological Review* 49:711–25 (December 1984).

10. V. Covello, D. von Winterfeldt, and P. Slovic, 1987.

11. R. E. Kasperson, and I. Palmlund, "Evaluating Risk Communication," Unpublished paper, CENTED, Clark University, January 1987.

12. Roger E. Kasperson, "Six Propositions on Public Participation and Their Relevance for Risk Communication," *Risk Analysis* 6(3):276 (September 1986).

Chapter 2

PESTICIDE RESIDUES IN FOOD: THE CASE OF EDB

Introduction

In environmental health policy, certain acronyms take on a powerful symbolic function and evoke lasting images. Ethylene dibromide (EDB), a pesticide that was widely used in agriculture, is one of these. This case study will explore the key questions for risk communication that derive from the public controversy over dietary exposures to residues of EDB. Although the actual history of the use and risks of EDB goes back to the 1930s, this study focuses on the EDB controversy that occurred during 1983 to 1984. Because the EDB controversy was deeply rooted in a set of broader concerns—the dynamics of regulating pesticides—this case study is relevant to continuing policy debates. Also, as a study of risk communication the EDB controversy presents a textbook example on how a national discourse on risk is connected to state and local responses to risk management.

In this study risk communication is broadly defined to include any message about EDB from any source to any receiver. Therefore, this analysis examines media reports, messages emanating from federal agencies to the general public, communication between federal agencies, information exchange between federal agencies and state regulatory bodies, communications from industry and/or trade associations, communications from public interest groups, and communications from lay citizens.

Within this complex web of information about EDB are many messages, some directly contradictory. Different groups have divergent purposes for risk communication: some messages are intended

13

to calm an agitated public; some are intended to stir up sentiment or raise the level of agitation. Some messages demonstrate a commitment to the detached objectivity of scientific reasoning, whereas others strive to express raw outrage. Independent of the particular form of the risk message, many communicators in the EDB case have an implicit political agenda.

In this analysis we do not play the role of a modern-day Diogenes looking for the truthful account. In contrast, this analysis presents the multiple accounts of the EDB crisis, the assumptions that underlie a particular communication of risk, and the structural factors (legal, political, economic) that determine the patterns and direction of risk communication. Given the level of scientific uncertainty surrounding EDB, there are few indisputable facts and even fewer unarguably clear policy inferences based on risk assessments. Risk messages are embedded in a social context, which will determine in large part how information is received.

The EDB crisis as a risk communication event was of brief duration but high intensity. It was also a risk communication event with both a national and regional context. Federal agencies with statutory responsibility to regulate different aspects of pesticide exposures (Environmental Protection Agency—EPA, United States Department of Agriculture—USDA, Occupational Safety and Health Administration—OSHA, and the Food and Drug Administration—FDA) were involved in national policy and, therefore, national risk communication. There is an important local and state regulatory context for pesticides as well. State departments of public health and agriculture assessed the risks of EDB and communicated those risks independently of federal risk communication. In some states this assessment and communication of EDB risk diverged substantially from the federal risk messages. The risk communication strategies of EPA on EDB were filtered through the local social context and the message was "translated" into terms consistent with the political and economic environment of each state. Much of the risk communication of interest groups, such as representatives of businesses that used or produced EDB (grain industry, citrus producers), trade associations (Grocery Manufacturers Association—GMA), and public interest organizations (Natural Resources Defense Fund, American Foundation for Science and Health), was directed to the states, in particular to those states that established aggressive regulatory approaches to EDB that diverged from federal policy.

In this case study we will briefly describe and characterize the national risk communication on the risks of EDB and its interaction with the state-based risk communication. The primary focus of the case is on two states with contrasting risk communication experiences during the EDB controversy, Massachusetts and Hawaii.

The former is a state with very little agricultural use of EDB and a minimal linkage of EDB to the state's economy. The latter had used EDB in a variety of agricultural applications since the 1940s and has a history of numerous threats to food and groundwater contamination from pesticides, including EDB. The analysis of how the risk discourse on EDB differed between these two states explores the role of the local context in risk communication and the structural limitations placed on federal-level risk communicators.

This case study explores the many different aspects of the EDB risk. On the technical health effects side, there are different types of exposures to EDB—through food residues, groundwater contamination, and occupational exposures. Each of these exposures was the subject of regulatory activities by multiple federal and state agencies. Congressional committee hearings, which evaluated the effectiveness of regulatory activities on EDB in protecting the public's health, provided a major forum for risk communication. Scientists who held different views on the relevance of laboratory tests of chemicals for setting policies for human exposures used the EDB controversy to demonstrate their views.

Environmentalists who had criticized the federal regulatory structure for pesticides (the Federal Insecticide, Fungicide and Rodenticide Act—FIFRA) found in the debate on EDB an opportunity to advance a policy argument. Industry and trade associations used the EDB issue to advance the argument for weighting economic implications in the regulation of hazardous chemicals. The media attempted to package this conflicting information in a way that would make exciting news stories and connect with what they perceived as the personal concerns of viewers.

The case study is based on a variety of sources. These include interviews with state and federal regulatory officials, trade association and industry representatives, congressional staff members who conducted hearings on EDB, journalists, public interest groups, and citizens. Media reports on EDB were reviewed from 1978 to 1986 in a variety of local and national sources, including print and television. Public hearings on EDB at both the federal and state levels provided an especially rich source of information. Risk assessment documents produced by government, industry, and interest groups were also reviewed.

Historical Context

The public discourse on the risks of EDB is rooted in a more general concern about the use of pesticides in American agriculture and the environmental and health effects of these chemicals. The chemical pesticide use problem has been a core issue in the development of environmental policy in the United States, and it has been a central axis for disputes about the acceptable risks of chemicals in the environment. There are two general frameworks for understanding the evolution of risk assessment and communication issues surrounding pesticides in general and EDB in particular. These positions are evident in articles and analyses that appeared during the "EDB crisis" of 1983–1984. William Havender, a toxicologist and scientific adviser to the American Council on Science and Health (an industry-oriented public interest group), framed the context in this way:

> *Just as individuals choose to take voluntary risks, society as a whole takes risks in order to provide the best possible standard of living for its populace. . . . [O]ne must evaluate the tradeoff between the risk and the offsetting benefits associated with that product's use. Nowhere does this apply more aptly than to the agricultural and health protection uses of pesticides. Because of the use of EDB and other pesticides, we in America have escaped the negative health consequences of eating uncontrolled amounts of insect fragments and mold toxins in our food.[1]*

Havender, whose testimony at congressional hearings and articles in major newspapers played an important role in EDB risk communication, represents the view that the EDB crisis is a natural consequence of making trade-offs between the risks and the obvious benefits to the public's health of pesticide use. No critical questions are raised about the status quo of pesticide regulation.

Jonathan Lash and Katherine Gillman from the environmental advocacy group Natural Resources Defense Council, who also figured prominently in EDB risk communication, framed the context in a different way:

> *American agriculture is hooked on chemical pesticides. . . . [H]undreds of species of insects have become immune to pesticides. Supporters of pesticides are visible and powerful. The network of advocates includes researchers at land-grant colleges, officials in state and federal departments of agriculture, . . . and the industry that manufactures pesticides. EDB is a symbol of health concerns taking a back seat to the needs of agribusiness.[2]*

Since the mid-1960s, especially since the publication of Rachel Carson's *Silent Spring*, there has been much debate between the two positions represented here on the effects of pesticides on the environment. Pesticide use was discovered as a factor in the contamination of watersheds and groundwater supplies and damage to food chains. A growing critique of the ability of the Department of

Agriculture to monitor pesticides led to EPA's becoming responsible for the regulation of pesticides in 1970. Congress passed a broad new federal pesticide law, FIFRA, in 1972, under which EPA regulates the use of these chemicals. There is a continuing debate as to whether FIFRA effectively protects the public against the health risks of pesticides.

The substance EDB has been in use since the 1920s. It was used first as a gasoline additive and later as a soil fumigant for nematode control; as a means of protecting stored wheat, corn, and other grains against destruction by insects and contamination by molds and fungi; as a treatment for fruit flies in citrus; and as a fumigant in grain milling machines. EDB was first registered as a pesticide in 1948. In 1955 tolerances were set for the presence of EDB metabolites in food. Thus, during this period, major structural elements of the EDB controversy were put in place.

In 1956 Dow Chemical, one of the major producers of EDB, petitioned FDA for an exemption from a tolerance in its use as a post-harvest fumigant for grains. The exemption was granted. The decision seemed to make good scientific sense at the time; EDB had well-known acute effects (bromide poisoning), but scientists thought that the compound would quickly dissipate in a fumigation application or convert to a relatively inactive substance. The *Federal Register* in July 1956 gave the following rationale:

> *When the fumigants are used according to directions proposed in the petitions, the residues, except inorganic bromide residues . . . will not be present in the finished food ready for human consumption; the cooking customarily given to these cereals before they are used as human food drives off organic residues.*[3]

To the farmers who used EDB it was a "miracle chemical," extremely effective in multiple types of applications and so volatile (but not a fire hazard) that it would not be present in final food products. From this time the use of EDB in agriculture greatly expanded so that by 1980 approximately 20 million pounds of EDB were applied to croplands each year.[4]

By the 1960s researchers had developed new technologies that were capable of measuring very small amounts of a pesticide residue, discerning traces at the hundreds of parts per billion level. Also, advances in mutagenic and carcinogenic research allowed scientists to measure many types of adverse outcomes in laboratory animals related to exposure to chemicals. The extrapolation of these results to humans became a controversial issue first within the scientific community and later in more public debates on the safety of chemicals.

Risk communications concerning EDB began to emanate from the scientific community as a result of adverse effects data from these animal experiments. By the early 1970s laboratory studies were linking EDB to genetic damage. In October 1974 the National Cancer Institute (NCI) issued a "Memorandum of Alert" describing a preliminary finding that EDB produced cancer in mice and rats.[5] Both EPA and USDA were informed through this document. At this point risk communication for EDB became a public and regulatory concern. In July 1975 the Environmental Defense Fund (EDF) petitioned EPA to investigate EDB on a priority basis. In November 1975 the EPA submitted EDB as a candidate for a process then known as Rebuttable Presumption Against Registration (RPAR). Thus began a review of data on the risks of EDB that, if compelling, would lead to a revocation of registrations of the pesticide issued thirty years previously. During the two years that EPA reviewed the data, the EDF petitioned the agency two more times. In 1977 EPA issued a preliminary risk assessment stating that EDB was likely to be carcinogenic in humans. Also in 1977 OSHA issued guidelines to limit occupational exposures to EDB by workers. EDB was emerging as a symbol for environmentalists of serious problems in the government's capacity to protect the public from the risks of pesticides.

From this point the sequence of EPA activity dominated risk communication on EDB, and this, in turn, was structured by the RPAR process, which involves the generation of position documents or PDs. PD1 is typically a statement of the rationale for review; PD2/3 reports the review findings, proposes regulatory action, and solicits public comment and scientific review by a scientific advisory panel; and PD4 summarizes the data, public commentary, and scientific review and makes a final ruling. PD1, issued in December 1977, presented the carcinogenic, mutagenic, and adverse reproductive effects attributed to EDB.

From the release of PD1 to the release of PD2/3 in December 1980, an increasing level of regulatory and public risk assessment and risk communication activities concerning EDB took place. The EPA received 105 rebuttal submissions; other studies reported on EDB residues in grain products; and lawsuits were filed by environmental advocacy organizations to speed up EPA decision making. The first national evening television news story on the carcinogenicity of EDB appeared during November 1978. In 1979 the National Institute on Occupational Safety and Health (NIOSH) completed a study showing EDB to be an animal carcinogen and recommended that permissible workplace exposures be reduced 200 times (from 20 ppm to 130 ppb). Labor groups such as the AFL/

CIO and the Teamsters began public efforts to reduce risks of EDB exposure in milling, trucking, and other industries. Also in 1979, DBCP, a chemical closely resembling EDB, was suspended by the EPA; this action heightened the concern of proponents advocating a ban on EDB. Thus, three years before the period conventionally identified as the EDB crisis, important risk communication activities were well under way, and the positions of the participants in the public debate on EDB were firming up.

In 1980 EPA released PD2/3, in which it announced its intent to cancel the use of EDB on stored grain, milling machinery, and felled logs. Cancellation of fumigation of citrus fruits was extended to 1983. The available scientific and risk benefit data were presented in this report. From 1980 to September 1983, further studies and deliberation by EPA and other federal agencies occurred, with increasing pressure from environmentalists to ban EDB based on the evidence presented in PD2/3. Allegations of inaction and footdragging on the part of federal officials were reported in a nationally syndicated column by Jack Anderson. In 1981 occupational exposures to EDB became an issue in California, and the state issued an emergency temporary standard for EDB exposure. Both the Teamsters and the AFL/CIO petitioned OSHA for an emergency standard. When OSHA denied the petitions, increased tensions arose between NIOSH and OSHA, which were reported in the national media and communicated quite dramatically in public hearings of the House Subcommittee on Labor Standards in 1983.

Groundwater residues of EDB, from soil injection in citrus fields, were detected first in Georgia in 1982 and later in California and Hawaii. This resulted in a great increase in state-level concern about the uses of EDB in general. In July of 1983, Florida banned EDB as a soil fumigant in eight counties in the state's citrus belt. Local media coverage of the EDB issue became more extensive at this point, although it was not yet a sustained national story. By this time states were requesting action from federal agencies on the EDB problem. Under pressure from local groups, Florida set an interim tolerance level for drinking water in the absence of an EPA tolerance on September 19, 1983.[6]

By now, EDB as a risk communication event was a more fully developed public crisis than previously. A Washington, D.C., news program presented a very detailed hour-long investigative report on EDB on September 12, 1983. This was followed by similar reports in both the electronic and the print media. A major theme of these reports was the aggressive state activity to protect the public's health in contrast to federal inactivity.

Reprinted by permission of Tribune Media Services.

Figure 2–1

More formal risk communication activity began on September 30, 1983, with the issuance of PD4 by EPA, which declared an emergency suspension on the use of EDB as a soil fumigant. This decision was announced at the first of three press conferences at which William Ruckelshaus, EPA administrator, presented the agency's position. National and local media attention increased, and public concern focused on other EDB residues, particularly in grain and citrus products. Florida began testing food products for EDB residues during October–December 1983 and on December 21 issued a stop-sale order and began to remove certain grain-based products from grocery shelves. This was the major event in the crisis; it instigated risk communication activities in other states and provided dramatic film footage for news broadcasts. The industry groups affected by the Florida ban began increased risk communication activities to protect the agricultural uses of EDB.

On February 3, 1984, after two months of deliberation and data gathering, EPA declared an emergency suspension (ban) on the use of EDB in fumigating grains. The EPA also set guidelines for tolerance levels of EDB in the food supply (900 ppb in grain, 150 ppb in flour/mixes, and 30 ppb in ready-to-eat food). Three days later Massachusetts adopted an emergency regulation and set a tolerance of 1 ppb. Over the next few months other states determined their own regulatory approaches for EDB residues in food products.

Widespread exposure to EDB throughout the population was communicated through many channels during February and March. Editorials, cartoons, and Op-Ed pieces began to appear in national print media (see Figures 2–1, 2–2, and 2–3). Scientists representing both environmentalist and industry groups began to speak out on the EDB issue, and EPA and other regulatory officials were put on the defensive, with much media attention focused on the delay in regulating the "most carcinogenic pesticide ever tested."

The EPA proposed to revoke the tolerance exception for EDB

Reprinted by permission of Tribune Media Services.

Figure 2–2

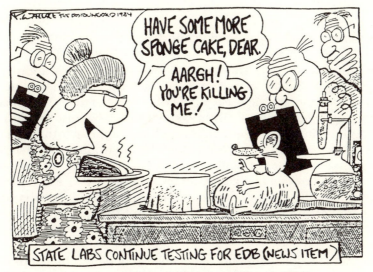

Source: Peter Wallace, *The Boston Herald*, February 17, 1984.

Figure 2–3

granted in 1956. On March 2, 1984, Ruckelshaus announced interim tolerances for citrus fruits and papayas at 200 ppb, of which no more than 30 ppb could be in the edible portion of the fruit.[7] By September 1, 1984, these tolerances expired, and any fruit with a detectable level of EDB residue, with the exception of imported mangos, was considered legally adulterated. Media attention by the time of this final Ruckelshaus announcement had waned. On April 3, 1984, FDA proposed action levels at the EPA-suggested tolerance levels, and these took effect on April 23. The last nationally televised evening news story on EDB was broadcast on October 12, 1984. State-level concern about EDB issues continued through the fall of 1986 when a congressional hearing was held on EDB pesticide disposal problems.

This historical summary indicates that EDB risk communication had a very long gestation period during which the essential morphology of the EDB crisis developed. By the time the issue became a public crisis with intense media attention, the major actors in risk communication activities had engaged in considerable interaction with the regulatory agencies. To understand the risk communication problems that occurred during the winter and spring of 1983–1984, one must place them in the context of EDB-related events of the previous ten years and the regulatory structures and risk assessments that shaped the EDB problem.

Regulatory and Legal Background

The complex and sometimes arcane logic of the federal approach to regulating pesticides is an important factor for understanding risk communication in the case of EDB. The ten-year sequence of events that led to the EDB crisis in 1983–1984 resulted in close public scrutiny of the pesticide regulatory approach through congressional hearings and lawsuits filed by environmental advocates. These investigations provided great detail on highly charged issues: inadequate interagency cooperation between EPA, FDA, USDA, and OSHA; allegations of the excessive influence of industry in the standard-setting process; failure to control the quality of scientific testing programs; and questionable practices of senior regulatory officials.

The four major regulatory agencies responded to the EDB problem in different ways and with distinctly different risk communication activities. Each agency contributed to the public perception of the EDB problem and, in turn, influenced the public perception of the other agencies' activities. Much of the media construction of the EDB crisis was a story of the inadequacy of the federal government to regulate exposures to dangerous chemicals, and the risk communication messages of each agency were received in this broader context.

OSHA

The Occupational Safety and Health Administration is responsible for the regulation of health and safety hazards in the workplace. Since the inception of the agency in 1970, there has been considerable controversy about how to implement this broad congressional mandate. Industry has vigorously maintained that the problem with OSHA has been too aggressive action that is disruptive to productivity; labor groups assert that the agency has inadequate resources to protect millions of American workers. The controversy has continued through the years of the Reagan administration. An enduring structural problem in regulating health and safety standards in the workplace is that while OSHA, in the Department of Labor, sets standards, the National Institute on Occupational Safety and Health (NIOSH), part of the Department of Health and Human Services, conducts scientific research. The two entities are often at odds, as was demonstrated in the EDB case.

In 1976 NIOSH declared that EDB exposures in the workplace were dangerous and recommended that the safety level be raised from 20 ppm to 130 ppb to protect the estimated 100,000 exposed workers. In 1981 California issued an emergency standard at the 130 ppb level, and labor groups petitioned OSHA for a national emergency standard. OSHA denied both this petition and a request by NIOSH to set a higher safety level. In a series of public statements, OSHA administrator Thorne Auchter maintained that there was not enough information to raise the safety level. After an industrial accident that killed two workers from acute exposure to high levels of EDB in 1982, extensive hearings were conducted in the fall of 1983 by the House Subcommittee on Labor Relations. Media coverage emphasized the delay in OSHA's response to scientific data when the agency proposed a higher safety level two weeks after the hearings.[8]

Thus, OSHA regulatory activities generated two key risk communication events that shaped the public discourse on EDB: Auchter's public statements on the lack of evidence on serious EDB exposures in the workplace, followed by the important symbolic event of EDB-related deaths and the hearings of the House subcommittee, which focused media attention on OSHA activity and provided extensive information that cast doubt on the objectivity of the agency's decision making.[9]

USDA

The United States Department of Agriculture is the leading federal agency on agricultural policy. Its regulatory focus is on food production and protection activities and to a lesser extent on health protection. The USDA was involved in the EDB case through its Federal Grain Inspection Service, which sets standards for and monitors the nation's grain supply, and through its Animal and Plant Health Inspection Service (APHIS), which has the responsibility to prevent the interstate and international spread of plant pests. These two responsibilities meant that the fumigation uses of EDB in grain and citrus were matters of concern to these agencies.

The congressional committees with direct oversight of USDA held no hearings on pesticide regulatory issues during the EDB controversy. Environmental activists considered this an indication that the primary role of USDA was to support farmers and agribusiness. The EPA was called to present testimony in hearings conducted by congressional committees with responsibility over EPA and FDA and concerned with the regulatory problems related to FIFRA. From the hearings (described in more detail under EPA, below) some important issues concerning USDA's EDB risk communication emerged.

A general finding of the 1984 January and March hearings was poor interagency coordination between USDA and the health-oriented regulatory agencies. Weak communications resulted in delays in the exchange of information about viable alternatives for EDB in citrus fumigation and the importance of EDB to the milling industry. From the hearings three specific issues emerged that became important in EDB risk communication and added to the perception of generic governmental problems in protecting the public against the risks of EDB. First, USDA challenged EPA's recommendations in PD2/3 by hiring a consulting firm to develop a risk assessment for the United States Fruit and Vegetable Association. The USDA agreed to pay for this study ($47,000) under a memo of understanding with the trade association to "maintain the registered use of EDB in quarantine fumigation."[10] Second, reports of "closed door" strategy meetings with industry were discussed in the hearing and reported in the press.[11] Last, the hearings brought out that APHIS manuals for the field use of EDB in soil injection allowed levels far in excess of EPA labeling requirements, which perhaps contributed to the groundwater contamination problem.[12] Although USDA officials were only minimally involved in the public debate about EDB, these critical events became important additional symbols of regulatory problems in the management of pesticides.

FDA and EPA

The agencies that played the principal roles in EDB risk communication were FDA and EPA, which share the regulatory responsibility for pesticide residues in food. The statutory basis for pesticide regulatory activities is FIFRA, which requires that products be registered by EPA before they can be sold or distributed in interstate commerce. A "registered pesticide" is one for which EPA has determined that its use will not involve "any unreasonable risk to man or the environment, taking into account the economic, social, and environmental costs and benefits."[13] For pesticides used in food production or storage, EPA established a tolerance (maximum allowable limit of pesticide residue) or an exemption from a tolerance for each food product on which it was to be used. When it was originally enacted in 1947 under USDA, FIFRA was oriented toward product efficacy and protecting pesticide users from the acute effects of the products. The act was amended in 1972 to incorporate broader health and environmental concerns and was transferred to EPA.

The key aspect of FIFRA for EDB risk communication is the reregistration provisions of the 1972 amendments, which charge EPA with evaluating the potential adverse health effects of the nearly 50,000 pesticide registrations that had been approved during the preceding thirty years. The initial congressional mandate was that this prodigious task be accomplished in four years. Much of the testing was to be done by the pesticide companies, which would supply the data to EPA. The process was inherently cumbersome, and the timeframe was eventually abolished. In 1975 EPA set up a "fast track" to remove or restrict a registered pesticide if new information suggested a potential hazard to public health or the environment. The process was called Rebuttable Presumption Against Registration (RPAR). (Mercifully, it is now known as "Special Review.") The series of studies, public calls for data, and the publication of the position documents (PD1, 2/3, 4) described previously form the core of the process. Congressional perceptions of undue delay in the special review process and the structural unsoundness of FIFRA in general were long-standing in the committees with oversight responsibilities for EPA and FDA (Government Operations Committee of the House of Representatives and two subcommittees: (1) Intergovernmental Relations and Human Resources and (2) Environment, Energy, and Natural Resources). In 1978 one member of Congress stated that the EPA standard-setting program is "abysmal and needs a complete overhaul."[14]

The statutory framework for FDA is the Federal Food Drug and Cosmetic Act (FFDCA), which authorized the establishment of tolerances and exemptions for pesticide residues in foods and raw agricultural products. There are no tolerances for food additives (the Delaney clause allows no known carcinogen in food additives). The FDA has the authority for enforcing tolerances established by EPA, monitoring the food supply, and seizing products that exceed established levels.

Even in the best of all possible worlds, with clear and effective communication between the agencies, EPA and FDA would still have had a difficult time implementing the special review program. The number of chemicals, the uncertainty of the scientific data, and the contradictory pressure from industry and environmental groups make the management and communication of this process inherently problematic. There were, however, other problems associated with both agencies' activities on pesticide registration in general and EDB in particular. Three congressional hearings on EDB produced more than 2,000 pages of testimony and supporting materials providing evidence of what Congressman Mike Synar called "bureaucratic footdragging by the lead agencies." The hearings were held because committee members and staff were concerned that a series of events concerning the special review of EDB "showed a general weakness in the program and . . . that the process had been politicized. *EDB was a prime example of the problems in the special review process.*"[15] EDB became a symbol for congressional concern with the broader statutory problems of FIFRA and the perception of serious biases in the administration of federal agencies charged with protecting the public's health.

These hearings were very important risk communication events in the developing EDB crisis. They provided a focus and a forum where the symbolic role of EDB became established. The essential terms of the public debate and the media presentation of the issues drew extensively from data first presented in the hearings. In all, five full days were devoted to careful scrutiny of the problem, and thirty people presented testimony and provided background documents. Fifty percent of these risk communicators were from the federal agencies (EPA, FDA, USDA), 20 percent from environmental and labor groups (NRDC, NCAMP, AFL/CIO, Teamsters), 16 percent from state governments (New York, Massachusetts, Florida, Texas), 7 percent from industry trade groups (GMA), and 7 percent were unaffiliated scientists. These accounts of the EDB problem became part of the basic lore of the EDB crisis. While much of this information had been available prior to the hearings and were part of early media reports, the hearings connected the disparate threads and wove together an integrated (if not objective) account. Timing was important: the September 26 hearing took place between the early communication of the EDB-contaminated groundwater problem in Florida and the EPA emergency ban on soil fumigation. The sequence gave the appearance that EPA actions were in response to public risk communications that the agency found embarrassing. While this may not be true, this inference was drawn in almost all media accounts on EDB in late September of 1983.

The key risk communication messages of the congressional hearings that played a significant role in subsequent media reports focused on the following:

- Undue delay in the special review process for political and economic reasons. A submessage was that EPA placed the concerns of industry above health considerations.
- The personal role of John Todhunter. Todhunter, a high-ranking EPA official, became established as the key villain in the disputed special review process. Principal symbols were the "lost" summary of the special review decision on EDB that postponed regulatory action for nearly a year, and his hiring of a business associate to provide a report that overstated EDB's importance in citrus fumigation. (Todhunter was the symbolic link to the previous administrator, Ann Burford, who resigned under allegations of excessive industry involvement in EPA decision making.)
- Secret meetings with industry officials and assurances to Florida officials that a ban on EDB use on citrus would be delayed: "all of the important decisions are made in private meetings where the public is not present."[16]
- Fraudulent data on pesticide registration testing submitted by the principal government-contracted laboratory.
- A critical reading of the risk assessment documents developed in the special review process highlighting the potency of EDB ("one of the most carcinogenic carcinogens ever reviewed by the Agency"); a message of "clear" scientific evidence to cancel EDB as early as 1969; FDA's inadequate monitoring of food residues and resistance to EPA's requests.

Risk Assessment

The risk assessment studies of EDB were conducted over a seven-year period, and they reflected the inherent uncertainty in estimating the risks and benefits of chemical hazards. The long period of EDB use in a variety of commercial contexts virtually ensured that there would be significant controversy about risk assessment studies, and those controversies played a significant role in EDB risk communication activities. In this section we will briefly describe the issues underlying the assessment of EDB risks and review three of the major risk assessments conducted by separate factions in the controversy.

As will become clear in this analysis, risk assessment provides a more formal statement of a group's underlying assumptions about the nature of environmental hazards and the level of economic/health trade-offs that is considered acceptable. While risk assessment is certainly part science, it is also part politics. The framework of a risk assessment and the particular pathways, exposures, and consequences that are considered important reflect the position of the assessor. Objectivity may be a goal of any risk assessor, but it was achieved in only varying degrees in the EDB case.

The most detailed and extensive risk assessments of EDB were conducted by EPA as a part of the series of special review studies discussed earlier. PD1 reviewed the toxicological data on the carcinogenic and adverse reproductive effects of EDB in animal tests.[17] PD2/3, issued in 1980, presented a full risk assessment including risk/benefit calculations as required under FIFRA. This is a quite complex analysis that required a review of the major uses of EDB (soil injection, post-harvest fumigation, fumigation of mills) and various minor uses. The commodities affected, estimated levels of EDB residues, dietary burden estimates, consumer exposures, and estimate of cancer risks related to exposure level are all components of the risk assessment in PD2/3.[18] Multiple layers of uncertainty are inherent in such a series of calculations; each estimate was subject to criticism, ranging from disagreements about the amount of grain product the average American eats to the specific exposure route used in the animal bioassay. The EPA concluded that EDB was a potent animal carcinogen and that a change in registration was warranted.

In PD4 EPA reconfirmed the risk and made a change in the risk assessment that would be criticized in the public communications of industry groups. The major change was in the mathematical model used to extrapolate risk from the NCI bioassays (themselves the subject of controversy in the scientific community). The EPA adopted a "one-hit model with Weibull timing," which was more sensitive to EDB exposure at very young and very old ages. The EPA also increased its estimate of the amount of EDB residue in grains.[19] EPA officials argued that the new model was more "conservative" and a response to critics that the assessment was biased against setting a strict tolerance for EDB. In fact, EPA received pointed criticism of the risk assessment from both industry groups and environmental groups in congressional hearings and in risk assessments developed by other groups.

A key data element communicated to the public in PD4 was that the estimate of cancer risk due to EDB exposure was 3.3 per 1,000. This figure loomed large in all subsequent risk communications and, depending on the perspective, was a symbol of a serious threat to public health or a symbol of poor science on the part of the EPA carcinogen assessment group. The other key communication was, of course, the guidelines for EDB residue tolerances in food products.

The major industry-oriented risk assessment was conducted for the GMA by Joseph Rodricks, a former FDA official who was a consultant with the Environ Corp. Rodricks criticized the EPA risk assessment from top to bottom. His report critiqued the NCI bioassay for EDB ("not particularly suited for assessing human risk"), the human exposure through grain consumption data (too high), and the use of the "Weibull timing" model ("is obviously absurd and may not apply at the extremes of age").[20]

The GMA risk assessment, using a different set of assumptions than were in the basis of PD4, estimated the EDB cancer risk at only 1 in 4 million for children and 1 in 12 million for adults. Rodricks stated that the risk assessment model used in PD4 was "not previously used in regulatory decisionmaking, and, moreover, has not been subject to peer review. . . . The [EPA] results are thus rejected pending complete scientific review."[21] The GMA risk assessment was distributed to health officials in every state and was cited in numerous national news stories on EDB by Rodricks and GMA officials. The document also was part of GMA discussions with EPA about PD4 and the tolerance guidelines.

New York State developed a risk assessment for EDB that differed from both the EPA and the GMA risk assessments. New York health officials led by Nancy Kim presented a broad risk assessment framework for EDB, considering all possible routes of exposure (grain, citrus, air, and water) and reproductive as well as carcinogenic effects.[22] While their mathematical model (GLOBAL) estimated lower cancer rates per exposure level, their risk assessment opted for a much greater margin of safety (greater than 1,000) and chose a child as the standard because children are the most susceptible population. Accordingly, the New York State guidelines for tolerances was 5 ppb, or 10 ppb if citrus use was banned.

Many states, including Massachusetts, based their assessments on the New York risk assessment model. There was also widespread media coverage of the New York risk assessment process as Governor Mario Cuomo held a press conference to discuss the results and the state's opinion that the federal standards were too lax.[23]

The risk assessments in the EDB controversy were all developed from technical data, and each stakeholder could point to a "legitimate" scientific basis for its position. Thus, in terms of the social and political dynamics of risk communication, no communicator was able to take the scientific highground and claim to be the only one using a science-based vs. an ideological approach in assessing the risks of EDB. Each account of the EDB crisis expressed beliefs in terms of data from a risk assessment. The problem was that in addition to the three major risk assessments described here, at least five other risk assessments were developed, utilizing but re-interpretating the data from one of the major assessments (for example, Massachusetts derived its risk estimates based on the New York data). In general, the risk assessment cited in a communication about the risk of EDB was chosen to fit the position of the communicator.

Risk Communication: The Mass Media and the Public Image of EDB Risk

For many persons the EDB risk communication crisis was the period of intense and confusing media coverage that occurred between September 1983 and April 1984. A lengthy analysis of the differences between EPA's intended risk messages and what the media reported can be found in *EDB: A Case Study in Communicating Risk*.[24] The basic assumption in this report was that "errors" occur in the translations of scientific "macro-risk" communications from EPA in the mass media. Media accounts, in this interpretation, focus on "micro risk": "What does this exposure mean to me?" The major public information problem for this analysis is the dichotomy between EPA's statistical analysis of a population-based problem and the personal concerns of individual citizens. The report includes extensive quotations from the media reports on EDB. We will not duplicate those efforts in this case study, but we have examined many of the same media documents and will present a different framework for interpreting the public image of EDB.

The history of public risk communications about EDB goes back to 1977. The first national evening television news story was on November 13, 1978; it reported the NCI findings of animal carcinogenicity and concluded "a government decision will be made in January." Occupational health risks from EDB were reported in newspapers from 1981 with coverage of OSHA decision making and emphasizing a message of federal inaction regarding a "grave danger." Many reports on EDB in particular and pesticides in general were published before the Florida groundwater controversy of September 1983. Also, and importantly, EPA had been presented as a "problem" in national news reports throughout 1981 and 1982. The resignation of the administrator, Ann Burford, was a lead story on national news broadcasts during this time. It is too limiting to restrict the EDB risk communication crisis to the period immediately surrounding the three EPA press conferences from September to March, as the media response was clearly conditioned by these previous events. For example, one Washington, D.C., television station reported on an EPA press release describing a proposed assessment of state data on EDB in food products in September 1983. This was immediately followed by an update on the legal problems of former EPA officials who had been forced to resign.[25]

A second issue is the lay public response to EDB risk communications. Although the response of many consumers interviewed at grocery stores when muffin mix was removed from the shelves was the "micro-risk" question, this does not explain the difference between the EPA message and state-level responses. As we will discuss in the two studies of local response to EDB that follow, the EPA message was received in a political and social context that varied from state to state. Elected officials and state regulators responded to EDB in very different ways, and these responses greatly influenced media coverage. Each local or regional newspaper placed the EDB story in a local context. The carefully crafted messages of the Ruckelshaus press conferences were embedded in the social question, "What does this mean to us, given our economic association with EDB?"

The third important factor in understanding the EPA message was that EPA was one of many message senders. Other federal agencies, environmental groups, and industry groups sent distinct and powerful messages also. While EPA was perceived as the sender most responsible for communicating EDB risks, it was not necessarily perceived as sending the most legitimate messages (see earlier discussion of congressional hearings). Editorials, interviews, and even cartoons indicate that EPA had to gain legitimacy the old-fashioned way in this risk debate; they had to earn it. As we will describe in a brief overview of the national media coverage, this was a continuing uphill battle.

To describe the national context of EDB risk communication, we analyzed forty-one news broadcasts from the EPA media archive and a sample of newspaper reports taken from the Dialogue data base. We also examined EPA press releases and videotapes of their press conferences. Press releases and other communications from the GMA, the most active industry group in the EDB debate, were also reviewed.

The national news story was framed in a very conventional way, according to journalistic procedures. Risk messages became news if they were dramatic, provided statements of clearly established conflicts between industry, environmentalists, and government, and addressed the reporter's perception of the audience's concerns. Television coverage was most extensive when dramatic visual images could be presented. The only period when the three major networks ran EDB stories on the evening news on two consecutive days was February 2 and 3, when vivid images of products being pulled from shelves, lists of tainted products, and interviews with fearful and confused consumers provided powerful visual images of a national crisis.[26] Newspapers sustained the EDB story more than television, with frequent and detailed accounts throughout the period. For example, the *New York Times* published fifty-eight news items or editorials about EDB between January 1983 and April 1984, whereas there were only twenty-three EDB stories on the three nightly television network news broadcasts.

The major symbolic events established for a national audience flowed through the media, but they should be viewed as a social construction of the EDB problem in which all interested parties played a role. The first media symbol was EDB as a chemical so hazardous that death could result from workplace exposures. There were many reports of occupational exposures to EDB, but two particular presentations characterize this image. *USA Today* (September 20, 1983) presented a full-page "opinion" section on "The Pesticide Threat." The editorial included the following comments on EDB: "worst carcinogenic chemical around," "virtually every worker exposed to legal limits could expect to die of cancer," "EPA and OSHA regulatory measures come late," "EDB must be banned. Pesticides were meant to kill pests not us." To "balance" the editorial, there were statements from Thorne Auchter of OSHA, an environmentalist, and a citrus industry-oriented writer for a Florida newspaper. Seven "Voices from Across the USA" presented a range of citizen responses. A large cartoon showed a disintegrating statue resembling Rodin's "The Thinker" with OSHA across his chest. The statue pondered a cannister of EDB emblazoned with a skull and crossbones (see Figure 2–4). Quotations ranging from John Todhunter of EPA ("We don't want to fool the public and take an action just for the sake of saying, 'I took an action' ") to Maureen Hinkle of the National Audubon Society ("EDB poses the highest risk of getting cancer of any chemical we know of") appeared in a "Quotelines" section of the page.

The television version of this message was a September 25 NBC "First Camera" report on EDB. This one-hour show stated that "EDB makes rats drop dead" and showed in excruciating detail the two California chemical workers who were exposed to a large amount of EDB. The viewer saw the two men being hosed off,

Source: David Seavey, *USA Today*, September 20, 1983.

Figure 2–4

rushed to an ambulance, and die. A particularly dramatic five-minute segment showed a retired fumigation worker who was exposed to EDB shaking uncontrollably from a nervous disorder related to his exposure. These images were merged with interviews of EPA and OSHA officials, a discussion of secret meetings and the delay in banning the chemical, and an allegation that government inactivity was criminal and like murder. This jumble of images provided a lasting symbolic context for the EDB crisis for the public and for reporters. The latter replicated versions of the NBC story in both print and electronic media, using the same film footage and interviews with the same respondents.

The three press conferences by William Ruckelshaus (who had

only been at the agency six months prior to the events of September 1983) and the series of press releases, appearances on network news programs, and media interviews that involved many EPA officials constituted the formal public risk communication activities. These activities were media-conscious activities—that is, they were communications intended to be mass messages. At each stage of the EDB crisis, EPA's message was a part of all national news reports we reviewed. For example, the February 3, 1984, nightly news broadcasts on the three major networks all presented the basic EPA message: "calm down, EDB is a long-term health risk which we are eliminating through the tolerance guidelines, and this is not an emergency." This message was allotted only about one-third of the time devoted to the story; the other messages were the seven-year delay and that workers had died from EDB exposure. Each network discussed the delay as a causal factor for the crisis. Two networks showed Ruckelshaus responding to a question about the delay: on ABC his response was a clearly frustrated, "You are asking me to give an answer I really can't give"; on CBS he declared, "My job is to be the Administrator of EPA while I'm there. My job is not to poke around in the past to see what happened."

The delay was the second central image presented in national news, and it was a result of the legacy of the previous EPA administrator. While the press conferences and materials were factual and clear (if cumbersome in the case of PD4) and Administrator Ruckelshaus was very good as a public communicator, EPA's message was placed in a historical context of failure to take appropriate action. In our review of the television reporting on EDB, delay or inaction on the part of the federal government was mentioned in almost 75 percent of the stories. The percentage is similar in newspaper stories.

A third national image was that of "dueling experts." Many media reports included statements from scientists, environmentalists, industry or trade group representatives, and EPA spokespersons. The most visible scientists were Samuel Epstein, an environmental health physician and author of two books highly critical of the government's policy on carcinogens; William Havender, a geneticist with the American Council on Science and Health who authored a report on EDB that downplayed the risks; and Bruce Ames, a biochemist from the University of California at Berkeley whose studies of "natural" dietary carcinogens were often cited in the EDB debate. Each of these visible scientists could be called on to give a position on EDB that was predictably supportive or unsupportive of stricter controls on EDB.

These scientists provided technical justifications for a confusing range of policy options for EDB, from a complete ban (Epstein) to no further regulation needed (Havender). Their comments were presented in news stories, editorials, and panel discussions. As we will describe in the following state-level cases, scientists and other dueling experts also played a role in local risk communication controversies on EDB.

National news reporting reached its peak activity during the first two weeks of February 1984, when reports of EDB exposures in grain products provided the most generalizable risk context. By the time of the third Ruckelshaus press conference in early March, national interest in the story had waned. There are many indicators of this decline, but one particularly clear one is that the time devoted to EDB in national evening television news programs averaged 15 seconds compared to the nearly five-minute segments following the February 2 news reports.

Risk Communication: Massachusetts

Massachusetts played a major role in risk communication activities concerning EDB. Following the first reports of food residue data from Florida in December of 1983, state public health officials began an aggressive testing protocol, which culminated in emergency regulations setting the strictest standards in the nation for EDB tolerances in food (1 ppb). Network news reports focused national attention on risk communication activities in this state. In this section we will analyze the risk management and risk communication activities of state officials as they developed a public message of a "crisis under control." We will review the public hearings, media reports, and industry communications that took place in Massachusetts to better understand how local politics and economic contexts determine the risk messages received by citizens.

Economic and Political Context

Agriculture does not play a major role in the Massachusetts economy. There are no grain production activities and no citrus production in the state and, thus, little use of EDB. The only group concerned with EDB during the years between EPA's issuance of PD1 and the events of the winter of 1983–1984 was the state Department of Public Health's Division of Food and Drugs, the key regulatory agency monitoring health risks in the food supply. The toxicologists and environmental scientists in this division were well aware of the growing uncertainty over the safety of EDB use as a pesticide. Nancy Ridley, the director of the division and a key player in the state's subsequent decision making on EDB, recounted:

> We had been aware of the issue for some time, particularly the problem as an occupational health issue. The post-harvest fumigation uses were of some concern, but since Massachusetts uses were minimal (not a grain processor, limited use on fruit trees, bee hives) we decided to put the issue on hold. PD4 indicated that food was not a serious problem but then the groundwater data raised our level of concern.[27]

The EDB issue had a history for these agency officials that was rooted in the scientific and regulatory disputes that occurred between 1975 and 1983. Although the events in Florida in December precipitated a much higher level of action by the state, from their perspective risk communication on EDB had been taking place for nearly ten years.

When Florida communicated the results of its testing on the levels of EDB residue in citrus and later in grain products during December of 1983, Massachusetts public health officials immediately recognized that EDB residues might present a potential threat to the public and quickly began a program to test grain-based food in Massachusetts. Yet, the Florida data did not result in similar actions in other states. What prompted this decision from Massachusetts officials? Three factors seem important: the analytic capacity of the health department's laboratory, the scientific position of the health department that does not recognize thresholds for carcinogens, and the political reaction of state officials to the delayed federal response to the EDB problem.

Massachusetts is well known in the public health field for the technical capacity of its laboratory. It is one of ten or so states that could mount an aggressive testing program in such a short time frame. In an interview, Ralph Temperi, director of the Food and Drug Laboratory, said that the trained personnel were in place and only the special glassware and supplies to set up the testing were needed.

The second factor, the implicit health department policy on thresholds for carcinogens, was even more important. The approach of the Massachusetts DPH followed from the belief that the only safe exposure to a carcinogen is no exposure. Every departmental official interviewed talked about the threshold issue. It was a major symbol that both guided their internal activities on EDB and provided a framework to explain their actions to the media and the public. Nancy Ridley stated it this way:

Based on the information presented in PD4, there were sufficient problems with EDB from a public health perspective that we could not tolerate any additional exposure if there were residues present in our food. We have in Massachusetts a departmental policy that does not recognize thresholds which is the basis for our action. The first time this became a really formalized policy was with EDB.[28]

The issue of thresholds and a policy for regulating environmental carcinogens was a leading component of the department's overall strategy. A center to investigate environmental causes of diseases had just been established in the department around the time of the EDB crisis. An important figure in this risk activity was the state's commissioner of public health, Bailus Walker, an environmental health specialist who was intent on implementing a policy of no thresholds where a substance could be eliminated.

The third condition that structured risk assessment and risk communication activities in this state was the response to the federal activities on regulating EDB. Commissioner Walker stated the position this way in early February 1984:

States have often looked to the federal government for leadership. I think that we would feel much better about the EDB problem had we seen some evidence that the EPA was acting on the issue. The agency was looking at the economic considerations rather than focusing in sharply on the health impact of EDB. It took state action to move them in the [right] direction.[29]

Other department officials echoed this lack of faith in the federal government to protect the public's health. All of the officials interviewed mentioned the "secret meetings" between EPA and industry officials. They were well aware of the discussions of these meetings in the congressional hearings on EDB during 1983, and had read about them in media reports and professional journals. As a result, it was with a serious lack of trust that department officials received federal responses to EDB. Stephen Havas, deputy commissioner of public health and the leader of the team assigned to develop a state response to the crisis, described the trust issue this way:

Looking at the scientific evidence EPA's actions did not look sufficient to protect the public's health. [The response] was overly cautious, to be quite candid. It was done more for the growers and the industry than the public. . . . The closed door meetings which continued created a perception in the public and state health departments that they were letting politics and business drive the issue instead of science.[30]

When the results of the laboratory tests became available in late February, state officials appeared to define their response in a context of low expectations of appropriate response from the federal government. The resulting sequence of risk communication activities in Massachusetts consistently operated out of this context.

The Discovery of EDB in Grain Products

During January of 1984 the results of the EDB testing program began to accumulate. There was clear evidence of the chemical in similar grain-based products, as was found in the Florida tests. Massachusetts officials had two options: to take no action and wait for EPA or FDA action, or to follow the lead of Florida and set emergency action levels for EDB in food and enforce this policy through product embargo.

On February 1, the first communication of EDB risk reflected the state's decision to follow the latter course. A press conference was held, and the results of the testing program were announced: levels of EDB exceeding 1 ppb were found in forty-six out of ninety-six foods tested. The state's communication of the risks of EDB was dramatic and bold: Governor Michael Dukakis called William Ruckelshaus to urge an immediate ban on all further pesticide uses of EDB, stating:

> This is a national problem which requires an immediate response by the federal government. . . . I find it extremely disturbing that the EPA has apparently been considering a ban on the use of EDB on food since 1981 and has not yet chosen to take final action to halt its use and eventual spread into the food chain.[31]

The state challenged the federal government to act first or an emergency response by Massachusetts officials was likely. Commissioner Walker advised the public to return tainted products to the place of purchase. A list of products with a detectable level of EDB was published in newspapers throughout the state with front-page coverage. The Rupert Murdock-owned *Boston Herald* ran the infamous headline, "Killer Muffins Found on Grocery Shelves." Film crews from local and national television recorded the removal of flour and cake mixes from the shelves of grocery stores. Representatives of the food industry (the Massachusetts Food Association—MFA) expressed opposition to "unilateral action by the state [because] we could have trouble getting supplies."[32]

The product list became, de facto, a major instrument of risk communication and set the context for much of the subsequent risk communication activities by industry and consumer groups in the state. Products were listed by brand name, lot number, and the

amount of EDB detected. The press conference and the media stories mentioned the long-term nature of the cancer risk and stressed
that consumers should not panic, but the potency of EDB as a
carcinogen ("the most potent cancer causing pesticide ever tested by
EPA") was a strong message in the early reports. In fact, in the
Boston Globe the list was accompanied by a summary of the risk of
EDB printed in large boldface letters. The undeniable consequence
of this message was the same as with the national coverage—fear.
The difference was that the state had taken a position to immediately
protect the public and ordered specific products off the shelves;
there was no call for further information or a phaseout in the Massachusetts message.

At the time, publishing the list was not considered a formal risk
communication. Havas reflected that "we found high values and
responded to calls by the media and reacted to public interest. If
you announce a testing program you will be asked what you found."
Ridley also recalled that the potential impact of the list as a risk
communication device was not considered prior to its release:

> We had little time for the public relations people to package the
> information. We did not even think of the packaging, that the names
> of products on the list were going to be given out as a press release
> and that those names were going to be on tomorrow morning's front
> page of the Globe. [33]

During the next two days (February 2 and 3) EDB was the
leading news story in Massachusetts in both print and electronic
media. Much of the reporting focused on the tainted products and
consumer and food industry responses to the state's announcement. The *Boston Globe* ran seven separate articles in these two
days, and other newspapers around the state had similar intensity
of coverage. The media accounts drew on local sources for information: DPH officials, representatives of grocery chains, and consumers. National statements were reported in a context of state risk
communications. The *Globe* reported a story leaked by the *Washington Post* that Ruckelshaus had decided to set standards for EDB
residues in grain and that this would result in the destruction of "as
much as 13 percent of grain-based food now on store shelves." In
addition, "the federal action would come as more than a dozen
states . . . appear on the verge of unilateral action."[34] The federal
government was presented as a reactor to the more assertive actions of the states. This was the context for the EPA risk messages
and the Ruckelshaus press conference that took place the next day.

In Massachusetts, the Ruckelshaus message on February 3
brought conflicting "official" statements of risk to the public. The

EPA action "put the agency at substantial variation with concerned state officials and leaves key practical questions to be resolved at the state level."[35] Picking up on Ruckelshaus's legal constraints to offer only "guidelines" to the states, Commissioner Walker said, "As Mr. Ruckelshaus said, the states are free to be more or less stringent."[36] Regional EPA officials stressed that EPA was legally required to weigh the risks of a chemical agent against its benefits. Massachusetts officials stressed that "there is no level of a safe carcinogen," and that EPA considered the economic impact on the agricultural industry. "We arrived at two separate conclusions," Walker stated.[37]

The EPA message of "calm down," and its assurance that the regulatory process was now in place to remove EDB from the food pipeline in a reasonable amount of time, did not deter Massachusetts public health officials from their course of action. The state's intent was to force the issue by developing as strict a set of tolerances as possible, approaching the no-detectable-level standard. An emergency meeting of the Public Health Council was called, at which DPH officials presented a proposal to set far more stringent limits for EDB than those suggested by the federal government. The logic was that an imminent hazard to the public's health existed and that the state would not permit any continued exposure. There was some discussion within the department on this course of action, but in one official's words, "prudent public health policy is to go too far rather than not far enough."

The risk assessment conducted by the state put EDB exposures in the context of "elevated cancer in the Commonwealth" and strengthened the belief that EPA guidelines were "unacceptably high and not adequate to protect the public's health." The document proposed an immediate standard of 10 ppb and a level of 1 ppb effective in thirty days. The message communicated to the Public Health Council and to the public was based on the no-thresholds argument and human cancer risk estimates derived from animal data. The human cancer risk data estimated in PD4 were cited and applied to the Massachusetts population:

> *Based on animal data, the EPA has estimated cancer risks for humans exposed to EDB at current average dietary levels for a lifetime. The estimated range of cancer risk from EDB is 3.2/10,000–15/ 10,000. Calculating from these rates, up to 8,700 Massachusetts residents now living could develop cancer if they consumed EDB at the levels currently present in food for a lifetime.*[38]

The emergency regulation was passed on February 6, 1984, by a split vote of the council, and the 10 ppb tolerance became effective immediately. Another list of an additional forty-two food products was distributed at the meeting and to the press. Voluntary compliance was expected from the industry in removing products from the shelves with residues of more than 10 ppb.

The response from industry was swift. Milton Segel, president of the Massachusetts Food Association, emphasized the problems with implementing a standard based on testing random lot numbers of products. He noted that different batches of products have widely varying levels of EDB and that grocers should remove only those products with the code numbers tested. The limitations of the testing protocol were also clear to DPH staff. Beth Altman of the Food and Drug Division received most of the phone calls from confused consumers. She recounted:

> *Most of the direct risk communication to the public was in response to people calling in to find out the risks of products on their home shelves. Callers did not understand the risk assessment; they could not understand the limitation of risk to particular lot numbers. To say that this lot is tainted but others might be safe was not an adequate communication to the public.*[39]

Most of the early calls were from highly educated health food eaters who were upset that even "natural" whole grains were tainted with EDB. These callers tended to be from the wealthy suburbs of Boston. The DPH prepared an information statement for consumers explaining the emergency action and answering a few standard questions about the level of EDB in product groups, which foods are "safe," and strategies to minimize EDB in the diet: "eat organic grains and grain products that require long cooking times."[40]

National industry groups had an immediate reaction to the Massachusetts actions. In fact, the Grocery Manufacturers Association of America had been in contact with Massachusetts officials since early January as a part of their general risk communication activities directed to the states. In a prepared press release, GMA's president, George Koch, took strong exception to the Massachusetts EDB guidelines:

> *This decision flies in the face of all scientific data compiled after months of study by the EPA. . . . For some reason, which is not related to public health, Massachusetts wants to set its own standard. . . . This action is a tremendous disservice to the consumers of Massachusetts. It will increase costs and waste tons of good food.*[41]

Other industry officials sounded similar harsh criticism of the emergency standards. They also focused on the divergence from the EPA guidelines (which industry groups now supported) and the refusal of the state to recognize the effect of heating to reduce EDB levels in intermediate food. Distributed through the wire services, these industry communications were cited in many media reports throughout the country.

The emergency standards strategy also generated internal DPH problems. District health officers were overwhelmed by the great effort required to monitor products. Numerous memoranda and updates were sent to local boards of health, and the risk message was far from clear. At the same time district health officers were to serve as "prime disseminators of information."[42] The risk communication infrastructure for effectively addressing the public's concerns was lacking. Some DPH staff felt that this massive effort, involving many hours of overtime, was intended to be a political communication for a national audience rather than to protect citizens of the Commonwealth. Deputy Commissioner Havas agreed that a national message was an explicit goal behind the Massachusetts actions:

> *We were absolutely sending a message to the nation. We were proud of that. . . . We did not set the strictest standards just to be the strictest but we set them because it was reasonable and would go a long way toward protecting the public's health. We felt any other action by us was irresponsible.[43]*

In this first stage of risk communication on EDB, Massachusetts officials transmitted a very clear message to both EPA and to industry. The state intended to emphasize the position of not recognizing any acceptable level of dietary exposure to a known cancer-causing chemical. EDB was a test case of a heretofore theoretical public health policy. The discovery of EDB residues in grain products in the state was viewed by officials as a legitimate public health crisis and an opportunity to implement a no-threshold policy toward a carcinogen that could be controlled.

Risk communication to consumers was not as clearly developed. There were two messages in the citizen-oriented communications: that EDB is a powerful carcinogen and should not be in any food, and that the state health department is more concerned about your welfare than is the federal government. Both messages combined to communicate a sense of powerlessness among consumers who could not understand the risk assessment/product sampling activities of the state and responded to the reports with a mix of skepticism, worry, confusion, and anger at the whole situation. They correctly perceived that they were not the principal players in the EDB drama, which was to focus increasingly on industry, scientists, public interest groups, and federal and state health officials.

Court Action

The courts are an important locus of risk communication activities in disputes concerning the legality of proposed actions (or inactions) by regulatory agencies regarding suspected hazards. Such communications occurred in Massachusetts only one day after the emergency regulation was passed by the Public Health Council. A suit was filed by the American Grain Products Processing Institute (AGPPI) on February 9, 1984, asking a superior court judge to grant an injunction suspending the regulation on the grounds that the state cannot impose a stricter standard for EDB residues than the federal government. A Boston attorney, representing the AGPPI, interpreted a state statute as limiting Massachusetts standards to the level of federal standards. The attorney estimated that emergency regulations could cost Massachusetts grocers $20 million and drive up grain prices by 15 percent. He also stated in the suit that the amount of EDB in food products did not present a danger to the public's health. A Massachusetts superior court judge blocked the emergency regulation on February 10, sending the issue to the state supreme court for a final decision.

The media carried the story of the lawsuit as front-page news in newspapers across the state and as lead stories on local television news. The story was picked up by the wire services and received some national attention. Few people had heard of the AGPPI before this legal action against the state—not surprisingly, because the institute had been incorporated by four of the nation's largest food companies for the specific purpose of fighting the Massachusetts emergency standards on EDB. In less than two days following the Massachusetts action, Procter & Gamble, General Mills, Quaker Oats, and Pillsbury set up the Institute as a nonprofit corporation, hired Hill & Knowlton (a large public relations firm), and retained a prestigious Boston law firm (Choate, Hall & Stewart). No mention of the four companies appeared in the papers filed in superior court; it took a *Boston Globe* investigation to discover that linkage. Stuart Pape, a Washington lawyer with long-standing food industry associations, was identified as the president of the institute.

The formation of a scientific-sounding institute allowed four large companies with a clear vested interest in the outcome of state-level EDB regulation to develop an excellent risk communication vehicle. It gave the appearance of another voice in opposition to the "reckless" regulatory approach of the Massachusetts DPH. It is an example of the increased leverage industry groups can gain in a risk communication controversy. In this case, however, the Massachusetts attorney general successfully argued in behalf of the emergency regulations, and the injunction was overturned on February 15.

The continuing strong support of the local media played a role during the temporary injunction. The *Globe* published an editorial on the morning the court was to decide the issue, urging that the EDB ban be enforced. Characterizing EDB as a "clear threat to public health," the editorial raised again the delay in a proper federal response, declaring that "the Reagan Administration has been dragging its feet for three years on this matter." The editorial went on to chastise the "cowboys of public relations, galloping in to rescue what remains of the reputation of their polluting, poisoning, ravaging employers." Obviously the *Globe* editors had not appreciated the intense industry lobbying pressure directed toward the media by AGPPI.

Implementation Measures

With the "strictest standards in the nation" scheduled to become law on March 7, Massachusetts public health officials continued to develop their risk management strategy to implement the 1 ppb tolerance level. Numerous meetings and exchanges of letters with the Massachusetts Food Association (MFA) produced a working agreement on the timing of the product testing/recall process. After initial opposition to the ban, the MFA developed a sound working relationship with the DPH. One memo from Milton Segel, president of the organization, suggested that he get a desk at the state laboratories to facilitate these joint planning efforts. The MFA was caught in the middle between the state actions and the local food distributors and retail stores that they represented.

For example, Cargill, the second largest flour miller in the United States, notified all Massachusetts customers that it would no longer ship flour unless Cargill was indemnified and held harmless from any damages and costs in connection with the use of the products. The mailgram went on to urge that the matter be brought to the attention of the governor.[44] Risk communications between food companies, the trade associations, and state officials were extremely important communication events. Establishing a shared framework for regulating EDB tolerances would prove to be crucial to the implementation plans of the DPH, and much internal activity was directed toward this end.

Another state communication strategy was to persuade the other states in the Northeast to support the strict tolerances. New York, Maine, New Hampshire, Vermont, Connecticut, New Jersey, and Massachusetts health officials met on two occasions to discuss a unified regional position for a lower tolerance than that permitted under EPA guidelines. Massachusetts was the only state that supported a 1 ppb level; the others accepted a 10 ppb level in ready-to-eat food. Havas of the Massachusetts DPH, who attended the meetings and argued the state's position, recalled:

> *Beyond Massachusetts, New York, and Maine, the other states were exceedingly conservative, most lacked expertise in environmental toxicology. . . . [T]hey did not fully understand the issue. The two states with the strongest [environmental] departments, New York and Maine, supported aggressive action. Having allies made us less the wild-eyed radical.*[45]

The states never did reach an agreement. Some of the states reacted negatively to New York Governor Cuomo's attempt to portray himself as the coalition leader, with the other states following his lead. Newspaper stories in the *New York Times* and local papers presented denials by other states that any agreement was reached. Massachusetts public health officials clung to the 1 ppb level, even though they had drawn on the New York risk assessment data in setting their standard. The no-threshold argument had become the important symbol in the Massachusetts crusade. It seemed a nonnegotiable position and the core of their risk communication messages.

The Massachusetts risk message provided the dominant framework for reporting the EDB issue in local media. The federal message would continue to be nested within the Massachusetts context. For example, the third Ruckelshaus press conference on EDB levels in citrus was reported as front-page news in both small and large Massachusetts newspapers. The EPA's statements, however, received less attention than the simultaneous announcement by the Massachusetts DPH of its third round of product testing. The subordinate position of the EPA message is exemplified by the relative number of column inches given to Massachusetts vs. federal risk statements in the *Globe* on March 3: 5 inches summarized the Ruckelshaus press conference, while 14 inches were given to the Massachusetts message and the results of the product tests.

Public Hearings

Public hearings were required by Massachusetts law to make the emergency regulations permanent. These were scheduled for

March 19, and they provided a high-profile forum for the different factions on EDB to present their versions of the risks. In this period Massachusetts had to defend its strict standards against the counterclaims of industry groups and scientists who viewed the risks of EDB as less significant. The various risk communication strategies and messages represent a microcosm of the national debate and put the terms of the differences in even sharper contrast because of the focus on the extremely low Massachusetts tolerances. The substance of the debate was as much the rights of states to preempt federal standards as the risks of EDB.

The food industry risk communication intensified at the time of the announcement of the public hearings. Led by GMA, the strategy was to enlist the support of scientists who had national reputations and were opposed to the Massachusetts standards on technical grounds. The GMA conducted three days of briefings, press conferences, and meetings with legislative officials to criticize the "ridiculous, unnecessary, and unworkable" standards set in Massachusetts. John Weisburger, who had conducted the early risk assessments of EDB for EPA, said that the state health officials' approach "was not relevant to public health."[46] Joseph Rodricks of Environ (and a key scientific spokesperson for the food industry position) criticized the use of animal test data in setting public health regulations. GMA spokespersons discussed the technical impossibility of meeting a 1 ppb standard and the dramatic economic effects such regulations would have on the local economy. The GMA panel delivered a fear message that thousands of jobs would be lost and innocent bakers and merchants would be viewed as toxic criminals. Commissioner Walker dismissed this activity as an industry-supported "road show." The state would present the testimony of other scientists later in the month.

The verbal and written testimony received for the series of public hearings on the EDB regulations followed a predictable pattern. Industry groups presented data on the potentially devastating economic effects of the proposed state standards. During this commentary period, GMA had developed numbers to embellish its cost-benefit arguments. They contracted with a Boston consulting firm, Temple, Barker & Sloane (TBS), to produce a report entitled *The Economics of Massachusetts EDB Regulations;* it was widely distributed and became a source of economic data for opponents of the Massachusetts tolerance levels. The GMA also submitted its own EDB risk assessment documents and presented testimony by Sherwin Gardner, its chief spokesperson on EDB. The stable of GMA legal and scientific consultants also presented testimony at the hearings. The key scientific presentation of the GMA position was given by Rodricks, who summarized the messages of other scientists who opposed the action:

- Residues of EDB likely to be present in grain products during the next three years do not pose a significant health risk.
- The EPA guidelines are adequate to protect consumers.
- From his own risk assessment "there was no need to place any limits on EDB residues."[47]

Ten other scientists presented similar testimony that discounted the theory of no-thresholds, rejected the extrapolation of animal data to humans without epidemiological data, and questioned the potency of EDB as a carcinogen compared to that of background dietary carcinogens like aflatoxin. Among these ten were nationally known visible scientists in the EDB debate: Elizabeth Whelan of the American Council on Science and Health; John Weisburger, former EPA toxicologist; and Bruce Ames.

The food company officials presented less alarmist written testimony than their trade groups. James Behnke, senior vice president of the Pillsbury Company, set a cautious but conciliatory tone in expressing general support for the EPA guidelines which reflected the state of the U.S. grain-based food supply on February 3, 1983. More importantly for Massachusetts officials, he stated that EPA could have set lower guidelines for future dates and guaranteed that by the fall of 1985 all Pillsbury products would contain less than 5 ppb EDB. This statement supported the technical feasibility of the Massachusetts tolerance levels.

Support for the Massachusetts regulations came from the statements of sixteen individual consumers, all of whom supported a zero level of exposure. Citizen risk communications in the hearings expressed concern over EPA delays, saw the problem as a moral issue (like the DPH position on nuclear war), and emphasized the hazards of pesticides in general ("a failure of federal policy").

Other supporting communications came from scientists who accepted the Massachusetts risk assessment position of no-thresholds, the potency of EDB, and the applicability of animal-test results to regulatory policy. Nine scientists posed these justifications for the 1 ppb level and complimented Massachusetts on its aggressive actions to protect the public's health. The key representative of this perspective was Samuel Epstein from the University of Illinois School of Public Health and a very visible scientist on national news broadcasts of the EDB crisis. He argued that the standard should be zero, all uses of EDB should be banned, and that the replacement of EDB by methyl bromide should be precluded by regulation. Again, the key images were of a federal government more concerned with

economics than health and of a food and chemical industry that had
recklessly disregarded the public's health. Other visible scientists
presenting this position were Nancy Kim of the New York State
Health Department and Marvin Legator, a well-known toxicologist
from the University of Texas.

In total, there were fifty-six verbal and written risk communications on EDB during the hearings and comment period; thirty
were supportive of the 1 ppb tolerance level, twenty-three were
against this approach, and three were cautiously equivocal about
the risks of EDB. Each group constructed its own version of the
overall message. For Massachusetts officials, the hearing provided
legitimation for their position. The *Boston Globe* labeled the hearings as "enthusiastically endorsing Massachusetts toughest-in-the-
nation" limits on EDB in food. (It is understandable why senior
Massachusetts health officials characterized the *Globe* coverage of
the EDB issue as very, very good.)

Industry officials, in particular those from GMA, saw the outcome
quite differently. The GMA produced a twenty-five-page document, culled from the Massachusetts testimony, entitled *EDB: Science and Public Policy*. In this report, which was distributed
throughout the nation, GMA published twelve of the twenty-three
negative responses to the 1 ppb tolerance and in support of the EPA
guidelines. The introduction calls this "the most extensive scientific
information yet compiled on the EDB issue."[48]

Final Approval of the Strict Standard

Massachusetts won the symbolic victory to remain the strictest
state in the nation in the regulation of EDB. The Public Health
Council accepted the risk management argument of DPH officials in a meeting on May 8, 1984. Along the way the state had to
pass new legislation to preempt the federal standards for EDB
residues set by FDA on April 23. Commissioner Walker appeared
tireless in his efforts to maintain the strict state standards for this
"super carcinogen." Many resources had been directed toward
achieving the regulatory strategy that he desired: the testing program had cost more than $100,000 in direct expenses for equip-
ment and laboratory procedures and perhaps even more than that
amount in overtime and additional work by DPH staff across
many divisions.

Final regulations proposed eventual levels of 1 ppb for ready-to-
eat food, 5 ppb for intermediate foods, and 0 ppb for baby food. A
key component of the risk management decision was a series of
communications from the Pillsbury Company, summarized by Dep-

uty Commissioner Havas in the Public Health Council hearings. These communications supported the technological feasibility of these action levels, and provided the cost-benefit basis for the regulations, as evidenced by the following passage:

> *In retrospect, if we could have known how effective the ban, combined with an ongoing controls program, would prove to be in reducing EDB levels . . . we would have spent less energy debating the numbers and placed greater emphasis on removing the EDB.*[49]

This response and communication by a major grain-based food company was absolutely essential to the Massachusetts approach and allowed state officials to dismiss the TBS economic forecast as "an irrational, typical consultant report." Massachusetts EDB team leader Havas summed up the meaning of this experience from the DPH perspective:

> *It was a success . . . particularly how quickly we got EDB out of the Massachusetts food supply. What we did drove EDB out of the food supply for the entire nation . . . not just for Massachusetts. . . . [E]verybody got the benefit.*[50]

The interpretation of this case depends on one's perspective: either Massachusetts was a crusading hero against a suspicious and cumbersome federal regulatory structure, or it developed an irrational policy based on the self-aggrandizing activities of its public officials. From the risk communication perspective, the messages, metaphors, and symbols were clear and used consistently throughout the episode. The principal players were industry, scientists, and government. There was little citizen involvement in this case—no citizen-organized groups developed in response to the EDB crisis. The essential nature of the risk debate might be described as ideological: public statements of the principal actors could be generally predicted by their positions on states' rights to preempt, the validity of animal tests, and their level of trust in federal regulatory agencies. EDB was an important symbol used in an essentially political discourse. Massachusetts had no structural political or economic constraints to advancing an ideological position. There was no significant agricultural lobby in the state, the governor was supportive of the actions, the media were quite supportive, and the citizens were extremely supportive. A poll of Massachusetts residents taken by the governor's office in 1985 found that the state's action against EDB was the most popular policy of the year. It is unlikely that such a response would occur in an agricultural state or in a state with a conservative Republican administration.

Risk Communication: Hawaii

Hawaii's situation during the EDB crisis of 1983–1984 differed from that of Massachusetts in that Hawaii was a heavy user of EDB. The chemical was widely used by Hawaiian pineapple growers as a soil fumigant and by the Hawaiian papaya industry on native papaya destined for export sale. By 1983, when EDB first became a national concern, Hawaii already had a history of EDB contamination. In 1980 residues of EDB and other pesticides were first detected in drinking water. In the following years pesticides were detected in island milk and water supplies. The risk communication context of EDB was dominated by a developing local concern over decades of extensive pesticide use in Hawaiian agriculture and a growing fear of widespread environmental contamination. Both local and national EDB risk communications were received within the context of an agricultural state attempting to balance its public health responsibilities with the economic concerns of growers.

Agricultural industries have historically played an important role in the Hawaiian economy, and decisions that affect agriculture are politically charged. A thriving agricultural sector is important to the state, and local government has protected the industry's interests, including the issuance of special permits to use federally restricted pesticides.

Prior to 1980 Hawaii had very little experience with environmental pollution in general, and there was scant concern about pesticide pollution. When Hawaii experienced a rash of pesticide contamination events between 1980 and 1983, this heightened both public and government concern. However, government agencies responsible for the regulation of pesticides and the monitoring and management of pesticide problems lacked adequate technical, monetary, and staff resources. Hawaii had traditionally looked to other states (such as California) and the federal government for technical assistance on pesticide risk assessment. Traditions of citizen organizing and public criticism of government policy were less developed in Hawaii as compared with the mainland. Hawaiian people almost took their environment for granted; there were few active environmental groups and even fewer interested in pesticide issues before 1980. Environmental groups had primarily focused on endangered species and land-use issues and were generally supportive of agriculture. After 1980 environmental and citizen groups became more active in pesticide control. For example, "Life of the Land" and the Legal Aid Society (of Maui) tried to stop the EPA exemption for pineapple growers to use dibromochloropropane (DBCP) in 1980.

By the fall of 1983, when EDB first became a national concern, the adverse effects of agricultural pesticides had become a serious local issue. Risk communication pertaining to underground drinking water was the first focus of Hawaiian concerns, and the symbol of "the poisoning of paradise" gained national significance in the EDB controversy.

Local Use of Pesticides

EDB had been a heavily used pesticide in Hawaii by both pineapple and papaya growers since the 1950s. Hawaii's tropical climate, which was ideal for growing pineapples, was also ideal for the propagation of many insect species. In the past, natural methods to control pest infestations had been used, such as field and crop rotations and biological controls. But as agricultural industries expanded and intensive methods of monocropping (planting one crop, year after year, on the same piece of land) were adopted, the insect problem was exacerbated, and synthetic pesticide use in Hawaii increased.

In the last comprehensive survey of pesticide usage, completed in 1977, approximately 4.2 million kilograms (or 9.3 million pounds) were used. Agricultural usage amounted to 82 percent of the total. In 1977 approximately 4.3 million pounds of nematicides were used to control the worms that attack the roots of the pineapple plant.[51]

EDB had been used extensively as a nematicide by the major Hawaiian pineapple industries, Dole Pineapple (a Castle and Cooke subsidiary) and Del Monte. The third pineapple company, Maui Land and Pineapple, used the nematicide dibromochloropropane (DBCP). Dole Pineapple abandoned its use of DBCP around the time that it was detected in groundwater in California (1979) in favor of EDB, whose residual effect was shorter but still adequate to kill the resistant breed of nematode. This switch from DBCP to EDB is ironic, as EDB is ten times more carcinogenic than DBCP.[52] EDB was also used by the Hawaiian papaya industry as a fumigant to control fruit flies in post-harvest papayas exported to the mainland and foreign countries.

Statewide usage of pesticides per square mile was quite high (local Hawaiian newspapers claimed that more pesticides were used in agriculture per acre in Hawaii than anywhere else in the United States). Hawaiian growers received many special exemptions over the years to use pesticides that had been banned elsewhere in the nation or that are registered under FIFRA for other uses. Such pesticides included heptachlor and DBCP, which continued to be used in Hawaii in the 1970s and early 1980s.

Many Hawaiian crops are categorized as minor or luxury crops and were not considered by EPA when they were licensing pesticides. Further, some of the pesticides used in Hawaii were never tested in a tropical climate, and their synergistic effects have received little study.

State Pesticide Regulation

The Hawaiian Department of Agriculture (DOA) is the lead agency responsible for pesticides management. The DOA manages the proper importation, licensing, production, sale, and use of pesticides as mandated by the Hawaiian Pesticide Law (HPL). Occupational exposure to pesticides is also regulated by DOA.

Through a cooperative agreement with EPA, DOA receives financial and technical assistance to implement FIFRA and HPL. Funds from EPA have steadily decreased over the years, and more responsibility for pesticide management has fallen on the state.

The Department of Health (DOH) is the lead agency in monitoring pesticide residues. This program includes statewide monitoring of fruits, vegetables, and dairy products; the frequency of testing specific commodities varies according to current issues and funding. The DOH also monitors for specific pesticides in drinking water.[53] The monetary and technical resources of the DOH available for the testing of food products are very limited. "The lack of a good laboratory facility has been a major bottleneck to effective management," asserted a state representative.[54] According to individuals both within DOH and other agencies, a shortage of both staff and money constrained the state's ability to handle problems of the magnitude of EDB and necessitated a reliance on other states and the federal government for technical support and guidance.

Some government officials and concerned citizens see a conflict of interest in the DOA's serving as both the leading advocate for agricultural policy and the regulatory agency concerned with pesticides. There have been recommendations that this responsibility be shifted either to the DOH or, better yet, to an independent environmental agency, such as a state EPA. This suggestion was made in February 1984 at the height of the EDB crisis and was widely discussed in the local press. For example, an editorial stated:

> At a time of tight budgets, the thought of adding another government office may not receive wide support. But given the possible extent of pesticide contamination here and the weaknesses in the present oversight system, development of a local EPA is worth considering.[55]

Both agricultural industry and DOA officials interviewed believe that the DOA can manage this area more efficiently than any other agency. They do not favor the establishment of a state EPA.

Pesticide Contamination in Hawaii

In the words of a state environmental epidemiologist, "the credibility of the DOH was already suspect, because of heptachlor, by the time EDB became a problem."[56] According to a state representative, "News of EDB dribbled into Hawaii amidst paranoia and fear."[57] Hawaii's perception and management of the EDB risk was directly influenced by the state's recent history of pesticide contamination and the way in which these earlier risks were handled.

The most dramatic contamination event in Hawaii's history was the presence of heptachlor in local milk supplies. This episode remains a critical part of Hawaiian risk communication folklore. In January 1982, samples from several dairies in Oahu were shown to contain extraordinarily high levels of heptachlor. Heptachlor residues on the green stalks of pineapple plants (commonly known as green chop and used as animal feed) had been ingested by dairy cows and eventually found their way into milk sold in supermarkets throughout the state. Heptachlor is known to cause liver damage and cancer in laboratory animals. The state DOH sent dairy samples to a federal laboratory in San Francisco, California, for verification and further analysis. During the next two months milk continued to be sold and the Hawaiian public was not informed of the problem. Fifty-seven days after the initial discovery of contamination, the first recall of milk was announced. Eleven successive dairy recalls followed.

This contamination crisis suggested negligence on the part of the pineapple industry and the state government. The presence of heptachlor in milk constituted only part of the problem; the subsequent mismanagement and information delay made this event a risk communication crisis. Questions such as, How could this have happened? Why didn't the state tell us? and How can we think this won't happen again? concerned many citizens. According to an EPA official, "the state was in shell shock as a result of the mishandling of the heptachlor problem."[58]

Lawsuits were initiated and settled, scientific and medical studies were undertaken, and the director of health resigned. The Hawaiian Heptachlor Research and Education Foundation was established with money from suit settlements to continue to monitor the adverse health effects of exposed individuals, including babies who drank tainted mother's milk.

This event greatly affected public confidence and trust in the government. Government actions were criticized, and state officials recognized a need to reestablish their reputation and correct past mistakes in pesticide management.

Hawaii's EDB Crisis

Local discoveries of EDB in drinking water wells created a state-based crisis in Hawaii, which preceded the national concern and dominated the experience of risk for residents. Although Hawaii continued to depend on EDB and other pesticides for use in local agriculture, citizens were increasingly fearful of detrimental side effects.

Detection of EDB residue in underground drinking water sources first occurred in 1980 in a well in Kunia, a Del Monte plantation village. The cause of contamination was attributed to a leaky storage tank, and the well was ordered closed by the Department of Health. In the following year, minute traces of DBCP were found in drinking water sources in central Oahu. A few other instances of the detection of EDB and DBCP in drinking water occurred during the next few years, with a cluster of contamination discoveries made in July 1983.

In June 1983 Stuart Cohen, a scientific adviser with the EPA in Washington, consulted with Lyle Wong, chief of the DOA's pesticide branch, about the protocol of "going out in the field" to test for EDB. Cohen had first become interested in the vulnerability of Hawaiian groundwater to contamination in 1979, and in 1981 stated that "there is nothing to prevent DBCP from leaching into Hawaiian groundwater." Cohen and Wong obtained the cooperation of the state DOH, and in July 1983 the Health Department began to run extensive tests on water. Contamination by EDB and DBCP was detected in eight wells throughout the state, in Waipahu, Mililani, and Maui. With new monitoring instruments, extremely low levels of the contaminant could be detected, down to 20 ppt.

No federal levels had been set for EDB in groundwater at this time, and the state DOH was uncertain of appropriate action. As a precautionary measure it ordered the closure of all wells with detectable levels of EDB. EPA's Stuart Cohen provided a risk assessment using the Wiebull Timing Method. The exact source of the contamination remained debatable, as groundwater even closer to fields treated with EDB was not contaminated. One theory advanced by DOA was that the contamination was gasoline-related.

Government Role

With aroused public concern about contamination of drinking water, the Hawaiian Department of Health was eager to resolve this crisis. Health Department Director Charles Clark publicly acknowledged the problem and quickly began an investigation. The wells were closed. Alternative drinking water sources were supplied, and in some cases carbon filtration was installed. The governor set up a task force consisting of local experts, agency officials, and researchers from the University of Hawaii to study the problem. Director Clark corresponded with the Office of Pesticides and Toxic Substances in Washington, seeking information that would be helpful in assessing the risk (the local EPA office could not provide much help as it is staffed only by clerks).

Recalling this period, Director Clark thought that the state health department handled the problem well. He felt that the DOH had gained the public's support, despite the fact that it had often been criticized for not being open with information. Another DOH official alluded to the environmentalists' desire that the state set a zero tolerance level but dismissed that as a marginal complaint. Articles that appeared in local newspapers during August 1983 stated that Clark was disappointed with the level of federal guidance during the groundwater contamination crisis.[59]

By this time the Department of Health was aware that establishing a good public image through timely risk communication was important. It hired a public spokesperson to coordinate information to the media and present a unified position. Although Director Clark stated that he did not allow his spokesperson to keep the press at a distance, reporters felt that the spokesperson was an obstacle to information.

The Media

The local news media avidly covered stories of EDB (and DBCP) contamination of groundwater. These events, which had begun in 1980 as isolated occurrences, had escalated into a major news story by July 1983. News reporters interviewed were in agreement that few citizens and newspeople took seriously the water contamination stories of 1980–1981, but by 1983 public interest in these environmental pollution stories was high, and the media gave them increased attention.

In the early stages of the EDB problem, the local print media served primarily as a channel through which the government could inform the public about its activities. The standard news story announced test findings and described what the government had done or planned to do about the problem, using the quotations of local governmental experts and EPA officials. Expert opinions provided a context in which to evaluate the present risk to the public and to describe regulatory options and strategies. Director Clark emphasized the importance of providing this risk information. His definition of a "good" news story was one where the level of con-

CORKY

YOU KNOW WE CANNOT JUST ARBITRARILY HALT THE SPRAYING
OF EDB AND DBCP. WE CAN ONLY ARBITRARILY ALLOW ITS USE...

Source: Corky Trinidad, *Honolulu Star Bulletin,* July 30, 1983.

Figure 2–5

tamination was announced, followed by a statement that such a level presented no danger (or other reassuring assessment). Longer articles on general pesticide issues also appeared in the press at this time, and these stories continued throughout the next few years.

Erosion of public trust in governmental officials was reflected in political cartoons that depicted an alliance between the government and agricultural industries (see Figures 2–5 and 2–6). This concern became more prominent in later months and years as the EDB crisis of 1983 became a turning point in citizen concern with pesticides.

The more critical reporting referred to the "poisoning of paradise" and the "chemical archipelago" and depicted lingering images of skull and crossbones: " 'Paradise' is being poisoned, claims new anti-pesticide group."[60] In January 1985, the *Honolulu Star Bulletin,* one of the two major Hawaiian newspapers, ran an in-depth exposé series called "Poisoning Paradise" nearly *three years* after the heptachlor incident and one-and-a-half years after the EDB crisis. The information contained within this series was primarily collected by a concerned citizen, who had investigated Hawaii's pesticide problems over many years in preparation for a book on the heptachlor problem.[61]

According to officials from DOA, DOH, and EPA, news report-

Source: Corky Trinidad, *Honolulu Star Bulletin*, July 8, 1983.

Figure 2–6

ing of environmental pollution in Hawaii was overreactive, sensational, and irresponsible. In their opinions, a pollution incident was likely to be a major headline, and reporters were inclined to turn a minor event into a front-page story. In some instances, news of water contamination problems appeared in the papers before they had been verified by officials, which required DOH to release a corrective statement. This type of reporting was more likely to appear in the smaller papers than in the major Hawaiian press.

Hawaiian reporters interviewed were in agreement that the state government provided timely information during the EDB problem. However, they continued to have trouble finding a reliable, knowledgeable source. Reporters stated that it was often necessary to contact or travel to Washington to get a complete story. Stuart Cohen of EPA supported this perception and said that the Hawaiian media were distinguished for the frequency of their calls. Only a few reporters from other states, interested in obtaining a more in-depth story, contacted him; the Hawaiian media called him often for detailed information.

Cohen gained the reputation of being a respected source of information. During his visit to Hawaii, he spoke to the press, briefed the governor, consulted with Director Clark, and was the lead speaker at a Mililani town meeting. Cohen said that he was completely surprised by the full auditorium of people desperate "to find a reliable source of information."[62] In this meeting he stated that EDB was ten times more carcinogenic than DBCP. This statement became not only a lead story in the media but also an enduring catch phrase that was often repeated by news reporters and citizen activists. "The chemical EDB that has forced the closure of four wells on Oahu is 'ten times more potent' than DBCP, another pesticide that has caused closure of two wells and restricted use of two other[s]."[63] And a month later, "Cohen has called EDB one of the 'riskier chemicals' the agency has dealt with. It is ten times more toxic than dibromochloropropane, or DBCP."[64] This statement would also appear in a letter to William Ruckelshaus by the town of Mililani: "EDB is a known carcinogen, and, according to the EPA, ten times more toxic than DBCP."

Cohen's statement provided a comparative risk context for EDB by contrasting it to a familiar and related risk. The dramatic impact of this statement supported the escalating public fear about pesticide contamination. Cohen believes that the newspapers actually reported his statement inaccurately.

One explanation for reporters' difficulty in obtaining information is that the state itself did not have it. The health director, though more forthcoming than the previous director, was not scientifically trained. His background was in labor and education, not medicine or public health. The state health department had limited knowledge and experience with trace amounts of pesticides in water and certainly did not have the facilities to run extensive tests.

In addition, the state was reluctant to report a problem when uncertainties existed. In the words of Director Clark, his job "was to disseminate any information which the department knew to be the *truth*."[65] This statement accurately describes much of Hawaiian officials' behavior in the dissemination of information about environmental pollution. The government chose not to release information until it was clear that the problem would not go away: "why alarm them if we don't have to?" This strategy of holding back information in an attempt to avoid public criticism and bad publicity contributed to the escalation of distrust in local and national regulatory agencies.

The Hawaiian government was also very interested in protecting the agricultural sector. Agricultural industries formed an effective lobby in the state, and government officials were responsive to this pressure. According to some officials interviewed, there was a feeling that the agricultural industries had already been damaged by international competition from developing nations and state officials didn't want to "kick them while they were down." Other officials feared that perhaps the pesticide contamination problem was an omen that the pineapple industry would leave Hawaii altogether. The Hawaiian government did not want to create undue public alarm and criticism focused on industry if the level of risk was uncertain.

Public Participation

Because Hawaii's EDB risk involved specific occupational and groundwater exposure, it aroused organized citizen concern. This is in contrast to states like Massachusetts where there was no organized citizen response to EDB residues in food. The most notable citizen action occurred in the town of Mililani. In July 1983, a municipal well, crucial to the town's water supply, was closed due to contamination with pesticide residues (EDB and DBCP) and the industrial chemical TCP (trichloropropane). According to State Representative Sam Lee, a native of this town, most people were initially hesitant to speak up and voice their concerns. Many of the town's residents were children of plantation workers and supportive of agriculture. Lee asserted that "the American tradition of speaking up is not as well developed in Hawaii," and citizens needed prompting and encouragement to organize a protest effort. With the benefit of Lee's leadership, and given the highly personal nature of this threat, the citizens of Mililani organized a remarkably effective effort. They forced corrective actions, including the delivery of alternative emergency water and supplies, and the installation of carbon filtration and water monitoring systems which are still in effect today. The town wrote a letter to William Ruckelshaus and another to the governor of Hawaii expressing their concern:

> *EDB is a known carcinogen and, according to the EPA, ten times more toxic than DBCP. It is also mutagenic and has reproductive effects. These findings are based on oral ingestion by animals. With regard to human beings, the acute toxicity of EDB is similar to that of DBCP. In other words, both are poisonous and cause several general toxic changes in major organs, resulting in death. [Excerpted from the letter to the Governor]*[66]

> *The Mililani . . . Neighborhood Board which represents 25,000 people of Mililani requests that the EPA, in the interest of preventing further contamination of the water supply, abolish the exemption of DBCP for pineapple cultivation and ban the use of EDB as an agricultural pesticide. [Excerpted from the letter to Ruckelshaus]*[67]

Mililani residents hired a lawyer who filed a billion-dollar class action suit against five chemical companies, a testing firm, and two pineapple companies. The coalition also requested participation in federal hearings on the EDB notice of cancellation. Within twenty-four hours of this request, the Pineapple Growers Association dropped its objections to the cancellation of EDB.

The organizing of Mililani was significant in that it encapsulated the dilemma of an agricultural state hesitant to speak out against agricultural interests but unwilling to accept the contamination of cherished resources. Mililani won the 1985–1986 All-America City Award for Citizen Action, Effective Organization and Community Improvement (co-sponsored by Citizens Forum on Self-Government National Municipal League and *USA Today*). It was one of nine cities to win this award.

Industry Opposition to EDB Control

The Hawaiian officials were pressured by the pineapple and papaya industries, which viewed EDB as essential to their production, to support them in opposition to the potential ban of EDB. Governor Ariyoshi intervened and requested continued use of this pesticide for Hawaiian growers.

The pineapple industry had supporters both in government and in the university. A health effects report (October 1983), written by the College of Tropical Agriculture of the University of Hawaii, was very supportive of industry's interests. It critiqued EPA's position document 2/3 and minimized the danger of EDB, asserting that the route of exposure and different metabolic rates of humans and laboratory animals challenged the validity of EPA's findings. It concluded with a statement that EDB in Hawaii's water at current levels may not be at all hazardous.

With the support of the state government, Hawaiian growers felt that their real battle was with EPA in Washington. In November 1983, the Pineapple Growers Association retained a Washington, D.C., law firm to file twelve objections to the notice of intent to cancel registration of EDB. The association also asked for a hearing on the notice of intent to cancel. According to Tony Hepton, director of agricultural research for Dole Pineapple, and Bob Souza from the papaya industry, this was an uphill battle against media hype and popular support for the suspension. Bob Souza described their defeat in the following way: "We found out with EDB that the public can kill you." On February 13, 1984, Hawaiian newspapers reported that the Pineapple Growers Association had withdrawn its objections. The agricultural companies chose to redirect their energy into the development of alternative pesticides, with the papaya industry having a more difficult time finding an effective substitute. Local media reported on their search for an alternative in a sympathetic tone.

Although agricultural industries fought the suspension, they were also interested in maintaining a good public image and used the media for this purpose. Dole Pineapple, for example, announced that it would immediately stop using EDB after EPA issued its emergency suspension on September 30, 1983. Even though growers were given permission to use existing stocks, Dole decided to return their stocks to the manufacturer. The growers continue to be interested in risk communication issues, and the local farm bureau has recently (1987) begun a project to study and improve risk communication.

EDB in Grain Products

Hawaii's EDB experience was dominated by its concern about contaminated drinking water and the effect of an EDB ban on local agriculture. The presence of EDB residue in certain grain products did not receive much public attention. This EDB story blended into the more immediate local concerns and added to the growing fear that pesticides were poisoning the Hawaiian environment.

The DOH, the lead agency responsible for monitoring pesticide residues in food, chose to test a variety of grain products in early 1984. Although it did not have the facilities to test an exhaustive list of products, it did test a small sampling of locally produced and imported grain products. The University of Hawaii also conducted its own testing of products. The DOH tested native papaya and pineapple for EDB residues beginning in late 1983. EDB residues did appear in local papaya, sometimes in significant quantities.

According to a DOH official, Hawaii had an advantage over other states in that it was familiar with this chemical, but testing was limited by the department's facilities and tight budget. The state health department looked to other states, such as California, for additional information regarding contaminated products and protocols of recall.

On February 1, prior to the announcement of federal standards, Procter & Gamble withdrew from Hawaiian groceries certain muffin mixes found to contain EDB in California tests. Although this action was described as "voluntary" by DOH, a Procter & Gamble spokesperson said the company had no choice because the products would have been embargoed by the state. A Procter & Gamble spokesperson was quoted in the *Honolulu Advertiser* as maintaining that state officials were "needlessly frightening Hawaii consumers." Director Clark decided to recall food found to contain levels of

EDB above federal guidelines. Further, Clark ordered all of Procter & Gamble's muffin mixes off the shelf when the company failed to comply immediately with his request for a list of all Duncan Hines products shipped to Hawaii. A few days later Procter & Gamble presented the list. Perhaps Clark's tangle with Procter & Gamble and assertive actions were in response to past instances of pesticide contamination when the department's risk communications were criticized. The presence of EDB in grain did not significantly affect local industries (only one locally milled flour was affected), and it was not politically difficult for Clark to take strong action. Clark said that he welcomed the EPA announcement of recommended tolerances and proposed to use them as interim standards. They had no reason or incentive to promulgate stricter standards.

In Hawaii, stories of local recalls were often accompanied by descriptions of how the potential ban might affect local agriculture. "EDB Rulings Have Extra Significance for Isle Economy," read the *Honolulu Star Bulletin* on February 2, 1984.[68] "EDB and papayas may be banned," read a small local newspaper.[69] The tone at this time was sympathetic to the growers' concern about the potential ban. However, individuals who were directly affected by water contamination, such as the citizens of Waipahu, were less supportive of industry's desire to continue using this pesticide.

Director Clark was proud of the way the state dealt with the EDB scare. He felt that the health department demonstrated strong leadership and adequately protected the health of the people. He maintained that the department was an open and effective communicator and kept the public up to date regarding recalls and federal action through daily press releases and its spokesperson Don Horio. Although Hawaii did take the initiative regarding testing and recalls, Clark as well as other members of DOH felt that because of its small size, it was very dependent on federal standards and intervention.

When the EDB suspension was under consideration, it was unclear if it would be applied to Hawaii as well. The strong agricultural sector was accustomed to gaining special exemptions: "EPA to Ban EDB on Mainland, Not on Hawaii Soil."[70] Once the suspension was announced, there was some confusion in Hawaii about whom it affected. The papaya industry was allowed to use EDB for a longer period until an adequate substitute was tested. Although they adopted a double hot-dip method, growers did not find this as effective as EDB. All use of EDB ceased by 1986.

Hawaii's response to EDB and other pesticide risks presents a trajectory of increasing risk perception from a low level of concern to a crisis over a five-year period. It is also a situation where the recognition of a risk had direct economic consequences within the state.

Risk communication on the island of Hawaii was characterized both by the state's loyalty to its agricultural industries and by its increasing concern about pesticide dangers that affect its citizens. However, state officials' ambivalence in restricting pesticide use incited public criticism. The effect of citizen activists in local pesticide decisions is still evident and has reshaped the public debate.

Hawaii, like many other states with a significant agricultural sector, has some very hard decisions to make regarding the future use of pesticides. It has already taken some positive steps, including mandating the Office of Environmental Quality Control to investigate problems with past pesticide management and to recommend improvements. Further changes will involve the areas of research, monitoring, enforcement, and education of pesticide users and the public. Many of these suggestions involve improving risk communication.

Conclusion

A number of lessons can be learned from the EDB case, both for improving risk communication and for understanding the factors that limit its possible effectiveness.

1. When there is a strong local context of risk communication, it will develop a dynamic of its own. A long local history with a hazard or an especially aggressive response by local officials will largely determine how "outside" risk communications are received.

2. The EPA is only one of many risk communicators. In the EDB case other federal agencies, the Congress, the affected industries, public interest groups, and state health departments were all recognized as legitimate communicators. The view that the crisis arose because EPA's message on EDB was misinterpreted is far too simple. There was an inherent pluralism of communicators in the EDB case, and the diversity of messages did not necessarily represent errors in interpretation.

3. Related to both of the above points, the mass media seem to reflect the diverse risk communication messages. The reporting on EDB drew on a social context of fear, suspicion, conflicting evidence, and dramatic actions first initiated by separate state public health departments. The media emphasized the drama and conflict but obviously did not create it. The EPA message was reported, but in a critical context. Clearly, there were examples of misinter-

pretation of toxicological data and, in particular, the intricacies of the RPAR process. But, our analysis indicates that these errors played a small part in the media representation of the EDB risk.

4. Many of the risk communication problems in the EDB case, in particular the question of unnecessary delay in EPA actions, were rooted in the statutory problems of FIFRA and the special review process. Moreover, risk communication was impeded by the legacy of secrecy and illegal actions by former EPA officials. The EPA risk communicators, Ruckelshaus included, were forced to deal as much with historical factors as with current problems.

5. The multitude of risk assessments available as a base for decision making complicated risk communications on EDB. All factions in the dispute could put forward defensible "scientific" positions. Although the PD2/3 and PD4 were the most exhaustive and elaborate risk assessments, some states chose to develop their own assessments and regulatory approaches to EDB. Without the statutory constraints of FIFRA and the special review process, states like Massachusetts could move more quickly and create a public image of decisiveness.

Risk communication efforts can only be as effective as the risk assessment and risk management activities that precede them. In the case of EDB, the long history of a problematic federal approach to regulating pesticides, the conflict between state and the federal governments over preemption issues, and the public's fear and anger about carcinogens in food and water created a highly unstable substratum for risk communication activities. More attention to the cultural and social contexts of risk assessment and management might have improved EDB risk communications.

Endnotes

1. William Havender, *Ethylene Dibromide (EDB): A Report by the American Council on Science and Health*, April 20, 1984, New York, pp. 1–2.
2. Jonathan Lash and Katherine Gillman, "The Politics of Pesticides: How the Reagan Administration Failed to Control EDB," *Boston Globe Magazine*, March 18, 1984, pp. 13, 46.
3. *Congressional Federal Register*, July 26, 1956, p. 5620.
4. National Coalition Against the Misuse of Pesticides, Ethylene Dibromide (EDB) Fact Sheet, January 1984, p. 1.
5. National Cancer Institute, Memorandum of Alert on EDB, October 16, 1974.
6. Stephen H. King, State Health Officer, Department of Health and Rehabilitation Services, Florida, Testimony before Committee on Government Operations, U.S. House of Representatives, March 5, 1984.
7. William D. Ruckelshaus, press conference, March 2, 1984.

8. CBS "Evening News," December 13, 1983.

9. Extensive coverage of OSHA issues includes a full-page feature entitled "Opinion, the Issue: Pesticide Threat," in *USA Today*, September 20, 1983.

10. Congressman Robert E. Wise (West Virginia), *Government Regulation of the Pesticide Ethylene Dibromide (EDB)*, joint hearings before certain subcommittees of the Committee on Government Operations, House of Representatives, 98th Congr., 2d sess., March 5 & 6, 1984, p. 1325.

11. Statement of Deborah Berkowitz (AFL-CIO), *Contamination from Ethylene Dibromide*, Hearing before the Subcommittee on Toxic Substances and Environmental Oversight of the Committee on Environment and Public Works, United States Senate, 98th Congr., 2d sess., January 27, 1984, p. 214.

12. United States Government Accounting Office, *Problems Plague Pesticide Regulation*, January 5, 1984, p. 13.

13. Ibid., p. 3.

14. Natural Resources Defense Council, Report on EPA Regulation of EDB, Washington, D.C., 1984, p. 42.

15. Staff Member, House Committee on Government Operations, off-the-record quotation from interview, June 24, 1987.

16. Natural Resources Defense Council, p. 48.

17. EPA, "Rebuttable Presumption against Registration and Continued Registration of Pesticide Products Containing Ethylene Dibromide (EDB)," *Federal Register* 42(240):63134–63161 (December 14, 1977).

18. EPA, "Pesticide Products Containing Ethylene Dibromide (EDB): Preliminary Notice of Determination Concluding the Rebuttable Presumption against Registration: Availability of Position Document 2/3," *Federal Register* 45(239): 81516–81524 (December 10, 1980).

19. D. M. Rains and J. W. Holder, "Ethylene Dibromide Residues in Biscuits and Commercial Flour," *Journal of the Association of Analytic Chemists* 64:1252–1254 (1981).

20. Joseph Rodricks, *Preliminary Assessment of Risks from Exposure to Ethylene Dibromide Residues in Grain Products*, (Washington, D.C.: Grocery Manufacturers of America, Inc. and Environ Corporation, January 20, 1984).

21. Ibid., p. 3.

22. Nancy K. Kim, *A Risk Assessment for Ethylene Dibromide*, New York State Department of Health, February 21, 1984.

23. "Cuomo May Pull Grain Products from Shelves," *New York Times*, February 8, 1984, p. B-1.

24. H. Sharlin, *EDB: A Case Study in Communicating Risk*. Report for U.S. EPA, O.P.A., January 9, 1985. For a published summary see H. Sharlin, "E.D.B.: A Case Study in Communicating Risk," *Risk Analysis* 6(1):61–68 (1986).

25. WTTG-TV Evening News, September 30, 1983.

26. Vanderbilt University Media Archives, T.V. Abstracts, February 2, 3, 1984.

27. Nancy Ridley, interview, June 1, 1987.

28. Ibid.

29. Quoted in B. Svetky, "EDB: The Politics of a Super-Poison," *Boston Phoenix*, February 7, 1984, p. 1.

30. Stephen Havas, interview, June 8, 1987.

31. Press release, Governor's Office, Boston, Massachusetts, February 1, 1984.
32. Press release, Massachusetts Food Association, February 2, 1984.
33. Nancy Ridley, interview, June 1, 1987.
34. "State Finds EDB in Food Products," *Boston Globe*, February 2, 1984, p. 1.
35. "EPA Chief Bans Use of EDB on Grain," *Boston Globe*, February 4, 1984, p. 1.
36. Ibid., p. 5.
37. Ibid., p. 5.
38. Massachusetts Department of Public Health, "Recommendations on EDB to the Public Health Council," February 6, 1984, p. 5.
39. Beth Altman, interview, June 2, 1987.
40. Office of Public Information, Massachusetts Department of Public Health, "Consumer Information on EDB," February 8, 198:.
41. Grocery Manufacturers Association, press release, February 7, 1984.
42. Massachusetts Department of Public Health, memo from Division of Food and Drugs, February 15, 1984, p. 1.
43. Stephen Havas, interview, June 8, 1987.
44. Cargill, mailgram to Deputy Commissioner Havas, Massachusetts Department of Public Health, February 17, 1984.
45. Stephen Havas, interview, June 8, 1987.
46. John Weisburger, testimony at the hearings before the Massachusetts Public Health Council, EDB Standards, March 19, 1984.
47. Joseph Rodricks, testimony at the hearings before the Massachusetts Public Health Council, EDB Standards, March 19, 1984.
48. Grocery Manufacturers Association, *EDB: Science and Public Policy*, April 1984, p. 2.
49. Letter from Pillsbury to Massachusetts Department of Public Health, December 1984, cited in the *Proceedings of the Massachusetts Public Health Council Meeting*, December 1985.
50. Stephen Havas, interview, June 8, 1987.
51. The following information was derived from *Regulating Pesticides in Hawaii*, written by the Department of Environmental Quality Control (DEQC) in Hawaii, January 1985. Pages 25–29 from the study *Pesticide Usage Patterns in Hawaii*, 1977, by Wataru Takashash. These data and figures are estimates, as there are no complete records of pesticide usage in Hawaii. Only records of restricted use of pesticide products sales are required under Hawaii Pesticide Law.
52. Stuart Cohen, press conference, August 17, 1983.
53. *Regulating Pesticides in Hawaii*, January 1985, Department of Environmental Quality Control (DEQC), page 31.
54. James T. Shon, interview, July 17, 1987.
55. *Honolulu Advertiser*, February 17, 1984.
56. Bruce Anderson, interview, July 18, 1987.
57. Shon, interview, July 17, 1987.
58. Stuart Cohen, interview, August 15, 1987.
59. "EPA Guidance on Wells Disappointing to Clark," *Honolulu Advertiser*, August 13, 1983.
60. *West Hawaii Today*, August 25, 1983.

61. Stuart Cohen, interview, July 25, 1987.
62. Ibid.
63. *Honolulu Advertiser*, July 16, 1983.
64. *Honolulu Star Bulletin*, August 22, 1983.
65. Charles Clark, interview, July 17, 1987.
66. Letter to Governor Ariyoshi, from Mililani citizens, August 12, 1983.
67. Letter to William Ruckelshaus, from Mililani citizens, August 12, 1983.
68. "ECB Rulings Have Extra Significance for Isle Economy," *Honolulu Star Bulletin*, February 2, 1984.
69. *Garden Isle*, September 16, 1983.
70. "EPA to Ban EDB on Mainland, Not on Hawaii Soil," *West Hawaii Today*, September 16, 1983.

Sources

STEVEN HAVAS, Deputy Comissioner, Massachusetts Department of Public Health, June 8, 1987.

NANCY RIDLEY, Director of Food and Drugs, Massachusetts Department of Public Health, June 1, 1987.

RALPH TEMPERI, State Labs, Massachusetts Department of Public Health, June 1, 1987.

DAVID GUTE, Assistant Comissioner, Massachusetts Department of Public Health, June 1, 1987.

BETH ALTMAN, Division of Food and Drugs, Massachusetts Department of Public Health, June 3, 1987.

RICK JOHNSON, EDB Team Leader, U.S. EPA, Washington, June 28, 1987.

RICK FELDMAN, National Association Against the Misuse of Pesticides, Washington, D.C. June 28, 1987.

DON GREY, Staff Director for Congressman Mike Synar, Washington, D.C., June 28, 1987.

CHARLES CLARK, Former Director, Hawaii Department of Health, July 28, 1987.

BRUCE ANDERSON, Epidemiologist, Hawaii Department of Health, July 29, 1987.

MAURICE TAMURI, Food and Drug Department, Hawaii Department of Health, July 29, 1987.

REPRESENTATIVE JAMES T. SHON, State Representative, Hawaii, July 30, 1987.

ANTHONY HEPTON, Castile and Cooke (Dole), Hawaii, July 29, 1987.

JEAN AMBROSE, Reporter, *The Honolulu Star Bullentin*, Hawaii, July 29, 1987.

HAZEL CUNNINGHAM, Citizen Activist, Maui, Hawaii, July 30, 1987.

STUART COHEN, U.S. EPA, July 20, 1987.

Chronology

1921

EDB is found to inhibit lead deposits when leaded gasoline is used in internal combustion engines.

1948

EDB is registered as a pesticide with the USDA.

1949

EDB is registered for use as a soil fumigant.

1956

EDB is registered as a stored-grain fumigant and as a fruits and vegetable fumigant.

1969

The Hawaii Department of Agriculture study, "Evaluation of Pesticides Problems in Hawaii," suggests the vulnerability of Oahu's drinking water to pesticide contamination. Hawaiian growers continue to depend heavily on EDB and other synthetic pesticides.

1970

President Nixon creates the Environmental Protection Agency and moves pesticide regulation to EPA from USDA under the Federal Insecticide and Rodenticide Act.

Congress enacts the Occupational Safety and Health Act.

1974

The National Cancer Institute issues a "Memorandum of Alert" describing a preliminary finding of EDB's carcinogenic activity in rats and mice (oral route).

1975

The Environmental Defense Fund petitions EPA to investigate EDB and to take suspension or cancellation action.

A NCI study shows EDB to be carcinogenic in both sexes of rats and mice.

1976

The National Institute for Occupational Safety and Health recommends that EDB be treated as a potential occupational carcinogen and recommends a safe workplace level at 130 ppb (200 times lower than current standard).

Studies show EDB to be mutagenic.

1977

EPA issues a Notice of Rebuttable Presumption Against Registration. EPA cites literature identifying EDB as a carcinogen. EPA begins a special review.

1979

DBCP (dibromochloropropane) turns up in California wells and the EPA asks five other states, including Hawaii, who also use this pesticide to test their drinking water. No DBCP found in Hawaiian wells.

DBCP is banned by EPA for all uses except Hawaiian pineapple. DBCP continues to be used by one of the three major pineapple companies, Maui Land and Pineapple.

Dole Pineapple abandons its use of DBCP and switches to EDB. Del Monte had been using EDB.

NIOSH completes an inhalation study that shows EDB to be an animal carcinogen.

1980

NCI completes an inhalation study that shows EDB to be an animal carcinogen.

EPA issues position document 2/3: Preliminary Notice of Determination Concluding the RPAR. The *Federal Register* notice announcing the availability of the position document states: "After reviewing all available information the Agency has concluded that the presumption for oncogenicity, mutagenicity and reproductive disorders have not been rebutted. . . . The Agency has determined that the risks to human health of certain uses of EDB are greater than the social, economic and environmental benefits of these uses and that risk reduction measures cannot reduce the risks to an acceptable level. . . . After evaluating the human health risks and the economic benefits the Agency has reached a preliminary decision to immediately cancel use of EDB on stored grain, flour mill machinery and felled logs; to cancel use of EDB for post-harvest fumigation of citrus and tropical fruits effective July 1, 1983; . . . *Residues in wheat and citrus are documented in this notice."*

June: DBCP and EDB are found in a well in Kunia, a Del Monte plantation village. They are traced to an EDB spill. Hawaiian Department of Health orders the well closed.

"Life of the Land" asks EPA to cancel the Hawaiian Pineapple industry's right to use DBCP and EDB.

1981

OSHA issues an advance notice of proposed rulemaking to review an exposure standard for EDB, noting the highest worker exposure levels relate to its use as a post-harvest fumigant for grain and citrus.

Health and Human Services, Public Health Service, includes EDB in its "Second Annual Report on Carcinogens" for the first time.

June: Minute traces of DBCP are found in the drinking water sources in central Oahu, Hawaii.

July: Traces of DBCP are found in a Mililani well and on Kauai.

October: California issues an emergency temporary standard for limiting workers' exposure to EDB to 130 ppb. The milling industry claims to be exempt. USDA, EPA, and OSHA meet with industry and labor representatives over EDB's use on citrus: they promise to set a tolerance level in food.

The Teamsters and AFL/CIO petition federal OSHA for a nationwide emergency standard on EDB. Federal OSHA denies the petition.

1982

January 21: Samples from several dairy farms in Oahu, Hawaii, are shown to contain extraordinarily high levels of the pesticide heptachlor (known to cause liver damage and cancer in laboratory animals). The milk continues to be sold.

February: NIOSH urges OSHA to reevaluate its position and to set an emergency standard.

March: The Hawaiian public is first informed of the heptachlor problem.

The Department of Health makes eleven successive recalls of milk and dairy products.

EPA is notified that EDB residues are detected in groundwater in Georgia.

1983

June: California groundwater monitoring finds EDB contamination in soil fumigation areas.

July 15: Four drinking water wells in Waipahu are closed following the discovery of small amounts of EDB. These wells were tested for EDB at the behest of EPA.

July 21: Florida Agriculture Commissioner announces that EDB traces are found in two wells.

July 30: Florida bans EDB for soil fumigation in eight counties in the state's citrus belt.

August: Hawaiian Health Department director calls EPA guidance on pesticides in wells "disappointing," and the Honolulu City Council accuses EPA of "dragging their feet."

September 11: Florida finds EDB traces in forty-five wells, but EPA has no tolerance standard for drinking water.

September 12: Food Chemical News reports that USDA and FDA will sample food products for residues of EDB. USDA monitoring is to include meat and poultry.

House Subcommittee on Labor Standards holds hearings on worker exposure to EDB. Representatives from the American Federation of Grain Millers testify on worker exposure 100–200 times safe levels recommended by NIOSH and CAL-OSHA.

OSHA is reported ready to impose a drastic reduction in worker exposure to EDB.

EPA is reported ready to propose a ban on some uses of EDB and a phaseout on others.

September 19: Florida public health officer states that EDB in drinking water can cause cancer and sets a tolerance level at 1 ppb.

September 30: EPA announces plans for immediate emergency suspension of soil treatment uses of EDB, an eleven-month phaseout of post-harvest fruit and vegetable uses, and cancellation of all other major uses.

October 3: Japanese health officials in Tokyo and Washington, D.C., express concern about the planned EPA action against EDB, saying tests on citrus shipped to Japan averaged 130 ppb, which they consider safe for consumption.

October 7: Florida's ban on sale, distribution, and use of EDB goes beyond EPA's emergency suspension, but allows post-harvest treatment of fruit for export. (EPA order phases out such use by September 1984.) Florida also files an order permanently banning use of EDB as a soil fumigant.

October 11: By *Federal Register* notice, EPA expresses its intent to cancel registrations of pesticide products containing EDB, except that post-harvest use as a citrus fumigant may continue until September 1, 1984, to allow time for alternatives to be found.

December 9: The Florida Agriculture Commission reports that residues of

EDB are in numerous grain-based products, and lists specific products that should be returned by consumers to place of purchase. A stop-sale order is issued.

December 20: Florida Agriculture Department issues a second stop-sale order involving twenty-six grain-based products.

December 22: Florida governor issues a letter to William Ruckelshaus at EPA, asking that a federal standard for EDB in foods be set immediately.

December 23: Grocery Manufacturers of America files a suit in federal district court in Tallahassee asking for a temporary restraining order on the stop-sale order. The request is denied.

EPA rejects Florida's request to immediately set a federal standard but indicates that action will be taken after the first of the year. Ruckelshaus says he sees no "imminent danger to public health."

December 28: The American Farm Bureau Federation president writes to EPA asking the agency not to make a "quick and quiet" decision to end agricultural use of EDB.

1984

January 5: EPA announces steps to take regarding EDB. It decides on an emergency suspension of EDB for grain fumigation and mill spot-fumigation while proposed registration termination of such uses is handled by an administrative law judge; it orders an emergency suspension of EDB use as a soil fumigant; it initiates the process of revoking an exemption issued by FDA in 1956 that prevents EPA from setting tolerances on residue limits for EDB grain fumigation; it continues active work to determine EDB residue levels in grains and grain-based food products that could be used by states as a guideline until the exemption is revoked.

Hawaiian DOH lab and the University of Hawaii test grain and grain products for EDB. Over one hundred products are tested in a few months.

January 6: Massachusetts state labs begin testing for EDB residues in food.

January 8: The American Farm Bureau Federation president, at the opening of the group's national convention in Orlando, says that EDB is not harmful to humans, it is needed by farmers, and it should not be regulated by the individual states.

January 11: A front-page *Washington Post* story reports on an emergency meeting the day before between USDA's Block and EPA's Ruckelshaus. The *Post* estimates that 50 percent of U.S. grain has EDB residues, some of which could be in the food system for up to four years.

January 13: EPA releases a letter to all governors asking for reports on any EDB test data, for help in gathering data, and for completion of this action by the end of the month.

February 1: In California's tests Procter & Gamble muffin mixes (4 code lots) are found to contain EDB. P&G withdraws these lots from Hawaiian supermarkets.

Results of EDB test are reported by Massachusetts Department of Public Health. Levels exceeding 1 ppb are found in forty-six out of ninety-six foods tested. The governor urges an immediate EPA ban on all further pesticide uses of EDB.

February 3: The EPA issues its guidelines for tolerance levels of EDB in the food supply. The three-tiered standards allow 900 ppb on raw grain intended for human consumption; 150 ppb on intermediate products, including flour mixes, soft cereals, and other products requiring cooking before eating; and 30 ppb on ready-to-eat products such as cold cereals, snack foods, bread and all baked goods.

Director of Hawaii's Department of Health welcomes EPA's announcement of the recommended tolerances, and he plans to use them as interim standards to recall locally produced and imported grain products.

February 6: Massachusetts Public Health Council adopts an emergency regulation stipulating food products with 1 ppb or more of EDB may not be sold or distributed effective March 7, 1984. During the interim transition period no food products of 10 ppb or above may be sold or distributed.

February 7: New York Governor Cuomo gives President Reagan ten days to adopt stricter permissible EDB levels, calling EPA guidelines deficient. After ten days New York orders all grain products to be certified as complying with state standards: 6 ppb for ready-to-eat food and 30 ppb for intermediate products.

The American Grain Products Processing Institute files a civil action in Massachusetts Superior Court to stop the state from enforcing an emergency regulation adopted February 6, 1984, establishing stricter EDB levels than EPA's.

Ruckelshaus compliments Clark for setting tolerance levels for EDB in water at lowest detectable level (20 ppt).

Mililani Neighborhood Board and citizen leader Sam Lee are given permission to participate in EDB cancellation hearings.

February 9: The Superior Court of Massachusetts enjoins the State Department of Public Health and the Public Health Council from enforcing the state's emergency regulation. The ruling results from a civil action filed by the American Grain Products Processing Institute on February 7, 1984.

February 10: Pineapple Growers Association of Hawaii withdraws objection to EPA's notice of intent to cancel EDB.

February 13: Massachusetts Supreme Judicial Court is asked by the state to uphold the emergency regulation on EDB levels permitted in food products. This overturns a temporary injunction granted February 9, 1984, by the state superior court.

February 14: The Massachusetts Supreme Judicial Court lifts a lower court order and allows the state to implement an emergency ban on grain products exceeding state standards for EDB.

February 22: EPA proposes to revoke the 1956 tolerance exemption in the use of EDB on grain.

March 2: EPA announces interim tolerances of 250 ppb for whole fruit and 30 ppb for edible fruit.

March 19: Massachusetts Department of Public Health holds a public hearing on the emergency EDB regulation.

April 13: FDA proposes EDB action levels of 150 ppb for intermediate use and 30 ppb for ready-to-eat grain products to take effect once the exemption is revoked.

April 23: Revocation of the EDB exemption takes effect.

May 3: Massachusetts Public Health Council passes final regulations on EDB for an eventual tolerance level of 1 ppb for ready-to-eat food, 5 ppb for intermediate foods, and 0 ppb for baby food to be implemented in January 1986.

June: The Hawaiian Office of Environmental Quality is mandated to prioritize Hawaiian pesticide problems and to lay the groundwork for an improved pesticide policy. Two years are spent on this task. (By 1987, funding has not yet been received for the implementation of these suggestions.)

January: The Honolulu *Star Bulletin* carries a three-part series about Hawaiian pesticide management, called "Poisoning Paradise."

1985

December 10: Massachusetts Public Health Council declares the state's EDB regulations to be "technologically feasible."

1986

January: Massachusetts establishes EDB action levels that are the strictest in the nation.

November 10: Hearings on EDB pesticide disposal problems are held before the House Subcommittee on Government Operations.

Acronyms

AFL/CIO	American Federation of Labor/Congress of Industrial Organizations
AGPPI	American Grain Products Processing Institute
APHIS	Animal and Plant Health Inspection Service
DBCP	dibromochloropropane
DOA	Department of Agriculture (Hawaii)
DOH	Department of Health (Hawaii)
DPH	Department of Public Health (Massachusetts)
EDB	ethylene dibromide
EDF	Environmental Defense Fund
EPA	Environmental Protection Agency
FDA	Food and Drug Administration (U.S.)
FFDCA	Federal Food Drug and Cosmetic Act
FIFRA	Federal Insecticide, Fungicide, and Rodenticide Act
GMA	Grocery Manufacturers Association
HPL	Hawaiian Pesticide Law
MFA	Massachusetts Food Association
NCI	National Cancer Institute
NCAMP	National Coalition Against the Misuse of Pesticides
NIOSH	National Institute on Occupational Safety and Health
NRDC	Natural Resources Defense Council
OSHA	Occupational Safety and Health Administration
PD	Position Documents
ppb	parts per billion
ppm	parts per million
ppt	parts per trillion
RPAR	Rebuttable Presumption Against Registration
TCP	trichloropropane
USDA	United States Department of Agriculture

Selected Bibliography

HARRISON, E. BRUCE. "Cancer from Cake Mix? The Public Demands Personal Answers." *Public Relations Journal* 41(1):32–33 (January 1985).

JOHNSON, F. R. *Does Risk Information Reduce Welfare? Evidence from the EDB Food Scare*. Paper #870130, Economics Department, U.S. Naval Academy, Annapolis, Maryland.

JOHNSON, F. R. "Economic Costs of Misinforming About Risk: The EDB Scare and the Media." *Risk Analysis* (forthcoming).

MAZUR, ALLAN. "Media Coverage and Public Opinion on Scientific Controversies." *Journal of Communication* 31(2):106–15 (Spring 1984).

MEDIA INSTITUTE. *Chemical Risks: Fears, Facts and the Media*. Washington D.C.: Media Institute, 1985.

SHARLIN, H. "EDB: A Case Study in the Communication of Health Risk." Unpublished manuscript commissioned by the Office of Policy Agency, U.S. Environmental Protection Agency, Washington, D.C., January 1985.

———. "EDB: A Case Study in Communicating Risk." *Risk Analysis* 6(1):61–8 (March 1986).

Chapter 3

THE RELEASE OF GENETICALLY ENGINEERED ORGANISMS INTO THE ENVIRONMENT: THE CASE OF ICE MINUS

Introduction

Proposals to test genetically modified bacteria in an open field were first reviewed by agencies of the federal government in 1982. Five years later—after several court challenges, grass-roots opposition, a steady stream of national and local reporting, and an unprecedented review process by the Environmental Protection Agency (EPA)—the genetically altered bacteria were released at two sites in California.

At issue was a bacterium with an excised gene. This modification of the organism reduced its ability to nucleate the formation of ice crystals in the presence of moisture and an ambient temperature below 0°C—thus the term *ice minus bacterium*. The parental organism is ubiquitous in nature; the debate was over the novelty of the genetically modified strain.

There was no dispute about the fact that the risks of releasing ice minus bacteria were hypothetical or conjectural. However, the concern about this issue is the legacy of a long historical debate about the safety of recombinant DNA (rDNA) research. (DNA is the abbreviation of deoxyribonucleic acid, the genetic material found in all living things.) The early history of the genetics debate set the stage for the public response to ice minus. Symbols of mutant organisms going amok in scientific laboratories were commonplace during the 1970s. The prospect of releasing a laboratory-engineered microbe, therefore, tapped into a reservoir of public and, to some extent, scientific anxiety.

This case study examines the forms and channels of risk communication that evolved over several years in anticipation of an event. The risks associated with the event—field testing ice minus—were highly speculative. Risk communication about the planned experiment was prevalent in the media years before the event took place. Debates were carried on at the national, state, and local level, among lay people, scientists, lawyers, and regulators. Ice minus created a stir in the small farm community of Tulelake, California, as well as in the halls of Congress. It brought headlines in major U.S. dailies and monopolized local coverage in several California communities for months at a time.

Within the general story line of releasing a microorganism from the laboratory, there were major subplots. The field test of ice minus was planned independently by two institutions, the University of California (UC) at Berkeley and Advanced Genetic Sciences (AGS) of Oakland. They interacted with different communities and chose different approaches to risk communication. However, other key players such as the Environmental Protection Agency, the California Department of Food and Agriculture (CDFA), and the Foundation on Economic Trends (FOET), a Washington D.C.-based public interest organization, provided a common link between the subplots.

This analytic study responds to the following issues:

1. How were the risks of field testing ice minus addressed by major parties and constituencies to the controversy?
2. How, by whom, and in what form were risks communicated?
3. What symbols and analogies served as frames of reference for addressing the technological uncertainties?
4. How were the risks presented and interpreted by the media?

In this case we shall be investigating what people were saying about ice minus and how they were saying it. We shall examine different constructions of the risk event. We want to know why some local citizens were not satisfied with assurances by federal and state agencies, including many experts, that ice minus was safe. What additional assurances were being sought? Alternatively, to what extent was the opposition to the field tests based on philosophical or ideological considerations? Both substance and symbols are important factors in understanding these events.

We will be looking at EPA's role as the lead regulatory agency and primary source of technical risk analysis. The study will focus in particular on how EPA communicated the results of its risk assessment and how its communications on ice minus were received. We will also discuss nontechnical sources of risk communication by analyzing the risk discourse of citizen activists and representatives of lay constituencies.

Awareness of the historical and regulatory context in which the risk event occurs is essential for understanding the problems associated with risk communication. This case begins with a review of the social history of the ice minus episode and its legal and regulatory setting. Next we describe the risk assessment process and its outcome. The core of the study is an analysis of the multifaceted risk and benefit communications that were promulgated by the print media, institutions promoting ice minus, the regulatory agencies, and various citizens groups. The conclusion provides some perspectives on risk communication that build on a distinction between technical and cultural rationality of risk.

Historical Context

The first public discussion of releasing genetically engineered organisms into the environment took place around 1980. In that year the U.S. Supreme Court ruled that life forms were patentable. The organism of contention was a strain of *Pseudomonas* that was capable of degrading crude oil. Using natural processes of DNA transfer, a General Electric scientist had moved several genetic components from diverse organisms into a specially chosen microbe. Each component was suited for degrading one of the key constituents of crude oil. According to the patent application, an inoculum of the modified strain of *Pseudomonas* would be impregnated into straw and dispersed over an oil spill.

After the court's patent decision, the oil-degrading organism dropped from the public eye; it was not field tested or commercially developed. Extensive public debate about the environmental release of genetically modified life forms did not take place until 1982 when two scientists from the University of California at Berkeley modified two bacterial strains and sought permission from federal agencies to field test the microbes.

Using modern techniques of gene splicing (recombinant DNA, or rDNA, technology), Berkeley scientists Steven E. Lindow and Nickolas J. Panopoulos deleted from *Pseudomonas syringae* and *Erwinia herbicola* the genes that synthesize a protein responsible for catalyzing the formation of ice crystals. Under laboratory conditions, the scientists demonstrated that certain plants sprayed with the genetically modified organisms exhibited less frost damage at temperatures several degrees below freezing.[1] The next step in the development of the product was to test it under field conditions. Under guidelines first established by the National Institutes of Health (NIH) in 1976, all rDNA research funded by the Department of Health, Education and Welfare (now the Department of Health and Human Services, or DHHS) must conform to safety standards and review procedures issued by the agency.[2] Initially, NIH proscribed all releases of rDNA-produced organisms into the environment, but by 1982 the prohibition against deliberate releases was removed.[3] In its place a multi-tiered review process was established that involved the NIH Recombinant DNA Advisory Committee (RAC), institutional biosafety committees, and the director of NIH.

Lindow and Panopoulos submitted a proposal to the RAC on September 1982 in what began a five-year history of the first open field test of a microbe that had been genetically modified by rDNA techniques. The period was highlighted by court challenges, community activism, risk assessments, and a complex network of communications flowing to, from, and among scientists and citizens.

The history of the ice minus episode can be described through several different axes. First, there is the proposed Lindow-Panopoulos test, which was evaluated by two federal bodies and debated in the test site community and in the courts. A second but related axis involves the role of a newly formed biotechnology company, Advanced Genetic Sciences (AGS), which developed a strain of ice minus for commercial applications. This company's regulatory trajectory for field testing the microbe paralleled that of the Berkeley scientists. A third axis consists of the risk perceptions and behaviors of residents of the communities that were considered as possible sites for the deliberate release experiment; county and city officials from Monterey, Contra Costa, Siskiyou, and San Benito counties in California were involved in discussions about the prospect of hosting the unprecedented test. There were variations in the response by local officials and residents to different forms of risk communication.

A fourth axis consists of the scientific and regulatory approaches to assessing the risks of the experiment. Each axis illuminates a different perspective on the history of the risk event; each has its own internal logic and must be understood within a social, political, and cognitive context appropriate to the institutions in question.

The Lindow-Panopoulos test was finally approved by NIH's RAC in April 1983, seven months after it was submitted for review. A suit filed against NIH by the Foundation on Economic Trends derailed the original test date. In March a similar request for field testing an ice minus derivative called Frostban was made by AGS. Following the only regulatory path open to it at the time, the company presented its proposal to RAC, a body with no statutory jurisdiction over private companies. AGS was not required to obtain NIH's approval prior to releasing the microbe, but like other companies that valued good public relations during a period of public anxiety over biotechnology, AGS brought its proposal to field test Frostban to NIH under the agency's Voluntary Compliance Program.

Two events altered the course of the regulatory review process for ice minus. The first event took place on May 16, 1984, when a U.S. district court judge enjoined NIH from approving field tests pending the outcome of a lawsuit filed by the Foundation on Economic Trends. The first field test scheduled for May 25 was postponed, and biotechnology firms became increasingly skeptical of NIH's ability to provide a stable regulatory environment for the industrial community.

The second event that shaped the course of the ice minus release was EPA's decision to extend its oversight review of genetically engineered microbial pesticides prior to application in the environment. On October 17, 1984, EPA issued an interim policy on small-scale field testing of microbial pesticides.[4] The EPA designated ice minus a microbial pesticide on grounds that it would be used to displace the natural and indigenous parental strain—ice plus—which was responsible for the frost damage. Prior to its 1984 policy statement, EPA did not normally require an Experimental Use Permit (EUP) for small-scale (less than 10 acres) field tests of pesticides—chemical or biological. The new interim policy, however, stipulated prior notification for such tests involving the use of genetically altered microorganisms in order for EPA to determine whether an EUP was required.

Commercial institutions that either had submitted or planned to submit proposals to NIH for field testing new genetically engineered microorganisms (GEMs) turned to EPA as the agency of jurisdiction. Although Lindow and Panopoulos had obtained NIH approval for the field test (June 1, 1983), fearing a bottleneck from the FOET litigation, they also sought EPA permission in November 1984 under the agency's recently promulgated policy.

The EPA set up a review and risk assessment process tailored to the ice minus test (see later section). Both AGS and UC-Berkeley were requited to obtain EUPs. Data relevant to health and environmental assessment of the organism were reviewed by EPA's Hazard Evaluation Division (HED) in the Office of Pesticide Programs (OPP). On December 5, 1985, EPA issued EUPs to AGS for conducting a field test at an undisclosed site in Monterey County. The media reported it as a milestone event—it would be the first federally sanctioned deliberate environmental release of a GEM.

AGS's Ice Minus History

The event that set off a lengthy period of community debate within Monterey County was a letter, dated January 3, 1986, sent to the Board of Supervisors and signed by seven residents of Watsonville and Salinas, California. The letter stated:

> *Numerous questions about this test remain unanswered. A primary concern is that there are no guarantees that the bacteria will not spread beyond its [sic] designated test site. . . . The possible side effects of Frostban on the environment are unknown factors that could affect rainfall patterns since the organism in its natural state is an important agent in cloud formation. Furthermore, few studies on the toxicity of Frostban have been completed, thus the possible impact on local crops remains unknown. Another concern is that local authorities do not need to be notified of the whereabouts of Frostban's application until twenty-four hours before its use.[5]*

The letter identified several subtle and complex issues associated with the test, and many lay citizens resonated with its message of skepticism toward biotechnology. Glenn Church, a local Christmas tree farmer, persisted in gathering information and bringing the debate first into the political arena of Monterey County and then to other counties. The letter asked for a county ordinance under the aegis of its land-use control powers to regulate deliberate releases of genetically altered organisms.

The motivation for this letter was the concern of the signatories themselves; they were not, as some have speculated, goaded by political or environmental interests outside Monterey. However, as a result of the NIH litigation, a climate of uncertainty and anxiety over genetic engineering was portrayed in the media. Also, once the initial action was taken by the Monterey residents and reported by the press, their actions were reinforced by individuals and organizations outside of the county, indeed outside the country.

Within weeks after the citizens' letter was sent, the Monterey Board of Supervisors put into motion a review process that included a public hearing, a forty-five-day moratorium on the field test (extended to a year on March 24, 1986), and a fact-finding effort. County officials eventually learned that the location of AGS's field test was in close proximity to residences. The AGS also publicly acknowledged injecting ice minus into trees situated on the roof of its facility, although it had not received an EUP for an open-air test. An initial EPA investigation of the episode concluded that AGS falsified EUP applications (the charge was subsequently reduced) and failed to perform tests in accordance with agency requirements. The EPA suspended the EUPs it had issued to AGS pending further investigation and a review of the test data. AGS scientist Trevor Suslow contended that EPA officials made statements to the press and filed a preliminary complaint containing allegations of falsification ignoring the results of a technical audit that drew contradictory conclusions.

These disclosures intensified public opposition to the test. County supervisors strengthened their resolve to proceed with caution despite assurances from regulators and scientists that the test was safe. The supervisors instructed the Planning Commission to incorporate guidelines for deliberate releases of GEMs in the land-use regulations. At the federal level, EPA reviewed its siting requirements in issuing EUPs.

Facing continued opposition to its field test from Monterey citizens, AGS sought a new site in San Benito and Contra Costa counties. This time the firm solicited community validation of its work. It was more forthcoming and proactive in communicating its plans to local communities; site disclosure and informed consent by the town formed an integral part of AGS's new risk communication strategy.

On September 4, 1986, EPA gave AGS conditional approval of the field test pending a site assessment. By February 1987, AGS had obtained endorsements from San Benito and Contra Costa counties and approval of three test sites from EPA and the California Department of Food and Agriculture. In March 1987, the Contra Costa County Board of Supervisors and the Brentwood City Council adopted resolutions supporting a field test of Frostban near Brentwood. Local media attention mounted as the first open field test of a GEM was once again weeks away.

A coalition of environmental and community groups, including Earth First, the Berkeley Greens, and the Foundation on Economic Trends, petitioned the California Superior Court for a preliminary injunction against the test. The petition was denied. Finally, on April 24, 1987, after an early morning effort by an anonymous party to uproot the plants, an AGS scientist sprayed Frostban on strawberry plants in a field near Brentwood, California. The event brought to a close the first stage of a protracted federal review process and ushered in a new era for biotechnology.

Several weeks later, on May 12, 1987, the Monterey Board of Supervisors passed an ordinance regulating the location and siting of experiments in which genetically engineered microorganisms are released into the open environment.

UC-Berkeley's Ice Minus History

During the time the publicity regarding AGS had intensified (in the spring of 1986) and in the aftermath of its unauthorized rooftop tests, the Lindow-Panopoulos EUP application was making its way through EPA. The designated site for the UC test was the state university's agricultural field station at Tulelake, California, near the Oregon border.

Steven Lindow had performed much of the path-breaking work on the relationships between ice nucleation active (INA) bacteria and frost formation on plants. He discovered how bacteria can limit the damaging effects of ice to frost-sensitive plants at temperatures above $-5°C$; he also showed that by reducing the number of INA bacteria, frost injury to field-grown plants could be mitigated.

As early as 1983, when the FOET suit was filed against NIH, Lindow described his work to the growers in Tulelake. No public concerns were evident in that community between 1983 and 1986. In the aftermath of the Monterey County episode, UC held a public information meeting at Tulelake to discuss ice minus. More UC and EPA people were in attendance than residents, and the low attendance was viewed by UC officials as tacit approval by the community of the field test. This paralleled the experience of AGS, which also made several presentations to grower groups and agriculture-related organizations—before the Monterey controversy flared up—and encountered no problems.

However, in Tulelake, as in Monterey County, a small group of residents made a substantial personal commitment to study and organize others around the issue. Two women spearheaded a grassroots campaign in Tulelake that eventually brought a coalition of groups into court with the University of California. Ava Edgar, a telephone company sales representative with an eight-year residency in the area, became inquisitive about the ice minus experiment when she learned about the UC informational meeting after it had taken place. Edgar started a networking and information-gathering process that brought her in telephone contact with Jeremy Rifkin. A week after the UC meeting she heard Glenn Church speak at the local Grange meeting hall about the AGS experience. Church's discussion of the risks was an important motivating influence on Edgar, who had concerns about the health effects of the ice minus organism on her infant child. After the Grange meeting, Edgar teamed up with social worker Djuanna Anderson.

Edgar and Anderson began gathering documents (such as the EUP) and constructing a series of queries. Frustrated by the responses, the women, along with a small core group of residents, formed the Concerned Citizens of Tulelake (CCT). They filed requests with local governments, collected signatures, petitioned the city council, lobbied various groups, sent out dozens of letters, contacted state and federal officials, and even organized their own forum on behalf of the city. By widely distributing flyers announcing a forum on the testing of mutant bacteria, they attracted between 250 and 300 people to the high school gymnasium—about one-quarter of the town's population.

According to UC's public information bureau, this was the first time they had heard of community concern about the field test. Moreover, UC had difficulty processing the questions raised by the Tulelake citizens, which they termed as "very nebulous."

This experience heightened CCT's skepticism about the field test. During an interview Edgar reported: "We read all the pieces of information that we received; anybody that was in the position to say this [field test] was OK obviously did not read the material."

The CCT's organizing efforts began to pay off. They gathered support from other groups, including farmers, local environmentalists, and the Tulelake Irrigation District. In June 1986, both the Modoc County and Siskiyou County Boards of Supervisors passed resolutions opposing the UC ice minus experiment. University officials declared that local boards lacked jurisdiction over the field tests and threatened a legal override of any local opposition. On July 23, 1986, UC issued a press release stating its intention to conduct the field test on August 6.

With support from FOET and Californians for Responsible Toxics Management (CRTM), CCT took legal action against UC. On August 6, the scheduled date of the test, the coalition won a temporary restraining order from the California Superior Court blocking the field test for nineteen days. Two weeks later, on August 19, a negotiated settlement was reached among the parties to the suit. Under the agreement UC would issue an environmental impact review (EIR) for the ice minus field test, and the litigants would withdraw the petition.

The draft EIR was completed in the late fall of 1986, and UC held hearings at Tulelake in January 1987 to obtain public comments. Community resistance to the test began to ease up. The Modoc County Board of Supervisors withdrew its categorical opposition and instead urged that alternative sites be "vigorously pursued and assessed." Similarly, the Siskiyou County Board of Supervisors acknowledged the potential benefits of the ice minus test but, like its neighbor, urged pursuit of alternative sites.

The final EIR for ice minus was issued by UC on February 20, 1987.[6] The negotiated settlement placed the final approval in the hands of the court. No further legal actions were taken against UC, and on April 29, 1987, five days after the AGS field test was completed in Brentwood, the UC-Berkeley scientists consummated a similar experiment at the Tulelake area.

Thus, nearly five years after two Berkeley scientists first submitted a proposal to release ice minus in an open field, the first sanctioned release took place. For UC the review process involved NIH, EPA, and two counties bordering the agricultural field station. Mirroring AGS's experience, county and local politics shaped the final outcome. Both institutions modified their communication strategies when the controversy heated up and the threat of a local ban was recognized.

The two histories of deliberate release follow an independent but related series of events within distinct political configurations. Tulelake was affected by the controversy in Monterey County. Both episodes grew out of an intense national media response to biotechnology in conjunction with actions taken by a few persistent individuals who brought the ice minus test to the attention of their respective communities. Uncertainties at the local level derived, in part, from the novelty of the technology and the lack of a mature and consistent regulatory response. The history of ice minus is a history of a nascent technology, emerging regulations, and complex public perception about applied genetics. Next, we turn to the evolution of regulations for genetically modified life forms.

The Legal and Regulatory Background

There are three main tributaries to the regulatory genealogy of the ice minus experiments. Initially, the sole authority rested with NIH. Eventually, EPA exercised its jurisdiction. Finally, the Office of Science and Technology Policy of the Executive Office of the President created a coordinated framework for regulating biotechnology, adding a third independent federal body with an interest and influence on environmental release issues.

First issued in 1976, the NIH guidelines were designed to govern rDNA experiments funded by DHHS. They exercised no authority over the private sector. Within a short time, other agencies of government such as the National Science Foundation

(NSF) and the U.S. Department of Agriculture (USDA) adopted the guidelines.

For the first half dozen years, the NIH guidelines prohibited the intentional release into the environment of rDNA organisms. Progressive relaxation of the containment requirements eventually led to the removal of the prohibition on deliberate release in 1982.[7] Proposals for releasing rDNA-modified organisms met a multitiered review process consisting of RAC, the director of NIH, and an institutional biosafety committee. The NIH approved deliberate releases if, in its view, no significant risks to health or the environment would result.

Between 1977 and 1979 Congress considered but failed to enact new legislation regulating biotechnology.[8] A regulatory gap existed because the NIH guidelines had no statutory authority over nongovernment-funded rDNA work. In recognition of this problem, NIH established a voluntary compliance program to accommodate private firms that sought the imprimatur of RAC for its genetic experiments and industrial scale-up activities.

The RAC approved three proposals for releasing genetically altered life forms; two involved plants (Zea mays or corn, August 7, 1981; tomato and tobacco plants, April 15, 1983), and a third from UC-Berkeley involved the ice minus bacterium (June 1, 1983).[9] Several months after NIH's approval of ice minus (September 14), Jeremy Rifkin, head of FOET, filed suit charging that NIH/DHHS violated the National Environmental Policy Act and the Council on Environmental Quality for failing to file environmental impact statements for deliberate release experiments. The U.S. District Court for the District of Columbia ruled in favor of the plaintiff. On January 21, 1985, NIH published an environmental assessment for the Lindow-Panopoulos experiment. This action of the court established a new formal requirement for NIH that slowed down but did not qualitatively change its review process.

Around 1982 Congress began exerting pressure on EPA to play a role in regulating genetically engineered products. Initially, EPA wavered, stating that its regulatory authority for nonpesticides was insufficient, but by August 1983, EPA formally declared that it would use two existing laws to regulate biogenetic products: the Federal Insecticide, Fungicide and Rodenticide Act (FIFRA) and the Toxic Substances Control Act (TSCA).

Minimally, TSCA requires that firms file premanufacture notices for new substances or substantially new uses of old substances. The EPA then decides whether there is sufficient information to make an assessment and if so whether the substance will pose an unreasonable risk to health or to the environment. The agency must demonstrate adverse health effects before it restricts the use of a product. FIFRA, on the other hand, imposes a greater burden of proof on the petitioner, who must demonstrate the safety of the pesticide prior to marketing and distribution.

Firms like AGS that initially had sought NIH approval for field testing their products filed new applications with EPA when EPA declared ice minus a pesticide and brought it under FIFRA review. Thus, EPA's active oversight of biogenetic products began in 1984 with the promulgation of its interim policy on small-scale field testing.

The entrance of EPA, in addition to USDA and the Food and Drug Administration (FDA), into the regulatory field of biotechnology increased the level of complexity. Each agency had its own body of law which did not conform exactly to the new technology. Agencies differed on how they viewed their responsibility to safeguard against adverse outcomes. There was no consensus on basic definitions of such terms as "biotechnology," "genetically engineered product," and "environmental release." In some cases there was confusion over regulatory jurisdiction and a lack of agency coordination.

Fearing a loss of American leadership in the development of biotechnology resulting from excessive or irrational regulations, the Office of Science and Technology Policy published a coordinated framework for the regulation of biotechnology.[10] The framework established the Biotechnology Science Coordinating Committee (BSCC),[11] a demarcation of agency roles, a set of policies on regulating products of biotechnology issued by each agency, and the definitions of several key concepts, representing an expectation that a shared vernacular could be developed.

Thus far, the coordinated framework has not resulted in uniform regulatory standards. The BSCC only offers recommendations to participating agencies. Each agency has constructed its policy on its preexisting laws, which do not lend themselves to a uniform set of definitions or approaches to regulation.[12]

EPA staffed new personnel within the existing structure of the Office of Pesticides and Toxic Substances to handle biotechnology. No new laws were enacted; no old laws were amended. The extant regulations within the established legal framework were interpreted to address the new biogenetic products. Each product, like ice minus, designated for field testing or industrial use was to be reviewed on a case-by-case basis. Risk communication requirements for EPA derive from general statutes like the Administrative Procedures and Freedom of Information Acts. Beyond these, the agency exercises its discretion with regard to balancing requests for proprietary information with the right of the public to know. The EPA has no statutory obligation under FIFRA or TSCA to communicate directly to distinct subpopulations, but the community debates regarding ice minus demonstrated a significant public interest in EPA's risk assessment process.

Risk Assessment

Two aspects of risk assessment for ice minus will be discussed: (1) the assessment process set up by the regulatory agency, including internal and external review and the opportunity for public input; and (2) the technical assessment of the product, including the questions posed, worst-case scenarios addressed, data required, and conclusions drawn. In another section we will examine how the technical risk assessment was expressed by the media and interpreted by the public.

The Assessment Process

Anticipating a possible legal challenge to the ice minus ruling, EPA put into place what it considered to be an exemplary assessment process, which incorporated multi-agency review, public comment, the establishment of an independent panel of scientists, and communication to the media. The process began for UC-Berkeley when Lindow and Panopoulos submitted a notification of intent to conduct small-scale field tests of ice minus (December 1984). Between January and March 1985, the proposal was reviewed by each of the branches of the Hazard Evaluation Division (HED) within the Office of Pesticide Programs (OPP). The Science Integration Staff of HED then developed a consolidated assessment of findings (called a preliminary assessment). The notification and preliminary assessment were sent to other EPA offices for their comments.

In March 1985, EPA convened a panel of outside experts (formally designated a subpanel of the FIFRA Scientific Advisory Panel) to review the potential risks of releasing ice minus. The six-member panel of academic scientists included two microbiologists, a meteorologist, a community ecologist, a microbial ecologist and a plant pathologist. A similar review, conducted for AGS, gave EPA and its advisory panel a first-round experience with ice minus. With recommendations of the Scientific Advisory Panel, EPA had to decide whether an EUP was required and whether more data were needed. The agency ruled affirmatively on both for the AGS and UC notifications.

Generally, when EPA receives EUP applications, scientists within OPP review them and develop a preliminary scientific assessment. The EPA then publishes a notice of receipt of the applications in the *Federal Register*. This is the first opportunity for public comment. Ordinarily, other agencies or ad hoc consultants are not involved in the review. In the ice minus case, however, EPA sought comments from groups outside OPP, including other EPA offices, USDA, NSF, and FDA.

There was also a notable difference in EPA's reviews of the AGS and UC-Berkeley field test proposals. In the former case, EPA did not inspect the initial site in Monterey County before issuing the EUP. However, once issued, the EUP provided for EPA inspection of the site prior to the test so that the agency could prepare for aerial monitoring. When the controversy in Monterey County erupted, public attention was drawn to EPA's failure to make a site visit prior to issuing the EUP. From AGS's perspective, a detailed site description was provided and the site was visited by the county agricultural commissioner's office. In contrast, EPA did undertake a pre-EUP site inspection at Tulelake in April 1986.

The AGS's books were inspected after the EUP application was issued in response to claims of fraudulent data by a former employee of the company. As part of the regulatory tightening that took place after the AGS technical audit, EPA also conducted a books and records inspection of the UC scientists' supporting documentation for the EUP application.

The final stage of the review process consists of EPA's decision on whether and under what experimental conditions to grant an EUP. Both AGS and UC-Berkeley eventually received the green light to field test the ice minus strains. The EPA's presence at the site with monitoring equipment was one of the conditions for approval. Overall, EPA created a distinctive assessment process that emphasized external and interagency review and careful documentation of each decision point, including the agency's response to public comments.

The Technical Risk Assessment

Since no confirmed hazards were linked to the ice minus bacterium, the risk assessment dealt with hypothetical or speculative risks. A framework developed by Martin Alexander of Cornell University and modified slightly by the authors provides a useful heuristic in discussing EPA's risk analysis. In this model, the risk assessment of ice minus is divided into four stages: formation, release, establishment, and effect (see Figure 3–1). Each stage is a necessary condi-

Figure 3–1 Stages of Risk Assessment for Genetically Engineered Microorganisms (GEMs)

I.	Formation of GEMs
	Techniques used
	DNA sequences added or deleted
	Parent organism/pathogenicity
II.	Release into the environment
	Deliberate release
	Site characteristics
III.	Establishment in the ecosystem
	Survival, multiplication, dissemination, transfer
IV.	Effect on the Environment
	Harm to humans or the ecology
	Pathogenicity

SOURCE: Adapted from Martin Alexander, "Ecological Consequences: Reducing the Uncertainties," *Issues in Science and Technology* 1 (3):57–68 (Spring 1985).

tion for any adverse consequence of a genetically engineered microbe. For example, there could not be a harmful effect of a release unless the organism established itself in the environment. Initially, EPA's risk assessment concluded that the ice minus field test experiment posed minimal risk, but that additional data were needed to substantiate this finding. Eventually, the agency's evaluation of ice minus involved reviewing data for each of the stages. No single effect or causal chain stood out prominently as the crucial test. However, a few hypothetical scenarios did attract special attention.

Formation of GEMs. The EPA examined information about the nature of the host organism, how it had been altered, whether genes (coding and noncoding sequences) had been added or deleted, the purity of the cultures, the genealogy of the organism (e.g., whether it or some related strain is pathogenic to another species), and its proposed use patterns. This analysis was intended to determine whether the organism or its genetic modification is associated with adverse effects to plants or animals. The EPA was satisfied that human/animal pathogenicity has not been associated with the parent strain. In addition, the agency was confident from the data that ice minus was not pathogenic to crops in northern California.

Release into the Environment. Would the release of ice minus in the test plot disseminate beyond the buffer zone? The EPA determined that releases beyond the test site and buffer zone could occur but at relatively low levels. The agency agreed to monitor the site to determine the flux of organisms upward and outside the test zone.

Establishment in the Environment.   On the presumption that ice minus will escape beyond the test zone, will it multiply, proliferate, and overrun the existing microbial flora (e.g., the parental strain, ice plus)? The EPA was persuaded by the studies indicating that wayward ice minus strains would not outcompete indigenous microorganisms. Moreover, there are naturally occurring strains of ice minus already in the environment. According to EPA, the relatively low level expected to be present outside of the test site was another reason it had confidence that ice minus would not displace the existing epiphytic (leaf-harboring) flora. At most, they might coexist. The agency reached the following conclusion about UC's proposal:

> *The applicants' data show that the INA⁻ deletion mutants have no competitive advantage over their parental or other INA⁺ strains when applied to potato leaves under a variety of dosage regimens except when INA⁻ deletion mutants are applied at higher concentrations than the INA⁺ strains.*[13]

Effects on the Environment.   The EPA addressed the following questions in its evaluation: (1) Is ice minus or its parental strain pathogenic to any plants or humans? (2) Will a release of ice minus in a field test alter precipitation patterns? (3) Will a release of ice minus affect the range and survival of frost-sensitive or frost-tolerant insects? (4) Will the release of ice minus produce unanticipated outcomes?

The test data on ice minus were derived from Lindow/Panopoulos and AGS. In its decision on the UC-Berkeley proposal, EPA stated that "these tests support the conclusion that neither INA products nor their parental strains are plant pathogens and that the deletion of a small portion of one gene had no effect on plant pathogenicity."[14]

The agency also indicated that data supported a no-effect outcome for points 2 and 3. The precipitation scenario was reviewed by one of the ad hoc scientific consultants—a meteorologist—who did not consider it a problem because of the amounts of ice minus available. The fourth point raises some epistemological issues: How can one assess the risks of unanticipated events? The EPA responded by shifting the burden of proof: "There is no scientific evidence or rationale for suspecting that the INA⁻ products may give rise to any different and/or unpredictable behavior and impacts on the environment."[15]

Despite its confidence in the safety of the field test, EPA placed some monitoring requirements on the EUP. Its purpose was to build a data base for future assessments. Bacteria were monitored above the site to determine upward dispersion, and vegetation and soil outside the site were sampled for colonization of ice minus. The Hazard Evaluation Division raised a question about the potential of the organism to colonize newly emerging vegetation; it concluded that only if the deletion mutant substantially displaced the indigenous strains, would there be a potential for "subtle, yet possibly significant effects on the survival and geographic range of certain plants and insects."[16]

UC-Berkeley's Ice Minus Greenhouse Studies. The EPA requested additional data from greenhouse studies on the colonization of ice minus on young emerging vegetation from low dose inoculation. These studies were conducted with both ice minus and their parental strains on 67 plant species. One deletion mutant was compared to its parental strain for 180 different phenotypic characteristics; another for 32 characteristics. The applicant claimed that the strains were identical in all respects except for their ability to nucleate ice (EPA actually found differences in 11 of these characteristics). The applicant tested ice minus and its parental strain for pathogenicity in 75 annual and perennial plant species, particularly those crops important to agriculture, and native plants in northern California, including all major crops in the Tulelake region.

Potential Adverse Effects. After reviewing the additional data supplied by the applicant, HED identified two types of potential adverse effects that "raise concerns about the proposed test": (1) on the survival and geographic range of certain plants and insects, and (2) on climatological patterns. The connection between ice nucleation bacteria and precipitation was speculative but could not be ruled out. For either of these effects to occur, according to HED, the ice minus would have to displace the ice plus strains over many plant species and over a wide geographic area. As previously indicated, four simultaneous conditions must be satisfied: (1) dissemination outside the area of the field test; (2) survival and replication; (3) successful competition with and displacement of ice plus; and (4) colonization.

Events 1 and 2 were expected to occur; the "no-significant-risk" outcome hinged on events 3 and 4. The HED maintained that the "likelihood of all four of these events resulting from Lindow's test is extremely remote."[17] Thus, it was the low likelihood of competition and colonization on which the case ultimately rested for EPA. These conclusions were based on small-scale tests and did not necessarily apply to large-scale uses of ice minus.

Next we will examine how the risks and benefits were understood and communicated by different groups and institutions, including the print media. What level of simplification, hyperbole, distortion, or skewing took place? What images and symbols of risk were used by different parties? What impediments prevented a clearer understanding of the different interests involved in the risk event?

Risk and Benefit Communications

The political and scientific discourse relating to the risks and benefits of ice minus took place in a variety of settings during the five-year period that plans for the release of the organism were considered. The following propositions will help frame the context of these discussions.

- The risk communications for ice minus preceded the first environmental release.
- The risks in question were conjectural or unknown; the benefits were explicit (saving money on frost damage to crops) but untested.
- The plan to release the ice minus organism evolved after years of public discussion about laboratory issues in biotechnology; thus the residual public apprehension about rDNA work had a new orientation in deliberate release.
- The channels of communication were complex and multi-directional; a considerable amount of the risk communication was carried out by nontechnical people (e.g., environmental litigants, community activists, and the media).
- Active involvement in communicating to the public by the regulatory, academic, and commercial institutions began in the aftermath of strong local activism and legal challenges to the ice minus release.

While the arousal of public interest to the ice minus test was driven by the potential risks, the benefits played an increasingly important role as the issue matured. Benefits were framed in terms of (1) increased productivity; (2) dollars saved by reducing frost-damaged crops; (3) an environmentally sound alternative to current methods of frost control; and (4) a harbinger of a new revolution in agricultural biotechnology.

We shall now describe how different parties to the ice minus debate constructed the issues and communicated their ideas to others.

EPA Communications

From the outset EPA assumed that its decision on ice minus would be the target of a legal challenge, and as a result it took several precautions. First, it created a review process with more external consultation and scientific input than was typical for the agency. Second, it maintained careful documentation of each stage in the process. Third, the agency appointed a public information officer who kept various environmental groups apprised of the events and responded to inquiries. Fourth, EPA took some active measures to communicate with the media. Beyond these considerations, the agency was prepared to treat ice minus in a manner similar to that of the fourteen other biological pesticides it had registered under FIFRA. But the issue was hardly typical in view of the FOET suit under Jeremy Rifkin, periodic oversight by congressional committees, and the insatiable appetite of the press for stories about new products of genetic technology.

The EPA announced approval of the first two EUPs for AGS at a Washington press briefing on November 14, 1985. The agency used the opportunity to assure the press corps that it had taken special precautions with ice minus since it was the first rDNA-produced microbial pesticide. The conference was videotaped, and we had an opportunity to analyze EPA's official message regarding ice minus and its response to queries from the press about the risks.

The general theme of the press conference was that EPA had produced a careful and technically sophisticated review of the product. Terms like "extensive scientific review inside and outside the agency" were used. Still struggling with the tarnished image left by the Ann Gorsuch administration, EPA tried to project the view that the decision to grant an EUP for ice minus was driven exclusively by science and that politics played no role. Aided by charts and graphs, EPA senior official John Moore cited four conclusions the agency derived from the data: (1) ice minus will not have a competitive advantage over its parental strain; (2) the upward flux of ice minus is expected to be low (therefore it will have an insignificant impact on atmospheric bacteria); (3) the genetically modified bacteria will not displace the indigenous strains outside the test site; and (4) the strains being tested are not pathogens.

Inevitably, members of the press corps inquired: "What is the risk?" "Is it zero risk?" "If the risk is so low, why the press conference?" The response was assertive, crisp, categorical, and confident. The test will not result in any "foreseeable adverse results to the environment." The conclusions represent "informed judgment from the best science." "Infallibility only resides in Rome." The risk is "extremely remote." The agency considered a "vast array of possibilities" and considers the field test safe. No commitment was made, however, to the safety of releasing ice minus on large tracts of land.

Some reporters, confused by the pace and content of the material presented, asked for further simplification. In response, they received a few summary sentences. Overall, the press briefing tried to leave the impression of an agency in control. It imparted the view that there were neither scientific nor logical grounds for delaying the field test. Ice minus was treated as a clear-cut case with no uncertainties or ambiguities about the science communicated and no divisions within the scientific community.

As EPA's experience with AGS and Monterey County intensified, several incidents gave the agency a fresh perspective on its public image. First, EPA was criticized at the Monterey hearings and subsequently by members of Congress for not visiting the initial site of AGS's proposed field test. Second, in keeping the location of the test site secret, AGS and EPA created a wave of public mistrust and indignation. Representatives from the Monterey County Board of Supervisors reported that when they visited EPA headquarters in Washington D.C., they were treated brusquely and told that the county had no jurisdiction over the site, including its disclosure. When asked why the test site was not being disclosed, an AGS scientist replied:

> *We have serious work to do—we don't want to be heckled or have tomatoes thrown at us. . . . We're concerned about sabotage and we don't want people undermining our efforts to confine bacteria to the test site by coming and traipsing around.*[18]

The disclosure in February 1986 that AGS carried out an unauthorized open-air test on its roof brought public embarrassment to EPA. The agency swiftly initiated an investigation, subsequently fined AGS, withdrew the EUP temporarily, and required a validation of test data. By the time the agency reviewed the UC-Berkeley test, it had drawn up a formal risk communication plan with three components.

Notification Plan. A number of individuals inside and outside EPA were to be notified just prior to the press release announcing the issuance of an EUP. The notification network included members of Congress, state officials, environmental groups, industry, groups that were potential users of the product (ten environmental groups; zero local groups; seven industry/university groups). Local groups were visibly missing.

Q&A Document on Ice Minus. The EPA issued a document with thirty-eight commonly asked questions and answers on the science and regulation of ice minus. The document, which filled sixteen single-spaced typewritten pages, provided a common framework for all EPA personnel who interacted with the media and the public. Clearly written with a minimum of scientific vernacular, the document was distributed by AGS and UC in their public information packets. The Monterey County experience demonstrated that something more than *Federal Register* notices and media reports was needed to apprise the public of EPA's decision process and its justification for approving the field test.

The issues discussed in the document represent EPA's rendering of citizen concerns at that period of time, and for this reason the document is an important indicator of how the agency conceptualized the public's educational needs. It stated that no new DNA was added to the naturally occurring bacteria, suggesting that the modified organism will not possess new properties. It discussed the rationale for using rDNA-produced ice minus rather than ice minus organisms found in nature. By describing the experiment in lay terms, EPA attempted to break down the mystique about genetic engineering, particularly for the informed lay reader and science reporter. Clear and unambiguous responses were given to the questions: Has EPA inspected the test sites? Have local officials and the local populace been informed? Will people be exposed to the products treated by ice minus? Is ice minus toxic to people, wildlife, or domestic animals?

While EPA affirmed its conclusion that the environmental effects of the field test would be insignificant, it also indicated that approval of the field test was not tantamount to pesticide registration. There were additional steps with more extensive requirements.

It was also reported that under FIFRA, EPA was obligated to consider the potential benefits of ice minus against the potential risks. The agency clearly put very little effort into communicating the benefits of the research—the document devoted a single paragraph to the subject. Benefits were framed in very general and qualitative terms (e.g., the products could lengthen the productivity of frost-sensitive plants). Two benefits *would* follow (EPA did not use the term "might"): economic benefits to consumers and replacement of chemical pesticides. Neither of these outcomes was definite, and yet they were stated with a confidence equal to that used in communicating the risk evaluation. By the time the issues reached the farmers at Tulelake, the question of benefits had become highly politicized.

Press Release. On March 13, 1986, EPA issued a press release on the approved EUPs to Steven Lindow. The two-page document emphasized "the extensive scientific review we have established within the agency," the peer review panel and input from sister agencies. In the aftermath of the AGS embarrassment, the release also stated: "As an added assurance, we have conducted a site visit

and examined the books and records of underlying studies that
support our approval of the experimental field tests."

The EPA avoided the use of analogies, symbols, or references to
relative risk. The rather hackneyed term "minimal risk to public
health and the environment" appeared in the release. The agency
also emphasized that it would have an "authorized observer" at the
site in addition to its monitoring personnel. Unlike the situation
with the AGS tests, no proprietary information was associated with
UC-Berkeley, and this placed fewer constraints on EPA's open
communication strategy.

In contrast to its usual procedures with products it reviews, the
EPA assessment of ice minus was regarded by agency personnel as
somewhat more exhaustive. One EPA scientist described it as a very
labor-intensive process designed to demonstrate that the agency
knows what it is doing and that something can go through the regulatory system. In our view, the agency's efforts at risk communication
for ice minus—a product with high media visibility—were linked to
regaining public confidence in its authority and its fairness.

Local Citizenry, Public Interest Groups, and County Officials

Concerned Citizens of Monterey.   A small number of citizens in
several California counties developed an extensive network of communications that raised questions and issued warnings about the
field test. It started with letter writing and petitions, and within a
short time their actions had a sizable influence on local press coverage of the field tests. Their risk discourse shaped the agenda in
their own terms. In several cases, newspaper opinion and editorial
pages became available to community activists. Glenn Church used
the opinion page of the *San Jose Mercury News* (March 23, 1986) to
frame out the issues. As the first local activist criticizing the field
tests, Church established his legitimacy to the media as a citizen
spokesperson and founder of the group he named Concerned Citizens of Monterey.

Church spoke of "overnight evolution"—an emotionally charged
oxymoron. He used phrases like "far-reaching consequences," spoke
of frost-sensitive plants surviving in new ecosystems and of rainfall
patterns being altered. He questioned EPA's wisdom in waiving
some toxicological studies, characterized EPA's regulatory structure
as primitive, and warned about the effects of other genetically engineered bacteria scheduled for release. Bacteria designed to clean up
oil spills could, he warned, devour petroleum resources; organisms
engineered to degrade wood pulp could "defoliate forests and con-

sume wood houses." Church compared a biogenetic accident to Three Mile Island, Love Canal, and Bhopal. His liberal use of risk simile, notwithstanding, Church contended that he never opposed biotechnology per se; rather his main goal was to insure that communities were involved in the siting process for open field tests.

Lay citizens were more likely than scientists and regulators to draw analogies between ice minus and other environmental events. The issue prompted references to the region's legacy of ecological mishaps. One letter in the *Salinas Californian* summed up the feelings of many nonactivists:

> *The new technology conjures images of science fiction thrillers where the creations of mad scientists threaten the delicate balance of nature and our ecosystem.[19]*

The writer used the dibromochloropropane (DBCP) pesticide case as a reference point for ice minus, despite the differences in their composition and mode of action:

> *The EPA has proven time and again to have a horrendous track record in regulating and testing pesticide products. It was the EPA-approved pesticide DBCP which caused cancer and sterility among male factory workers and which has contaminated the drinking water of thousands of people in San Joaquin Valley.[20]*

Local citizenry often bring historical recollections to a risk event that are of little or no significance to regulators or scientific reviewers. But the relevance of historical precedent, as interpreted by popular culture, sets a different set of boundary conditions on risk communication. As the letter reveals, the cultural standards of relevance may be quite distinct from the technical standards.

International Communications. Within two weeks of the time Monterey County officials were petitioned by local citizenry about the field test, the issue had drawn international attention. Telegrams to the Board of Supervisors from twenty-six members of the European Parliament and twenty-seven members of the West German Parliament (Green Party) warned of the grave dangers of allowing ice minus to be released into the environment. The language was powerful, unambiguous, and dramatic.

The Green Party letter stated that there was insufficient research to ensure safety of the test—a direct contradiction of EPA's position: "Many scientists have asserted that no such methods currently exist for analyzing the environmental impact of genetically-engineered organisms." The letter indicated that Monterey is acting for the world in reviewing this "irreversible experiment" and informed the supervisors that West Germany currently has a regulation prohibiting such releases.

The telegram from the European Parliament members warned of unforeseen and irreversible consequences that could affect the "beautiful landscape of Monterey County" and the rainfall patterns of the region, concluding: "Man has no right to take such grave risks with our nature and our lives." A local Green Party translated the message of its European counterpart into political action by forming coalitions with community activists like Glenn Church.

The Rifkin Influence. Jeremy Rifkin's role in bringing public attention to ice minus stems from his 1983 court challenge to NIH for approving releases of genetically engineered plants and organisms without issuing environmental impact statements. His litigation was reported in the major dailies and wire services. Rifkin's organization was active in supporting the challenges made by Monterey and Tulelake citizens. He provided a body of technical critique of EPA's risk analysis that was cited liberally by the media and used by local groups in their challenge of the ice minus field test.

Rifkin solicited affidavits from ecologists that raised serious questions about ice minus and EPA's review process. These included unresolved questions about the pathogenicity of the organisms and the site selection for the test. With an excellent compilation of internal documents from EPA, Rifkin reported that EPA's own advisers recommended that the tests be carried out in a remote area. He also publicized that certain data requirements on the "environmental fate" of ice minus had been waived, despite recommendations to the contrary by members of EPA's science advisory panel. The EPA spent considerable time responding to Rifkin's legal challenges, such as the February 1986 petition for a preliminary injunction on the AGS field test—eventually denied by a federal court.

The message that came from Rifkin's group was of an agency that disregarded its own advisers, of an experiment that was troubling to eminent scientists, of a review process that neglected some essential questions, and of a test that could have catastrophic consequences for the environment. The mounting collection of lawsuits, agency petitions, and press releases emanating from FOET was the most important single factor shaping the media's reporting of genetic engineering risks. Rifkin crafted the form and content of an anti-establishment risk communication program. Despite his low credibility and *bete noire* status among many leading scientists, with his effective litigation and skillful use of the media, Rifkin shaped the biotechnology debate on his own terms and emerged as the foremost antagonist of the regulatory and scientific sectors.

Concerned Citizens of Tulelake. Two women at Tulelake organized a small grass-roots organization (Concerned Citizens of Tulelake) that opposed the University of California's field test of ice minus. Ava Edgar and Djuanna Anderson, whose involvement began after Glenn Church spoke in Tulelake about his experience with AGS in Monterey County, became consumed in the ice minus issue. They established an extensive network of communications to obtain answers to their questions, and challenged authorities to produce a stronger justification for the proposed field experiment. When they failed to receive satisfactory responses, they lobbied to block the test. They engaged in an extensive letter-writing campaign, attempting to gain recognition and support from the local and national press, elected officials from the city to the White House, regulators, farmers and environmental groups (see Figure 3–2).

With meager resources at their disposal, Edgar and Anderson outlined their concerns to anyone who would listen and to others they believed had an obligation to listen. The letters they sent to EPA assistant administrator John Moore and Senator Alan Cranston provide some of the principal grounds for their opposition to the test.[21]

1. Tulelake, Edgar and Anderson maintained, is situated on one of the richest wildlife areas in the country. When they learned that EPA was not requiring toxicity tests for humans and animals, they responded: "Because of the vast variety of animal and plant life and their interdependency, more pathogenicity tests must be made to protect the overall environment." They believed there was no justification for any pathogenicity test to be waived.

2. Potatoes and grain are the two major crops of Tulelake. Edgar and Anderson suggested that the public might reject Tulelake potatoes when it learned that they had been sprayed by mutant bacteria. By exploiting this possibility, CCT piqued the interest of local farmers who expressed concern that competitors could use information about ice minus to discredit their product. This was the first instance in the public dialogue about ice minus that a negative economic argument gained some currency. Local farmers referred to Temik-contaminated watermelons and the radiation from the Chernobyl nuclear power plant accident in the Soviet Union.

3. Edgar and Anderson claimed that the university did not have adequate liability coverage for the product. (This issue was initially raised by Jeremy Rifkin.)

4. They also claimed that monitoring of the site was inadequate; EPA had no plan to monitor the surrounding area. There was a chance that the airborne bacteria could change the precipitation patterns.

Tulelake Citizens' Communications Network

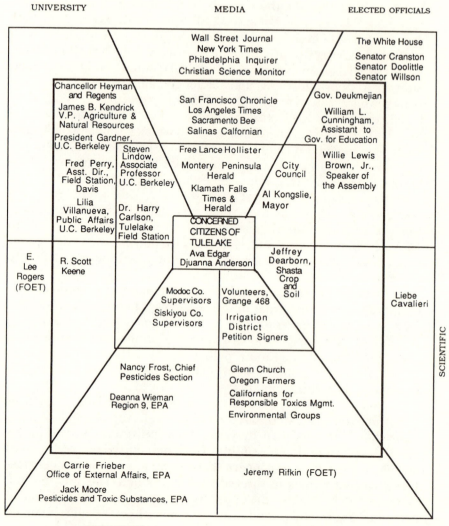

Source: Eileen Schell.

Figure 3–2

Sometimes risks are communicated in unexpected ways. Both CCM and CCT organizers obtained a posting label that EPA required EUP applicants to submit and display at the test site. The notice listed some physical properties of the bacteria, testing parameters, and several warnings, including the following:

- KEEP OUT OF REACH OF CHILDREN
- CAUTION: TOXICOLOGICAL PROPERTIES NOT FULLY INVESTIGATED
- DO NOT ENTER TREATED AREAS UNPROTECTED FOR 12 HOURS AFTER APPLICATION
- APPLICATORS SHOULD WEAR FULL PROTECTIVE CLOTHING INCLUDING GOGGLES AND RESPIRATOR
- APPLY ONLY DURING CALM WEATHER TO AVOID DRIFT

To an average person reading these warnings, the message suggests that a hazardous product is being applied. That is how community organizers interpreted and exploited the message, although that was not the intention of EPA, or UC and AGS. Edgar and Anderson developed their own risk communications, which became the instruments of their protest. They described UC's response to their inquiry as arrogant and imperious. Edgar commented in an interview:

> *The university (UC) is our employee. They should not put us off or belittle us. I recognize "bull" when they try to pass it off. It makes us mad. The university is arrogant. They say, "we want to help." Why then did we need a lawsuit and to take them to court?*[22]

With CCT skeptical of reaching some accord with UC, they pursued support from the farm community.

Tulelake Farmers. The early opposition to ice minus grew out of environmental concerns. However, later opposition expressed by the Tulelake farmers was based on the product's economic liability. Two economic risks were discussed among local farmers. The first had to do with the overproduction of crops as illustrated by the following citation from the *Los Angeles Times* (August 17, 1986).

> *Several growers, pinched between bumper crops and declining demand, have asked whether there is any need for a product that could mean even larger potato harvests in the future, particularly if it alienates consumers.*[23]

Siskiyou County supervisor and potato grower Norma Frey framed the economic risks of ice minus in quantitative terms:

> *Potatoes sold for $5 per 100-pound sack in 1984 but only $2.50 a hundredweight last year and as little as $2.00 this year. Production costs average $4.50 per hundredweight. Agriculture is already in such a backslide. We don't need something else to give us even bigger crops.*[24]

The president of the Tulelake Growers Association challenged the value of the product: "Nobody thinks they need [ice minus] now.

We already have good frost protection. . . . If it was available on the market today, there would be very few people who would use it in this area."[25] A representative of the Grange commented that his organization objected to the test because "the Board of Regents declined to accept full responsibility for any adverse environmental health, or economic impacts."[26]

The second economic risk factor was the possibility of a consumer boycott of Tulelake potatoes. Watermelons contaminated with the pesticide Temik had been withdrawn from the market, and farmers saw themselves facing a similar situation with their potato crop. Another example cited was the radiation-contaminated crops from the Chernobyl nuclear power plant accident. While Tulelake farmers expressed confidence in the UC Agricultural Field Station and the associated scientists, they feared a consumer backlash against what were being heralded as "mutant potatoes." As we shall see, the term "mutant" frequently appeared in the local press to describe ice minus. The CCT promoted use of the term and thereby put forward an emotion-laden risk message. Nevertheless, the farmers were not persuaded by CCT's framing of the ecological risks. Instead, they introduced another axis of contention based upon their economic self-interests.

Print Media.   Local and regional papers of the affected counties in California were the predominant news channels to local residents. How did these papers report the events? What level of risk information was presented? Which voices carried the risk information? What symbols, analogies, and referents were found in these stories?

We have selected three events as reference points for the analysis of the press coverage: (1) the birth of the Monterey County controversy (February 1986); (2) the Tulelake episode (May 1986); and (3) the period just preceding and up to the first field test of ice minus (April 1987). Each event is associated with a peak of media coverage. From that coverage, we chose a cluster of articles for analysis. We looked specifically at the headlines, sentences devoted to risk, individuals quoted on risk, symbols and analogies used, emotive terms, benefits cited, and depth of the risk discussion (0 = low, 3 = high). The data are presented in the appendices to Chapter 3.

Overall, the language and tone of the headlines are not emotionally charged, with the exception of a series of articles on the Tulelake case in local papers where the emotionally loaded term "mutant" appears frequently both in the headlines and in the text. The term was absent in the reporting of the national press and the major California dailies. They used the more neutral terms "genetically altered," "man-made," or "genetically modified" bacteria. The term "mutant bacteria" was used strategically by CCT to arouse a public response to the field test.

In the average story, relatively few column inches were devoted to a discussion of risk. Also, with few exceptions, there was very little depth to the risk discussion; the major dailies with their science writers did not cover risk any more systematically than did the small newspapers. Some local papers gave substantial space to the perceived risks by incorporating extended quotes from letters or reports.

With few exceptions, the primary bearers of risk information to both local and national media formed dichotomous groups: Lay individuals brought the warnings, while scientists carried the "all clear" signal. The local media sought a balance in the reporting of risks by treating community and national lay activists on a par with scientists. Experts cited in response to critics were associated with UC-Berkeley, AGS, or federal agencies. Rarely were independent experts referred to in the stories.

As reported by the print media, the controversy was driven almost exclusively by a discussion of risks. Except for passing reference to the function of ice minus as a frost inhibitor, very little attention was given to benefits. Of the twenty-five articles cited in the January 1986 period, one referred to "huge sums lost to frost damage," and another indicated that Frostban would reduce agriculture's dependency on synthetic chemicals. In the second period (May 1986), the benefits were still downplayed in comparison to the risks, although dollar figures were more visible. Estimates of frost damage to crops cited in press accounts varied enormously: in the United States, from hundreds of millions to $14 billion; worldwide, up to $20 billion. These figures were never referenced, and there was no indication in the reports of dollar savings to farmers or consumers from the use of ice minus. In the month preceding the first field test (April 1987) the balance in reporting began to shift from risks to benefits. Ice minus was cited as an alternative to environmentally unsound or expensive frost protection practices.

Just as the risk discourse dominated the media's presentation of ice minus, it also was the wellspring for most of the symbols and comparisons. We have already discussed the use of the term "mutant" in the Tulelake media. The most common analogies were Chernobyl, radiation-contaminated food, the *Challenger* disaster, and past pesticide problems such as DBCP and Temik. Alternatively, promoters of ice minus compared the bacterial introduction to adding a teaspoon of water into the ocean. Another analogy compared the changes in *Pseudomonas* to removing a key from a piano. Just as the removed key would not change the piano into a trumpet, the deleted gene from *Pseudomonas* would not make it into a new organism.

The recurrent themes in the risk discourse fall into two categories: (1) broad and unspecific images: unforeseen hazards, catastrophic effects, environmental roulette, interference with natural ecosystems, blind faith in technology; and (2) specific scenarios: endanger wild strawberries, alter rainfall patterns, toxic to humans and animals, stigmatize Tulelake potatoes, overproduction of potatoes.

The change in the symbols reported in the press between January 1986 and May 1987 reflects changes in the nature of citizen opposition. While Rifkin was a constant, whatever philosophical critique he held was in the shadow of his litigation strategy, which focused on risk assessment deficiencies. By April 1987, representatives of Earth First and local Greens (U.S. counterparts to the German Greens) were explicit in their categorical imperative "Don't fool with Mother Nature." In response to the new wave of environmental activism, the symbols of humans interfering with the natural order became more prevalent in the media.

Public Relations of AGS and UC. The AGS was not prepared for the hostile public reaction it received to its proposed field test in Monterey County. The company's public credibility was seriously tarnished by the publicity surrounding its unauthorized rooftop tests revealed in the early months of 1986. When AGS approached other communities about siting its field test, it adopted a new public relations strategy. The company developed a press packet with questions and answers about Frostban and introduced a videotape describing the product and the field test. The video was scripted, laid out, and developed by AGS scientists with the help of professional editors and graphic artists. Also, AGS opened up its channels of communication and stated publicly that it would not test its products in any community opposed to the experiment.

AGS's video, prepared before a second site was selected, exemplifies its risk communication strategy. The following is a summary of its message:

- Frostban is a natural approach to frost protection that employs nature's own methods. It is an alternative to synthetic chemicals which are indiscriminate and ecologically destructive. Chemicals not only kill disease-causing bacteria but also eliminate beneficial and benign ones.
- Frostban is a new solution to an age-old problem—frost damage to crops. It is far more desirable on energy and environmental grounds than older methods, i.e., smudge pots, sprinkler irrigation, wind fans, helicopters, biocides.

- Frostban is a model for combating diseases in a wide variety of plants. It will help reduce the buildup of toxic chemicals in the environment.

The video highlights scientists and farmers extolling the virtues and safety of ice minus. It states that AGS conducted 200 tests on Frostban and that there has never been a product as extensively reviewed prior to its field trial. The video portrayal of Frostban was designed to evoke a positive environmental image. It does not defend itself against critics but takes an offensive approach. The message that Frostban is anti-chemical and pro-nature captures the rhetoric of the firm's critics and turns their criticism into an asset.

Public relations for the UC-Berkeley test was initially handled by the superintendent of the Agricultural Field Station at Tulelake and the Berkeley Office of Public Information. News releases from this office were issued from 1983 to 1985. When local opposition arose, public and media relations were transferred to the systemwide news office of the University of California. Periodic advisories to the media on ice minus were issued from this office from May 1986 through April 1987. The centralization of the public relations activities to the news office of the UC president signified a departure from the conventional methods of handling community relations on the various UC campuses. Personnel at the Tulelake Agricultural Station were now receiving directions from the central administration.

The central news office issued press packets that contained EPA's Q&A document, past news releases, selected articles on biotechnology with alternative viewpoints, details about the organism, and EPA's monitoring program. A videotape was also produced and made available to Tulelake residents. In contrast to AGS's sophisticated public relations video production, UC used a cinema verite approach by focusing a camera on Steven Lindow and letting him talk for 45 minutes. In a down-home style, Lindow presented a mixture of scientific jargon and agricultural images. The technical details of the talk probably were in excess of what most lay people could comfortably understand. Logarithmic and three-dimensional graphs, the rapid fire of the monologue, and the sheer density of new information made this an unlikely prospect for effective risk communication. The tape was cut substantially in a second version. The image projected in the tape is of the quintessential agricultural scientist who emerged out of the farming community to engage in scientific studies that improve the lot of farmers, but the detailed science obfuscated some of the intended message. The UC news releases carried the viewpoints of the test's supporters but not necessarily those of members of the UC community with alternative positions. For example, a 1984 news release stated:

The test organism is not something "exotic" or some unknown biological entity that is being introduced into the environment for the first time. . . . Rather it is a natural and native common bacterium that has been "disabled" by having one of its genes taken away.

The image presented by UC's risk communication is that of a natural pest (ice plus) being transformed by scientists into a benign organism (ice minus). The news releases tried to capitalize on the status of the university in stressing the responsibility of its scientists while also emphasizing EPA's oversight of the experiment.

In April 1986, UC notified state and local agencies and the communities surrounding Tulelake of the planned test. A public forum was held on April 17 as part of UC's response to EPA's request that the community be notified before an EUP was issued. The university's legal counsel described their efforts to inform the public:

Throughout these contacts on informational efforts, including the sparsely-attended public forum on April 17, 1986, we have seen no indication of any serious or significant local controversy over the proposed field trial.[27]

Within weeks after the public forum was held, the organizing efforts of residents Edgar and Anderson produced a groundswell of opposition. Their own forum held at the local high school drew about one-quarter of the Tulelake residents. As a result of their petitions, two county boards of supervisors passed resolutions opposing the release. The community's skepticism about the safety of the test eventually was channeled into a lawsuit. Finally, a negotiated settlement between UC and the plaintiffs was reached. What appeared to the university to be an effective direct communication to the community proved to be a failure.

Low public attendance at a forum cannot be interpreted as lack of public concern. People need time to process information about new technologies. Also, local controversies do not just happen through some cognitive process. They are socially constructed. The community's skepticism toward some event or process may lie dormant until social catalysts activate these concerns. In the concluding section, we shall elaborate on a cultural framework for understanding the role of communication during the ice minus episode.

Conclusion

This case study reveals alternative conceptions and expectations of risk analysis and risk communication held by different parties that participated in shaping the political discourse of deliberate release.

In summing up this case analysis, we shall distinguish between scientific and cultural rationality of risk and show how the distinction clarifies the social processes of risk communication in the ice minus case. Finally, we offer some perspective on risk communication based on the empirical observations and a multi-cultural approach to the rationality of risk.

There were clearly some instances in this episode where open, timely, and generous presentation of information could have mitigated conflict and misunderstanding. But that is only part of the story. The development of ice minus arose out of the laboratory from one technical group; its release into the environment was reviewed by a different technical group. These two spheres communicated reasonably well. They could define the required information and process it according to their own needs. The innovation and regulatory spheres share a scientific culture. Of course, it did not always work so smoothly. Ecologists framed the risk issues differently than molecular geneticists. Nevertheless, these groups reached a consensus on ice minus far more readily than the technical and nontechnical sectors.

What factors divided the technical sphere and the popular culture on this issue? Why did some communities support the field tests and some oppose them? What role did risk communication play and what form did it take in this debate?

There are differences between technical and cultural rationality with respect to risk and risk communication. At different nodes in the social history of the ice minus episode, there was information exchange and dialogue but not understanding. Both the technical and cultural spheres depended on symbolic discourse to reveal their level of frustration on failing to reach the other side. Technical rationality views risks according to a set of accepted norms within a disciplinary context. The analysis is narrowly directed—that is, to the product in question. Vague, ambiguous, or philosophical queries have no place in the analytical heuristic. Technical risk is framed in terms of probability and hazard. The cultural context of risk perception is irrelevant, as is the institutional and historical background of the event and analogous risk management efforts.

In the popular culture, the approach to risk is expansive. It searches for historical examples with even mild resemblances to the event in question. It poses questions alien to the technical sphere such as, "Why do we need this product?" Cultural rationality does not defer to the principle of authority or the hierarchy of expertise. There is an appeal to folk wisdom, peer groups, and traditions. Any scientist is an expert; therefore, when scientists disagree, it is reasonable to doubt the optimists. The popular culture embraces many factors relevant to risk that play no role in technical risk analysis, including benefits, protection of proprietary information, historical analogies, and institutional trust.

Expectations about risk communication are intimately related to the forms of rationality related to risk. The technical sphere focuses on the transfer of information relevant to the risk analysis. Generally, this information is shared on demand, since it is rare that scientists or regulators consider it their responsibility to educate.

For the communities involved in ice minus, risk communication symbolized more than "access to information." It represented a demand for informed consent. The political rhetoric of the popular culture stressed control over its environment. Another factor frequently neglected in technical approaches to risk communication relates to processing. When lay people are faced with new technologies, a period of time is required to process information. The processing interval involves peer group discussion and confidence building.

With the distinction between cultural and technical rationality before us, we can illuminate some of the differences in risk information requirements for different actors in the ice minus episode.

Monterey County citizens and public officials were irate when they were told that the site location was secret and they had no authority over the tests. Risk communication to them meant disclosure and consent. It also meant consistency and adherence to principles of common sense. Local citizenry criticized EPA for its aloofness (the distant overseer); for not visiting the first site before an EUP was issued; and for approving a site in a residential area, against the advice of its advisers.

Both Monterey and Tulelake residents questioned the norms that guided the risk assessment. While they received considerable information about the affirmative steps taken in the risk assessment, some citizens were not satisfied until they understood the limits of the data and until they became familiar with norms such as "sufficient evidence." Here are some examples.

The following question was raised on many occasions: Why had EPA waived toxicological tests? Communication was not the issue in this case but rather scientific and regulatory norms. A response to the query by the California Department of Food and Agriculture establishes the point:

> *The conclusion that genetically altered ice minus bacteria are not pathogenic or harmful to man and animals was based on the absence of reported incidents and the fact that these bacteria do not grow at body temperature, rather than on definitive results of actual systemic infectivity and acute toxicological studies.*[28]

Lay people, unschooled in scientific and regulatory tradition, entered a new situation and questioned the rationale of "sufficient evidence." If this is a new and untested technology, they reasoned, why waive any of the tests?

The environmental impact report issued by the University of California was the closest that any of the involved institutions came to responding directly to the concerns of the local community. Many of the questions were not raised in any previous risk assessment. This document produced an inventory of citizen-derived questions with responses by UC. This inventory corroborates the distinction between technical and cultural rationality.

Tulelake citizens pressed for a worst-case scenario. The university's response was that the California Environmental Quality Act (CEQT), unlike the National Environmental Policy Act (NEPA), does not require a worst-case scenario. Thus, we have an unmet demand for information. Tulelake farmers requested information about the economic impacts of ice minus: In what way is it an improvement over existing methods for dealing with frost damage? Who will accept responsibility if there are market losses for Tulelake potatoes? Here again, the demands for information were largely neglected.

The technical analysis focused on the .5 acre test site, whereas citizens kept pushing for broader impact studies: for example, what would be the effect on precipitation if the atmosphere were filled with ice minus? Many seemed frustrated by the constraints placed on the so-called viable questions. The Congressional Office of Technology Assessment (OTA) contracted out two worst-case analytical studies of ice minus on precipitation patterns. Their report, published in the spring of 1988, indicated that "it is unrealistic to expect any significant negative impact on global climatological patterns from large-scale agricultural applications of ice-minus bacteria."[29]

A recurrent dilemma in communication between technical and nontechnical people arises from the language of confidence. It is well known that scientists couch predictions in terms of probability cognates. Citizens responded to the EIR with frustration; they were looking for an explicit "is safe" response with no shades of gray, but they met terms like "unlikely" and "no foreseeable risk." The culture gap reflected in the language of risk could not be more explicit. UC's response to a request for categorical answers was to state that "to date there has been no indication of any negative potential of the experiment to cause harm or have a negative effect."

Scientists and regulators made every effort to focus exclusively on the ice minus field test. When some respondents wished to discuss the wider implications of the test, risk communications within the conventional formats were ineffective. Representatives of the Berkeley Greens, Earth First, FOET, and CCT framed risks in terms of broad philosophical issues. Ice minus was a symbol to them of a new stage in human exploitation of nature. They brought nuclear power and petro-chemicals into the equation. Except for AGS's videotape, little attention was given to the broad ecological implications of the new genetic technologies.

Finally, this case draws our attention to the needs of humans processing information of risk events. People need time to assimilate complex issues. As our analysis indicates, newspapers provide a superficial view of the problem. Moreover, people gain confidence in their own ability to make decisions when they interact with others, both experts and peers, over an extended period. The final EIR issued by UC proved to be a useful means for bringing citizen queries to the attention of both the technosphere and other concerned individuals. The UC responses in the EIR were reasonably accessible to a lay audience. Ideally, once the clearest articulation of the issues has been achieved, the decision process should begin. In fact, the final impact statement came out on February 20, 1987. The release of ice minus took place barely two months later, which provided very little time for the public to evaluate the intricate responses by UC to many concerns. Here again, the multi-cultural model explains the role of the EIR as a medium of risk communication. For UC, the EIR was a means of discharging its responsibility under CEQA. It was the finale. For the citizens it provided the first opportunity for a reasonably full accounting of the issues—which some might view as the opening scene.

Endnotes

1. Steven E. Lindow, "The Role of Bacterial Ice Nucleation in Frost Injury to Plants," *Annual Review of Phytopathology* 21:363–384 (1983).
2. Department of Health, Education and Welfare, National Institutes of Health, *Guidelines for Research Involving Recombinant DNA Molecules*, June 23, 1976.
3. Department of Health and Human Services (DHHS), National Institutes of Health (NIH), "Guidelines for Research Involving Recombinant DNA Molecules," *Federal Register* 47(77):17180–17198 (April 21, 1982).
4. Environmental Protection Agency, "Microbial Pesticides: Interim Policy on Small Scale Field Testing," *Federal Register* 49(202):40659–40661 (October 17, 1984).
5. Glenn Church et al., letter to Monterey Board of Supervisors, January 3, 1986.

6. University of California, Division of Agriculture and Natural Resources, *Ice Nucleation Minus Research Field Test. Final Environmental Impact Report,* February 1987, 181 pp.

7. DHHS, NIH, "Guidelines for Research Involving Recombinant DNA Molecules."

8. Sheldon Krimsky, *Genetic Alchemy* (Cambridge, Mass.: MIT Press, 1982).

9. E. Milewski and B. Talbot, "Proposals Involving Field Testing of Recombinant DNA Containing Organisms," *Recombinant DNA Technical Bulletin* 6(4):141–145 (December 1983).

10. Office of Science and Technology Policy, Executive Office of the President, "Coordinated Framework for the Regulation of Biotechnology," *Federal Register* 51(123):23302–23392 (June 26, 1985).

11. Office of Science and Technology Policy, Executive Office of the President, "Coordinated Framework for the Regulation of Biotechnology: Establishment of the Biotechnology Science Coordinating Committee," *Federal Register* 50(220):47174–47176 (November 14, 1985).

12. Sheldon Krimsky, "Gene Splicing Enters the Environment: The Socio-Historical Context of the Debate over Deliberate Release," in John R. Fowle III (ed.), *Application of Biotechnology: Environmental and Policy Issues* (Boulder, Col.: Westview, 1987).

13. Frederick S. Betz, U.S. EPA Hazard Evaluation Division, Science Integration Staff, memorandum to Tom Ellwanger, Technical Support Section, Registration Division. Hazard Evaluation Division position on Lindow EUP applications, March 27, 1985.

14. Steven Schatzow, Director, Office of Pesticide Programs. Memorandum to John A. Moore, Assistant Administrator for Pesticides and Toxic Substances, May 12, 1986.

15. Ibid.

16. EPA, Hazard Evaluation Division, "HED Scientific Position on University of California (Lindow) Notification of Intent to Field Test a Genetically Engineered Microbial Pesticide," March 15, 1985.

17. EPA, Hazard Evaluation Division, "HED Position on Lindow EUP Applications," March 27, 1986.

18. Steven Cull, AGS plant pathologist, quoted in "Lawsuit Filed to Stop Bacteria Experiment," *Monterey Peninsula Herald,* January 19, 1986.

19. William W. Manning, "Concerned Justified about Frostban" (letter), *Salinas Californian,* January 22, 1986.

20. Ibid.

21. Ava L. Edgar, Concerned Citizens of Tulelake, letter to Jack [John] Moore, EPA Assistant Administrator for Pesticides and Toxic Substances, May 9, 1986; Ava L. Edgar, letter to Senator Alan Cranston, May 21, 1986.

22. Ava L. Edgar, interview, March 20, 1987.

23. Mark A. Stein, "The Worries Build over the Release of Bacteria Altered in the Laboratory," *Los-Angeles-Times,* August 17, 1986.

24. Ibid.

25. Elliot Diringer, " 'Ice-minus' Creator Has His Day," *San Francisco Chronicle,* April 30, 1987.

26. University of California, Division of Agriculture and Natural Resources, *Ice*

Nucleation Minus Research Field Test. Final Environmental Impact Report, February 1987. Transcript of public hearing, January 10, 1987, p. 146.

27. William A. Anderson, II, letter to Information Service Section, Office of Pesticide Programs, EPA, April 21, 1986.

28. Biotechnology Working Group, Division of Pest Management, California Department of Food and Agriculture, memorandum to Tobi Jones, Program Supervisor, Pesticide Registration Branch, CDFA, January 28, 1987.

29. U.S. Congress, Office of Technology Assessment, *New Developments in Biotechnology 3. Field Testing Engineered Organisms: Genetic and Ecological Issues* (Washington, D.C.: Government Printing Office, May 1988), p. 95.

Sources

Interviews

DJUANNA ANDERSON, Concerned Citizens of Tulelake, March 20, 1987.

FRED BETZ, EPA, Hazards Evaluation Division, May 25, 1987.

HARRY CARLSON, University of California Agricultural Field Station at Tulelake, February 3, 1987.

DAVID CHRISTY, *Herald & News* (Klamath Falls, Oregon), March 25, 1987.

GLENN CHURCH, Concerned Citizens of Monterey County, February 19, 1987.

EDWARD J. DANOSKY, Secretary, Tulelake Irrigation District, March 20, 1987.

AVA EDGAR, Concerned Citizens of Tulelake, February 25, 1987; March 20, 1987.

SAM KARAS, Monterey Board of Supervisors, March 23, 1987.

BRAD KEEFER, Freelance (Hollister), February 23, 1987.

DAVID KINGSBURY, Biotechnology Science Coordinating Committee (taped talk), April 19, 1987.

JULIANNE LINDEMANN, Advanced Genetic Sciences, April 22, 1987.

STEVEN LINDOW, UC-Berkeley, April 23, 1987.

WALLACE RAVVEN, UC-Berkeley, Public Relations, April 22, 1987.

JANE RISSLER, EPA, OPTS, May 21, 1987.

PATRICIA ROBERTS, EPA Legal Affairs, May 1, 1987.

TREVOR SUSLOW, Advanced Genetic Sciences, April 22, 1987.

LILIA VILLANUEVA, University of California Public Relations March 11, 1987.

WALTER WONG, Monterey Department of Environmental Health, March 24, 1987.

Videotapes

UNIVERSITY OF CALIFORNIA, *Biological Frost Control Experiment, Tulelake Field Station, Tulelake, CA.*

ADVANCED GENETIC SCIENCES, *A Natural Approach to Frost Protection*, circa 1986.

KGO-TV NEWS (SAN FRANCISCO): News coverage of the ice minus field test, April 24, 1987.

KQED-TV, "Express: Genetic Engineering," March 19, 1986.

U.S. ENVIRONMENTAL PROTECTION AGENCY, *Should Genetically Modified Bacteria Be Released into the Environment?* May 14, 1984.
U.S. ENVIRONMENTAL PROTECTION AGENCY, "Announcement approving the testing of frost damage preventing bacteria," Press briefing. November 14, 1985.
PBS-TV, McNeil/Lehrer, "Deliberate Release Experiment," May 4, 1984.
KRON-TV NEWS (SAN FRANCISCO), News coverage of the ice minus field test, April 24, 1987.

Chronology

1982

September 17: Steven E. Lindow and Nickolas J. Panopoulos submit a proposal to the Recombinant DNA Advisory Committee (RAC) of the National Institutes of Health (NIH) to field test strains of *Pseudomonas syringae* and *Erwinia herbicola* derived by rDNA techniques. The organisms are called ice minus to signify the deletion of the ice nucleation gene.

October 25: RAC reviews Lindow-Panopoulos field test proposal for ice minus; proposal is approved with split vote and forwarded to NIH director.

1983

January 10: NIH director rejects the Lindow-Panopoulos proposal pending additional data.

March 8: Lindow and Panopoulos submit to RAC a revised proposal for testing ice minus.

April 11: RAC approves the Lindow-Panopoulos ice minus field test with 19 affirmative, 0 negative, and 0 abstentions.

September 14: Jeremy Rifkin, Foundation for Economic Trends (FOET), files suit against NIH, charging violations of the National Environmental Policy Act and the Council on Environmental Quality in the agency's approval process for ice minus field tests.

1984

February 6: Rifkin presents a petition to RAC requesting disapproval of field tests of ice minus. Affidavits from ecologists support the petition.

March 22: Advanced Genetic Sciences (AGS) submits field test proposal to NIH/RAC; the firm is not legally required to seek NIH approval.

May 16: Judge John Sirica, U.S. District Court for the District of Columbia, enjoins NIH from approving UC field tests pending the outcome of the Rifkin suit; the Lindow-Panopoulos field test scheduled for May 25 is postponed.

June 1: RAC meets to evaluate the AGS proposal.

June–October: RAC working group reviews AGS submission.

October 17: EPA interim policy statement entitled "Microbial Pesticides; Interim Policy on Small Scale Field Testing" is published in the *Federal Register*.

October 18: NIH notifies AGS of field test approval pending further data and monitoring plan.

October 31: AGS submits a description of proposed ice minus field trial to EPA.

November 5: AGS notifies EPA of its field test proposal.

November 15: AGS submits additional data to NIH/RAC.

November 1984–January 1985: Hazard Evaluation Division & Scientific Advisory Subpanel of EPA and the Intra-Agency Working Group on Biotechnology review AGS notification to EPA.

December 31: Office of Science & Technology Policy publishes a coordinated framework for the regulation of biotechnology.

1985

January 21: NIH issues Environmental Assessment (EA) and a finding of No Significant Impact (FONSI) for the Lindow-Panopoulos field test experiments with ice minus.

January 10: AGS furnishes additional data to EPA.

January 22: EPA's Scientific Advisory Subpanel meets.

January 30: Hazard Evaluation Division of EPA completes its assessment of AGS's proposed experiments based on a review of additional data.

February 1: EPA notifies AGS that an Experimental Use Permit (EUP) is required prior to conducting field tests for genetically modified organisms. EPA outlines additional data requirements.

February 15: The Hazard Evaluation Division (HED) of the EPA rules on the Lindow-Panopoulos experiment: the risks are slight, but the evidence is insufficient to proceed.

February–July: AGS conducts additional research to provide EPA with requested information.

February 27: U.S. Court of Appeals for the District of Columbia rules that NIH may approve experiments releasing genetically engineered microbes, but only after formal environmental assessments are conducted.

March 15: HED recommends requiring EUPs for the Lindow-Panopoulos experiment.

April 15: NIH publishes *Federal Register* notice of the availability of the agency's Environmental Assessment of the Lindow-Panopoulos experiment; EPA requests public comment and the evaluation of the need for a programmatic EIS.

April 4–May 17: NIH receives written comments about NIH's Environment Assessment; of 14 letters received, 13 support the EA FONSI and 1 criticizes it.

July 2: AGS submits two EUP applications to EPA with additional supporting data for field tests of ice minus.

August 21: EPA publishes *Federal Register* notice stating receipt of EUP's from AGS and soliciting public comment.

August 27: EPA's Hazard Evaluation Division releases its preliminary scientific position on the AGS field test.

September–October: USDA, FDA, NIH, SAP subpanel, and Intra-Agency Working Group on Biotechnology review EUP materials on AGS test.

November 5: HED responds to comments from various review panels and the public about the AGS EUP.

November 14: EPA holds press briefing announcing that AGS will be granted two EUPs for the Monterey County test site.

December 4: Congressional hearings focusing on planned releases of genetically altered organisms are held before the Subcommittee on Investigations and Oversight, Committee on Science and Technology, U.S. House of Representatives.

December 5: EPA issues EUPs to AGS for conducting ice minus field tests.

December 17: At a legal status conference, NIH agrees to follow NEPA for recombinant DNA and the University of California (UC) agrees not to challenge the EPA's decision to require an EUP; UC also agrees to give the plaintiffs 30 days notice prior to conducting the experiment.

December 30: Lindow-Panopoulos submit EUP applications to EPA.

1986

January 3: Monterey County residents ask Board of Supervisors to request that AGS obtain a county permit for the ice minus field test.

January 14: AGS field test is discussed before the Monterey County Board of Supervisors.

January 15: Monterey County Board of Supervisors receives a communication from 27 members of the West German Parliament opposing the field test of ice minus.

January 27: The Monterey Board of Supervisors holds a public hearing to discuss the AGS test. AGS voluntarily delays the test 30 days to allow the county to study the issues.

February 11: Monterey County establishes a 45-day moratorium for the AGS field test.

February 20: Monterey County officials meet in Sacramento with Assemblyman Sam Farr and officials of the State Departments of Commerce, Health Services, Food and Agriculture and the Governor's Interagency Task Force on the Biotechnology Industry.

February 26: Allegations of unauthorized rooftop testing of ice minus organisms are disclosed by Jeremy Rifkin to the *Washington Post* and *New York Times*.

AGS acknowledges testing a genetically engineered microbial pesticide in the open air in February 1985 without experimental use permits from EPA.

Rifkin petitions a federal court for a preliminary injunction against EPA approval of AGS's open field test.

February 28: EPA Region IX initiates an investigation of AGS to determine and document the facts related to the outdoor testing, greenhouse testing, and proposed field testing of two strains of bacteria whose ice-nucleating genes were removed.

March 4: Congressional hearings are held about ice minus by the Subcommittee on Oversight and Investigations, Committee on Science and Technology, U.S. House of Representatives.

An EPA headquarters inspection team initiates a facilities inspection and data audit of AGS.

March 6: EPA Region IX inspectors and EPA Headquarters inspectors complete their investigations of AGS.

A federal judge rejects Rifkin's petition for a preliminary injunction against AGS's field test.

March 8: Monterey County officials discover the undisclosed location of AGS's test site.

March 14: AGS volunteers to repeat tree pathogenicity tests.

March 18: Monterey County extends the moratorium on ice minus field tests.

March 21: EPA completes the AGS enforcement case investigation. The review finds (1) falsification of EUP applications; (2) failure to test in accordance with EUP requirements.

March 24: EPA suspends EUPs of AGS and issues an administrative civil complaint to the firm. *Federal Register* notice appears for Lindow-Panopoulos EUP application.

Monterey County Board of Supervisors pass interim one-year moratorium on field testing Frostban (AGS's trade name for ice minus), part of a temporary land-use ordinance restricting any field testing of genetically engineered substances.

March 25: California Department of Food and Agriculture (CDFA) suspends AGS's field test permit.

April 1: Monterey Country Supervisors instruct the county planning commission to amend the land-use regulations to include experiments involving genetically altered organisms.

April 10: EPA completes review of repeated fruit and nut tree pathogenicity tests by AGS.

April 17: EPA personnel inspect the UC Agricultural Field Station at Tulelake, Calif.; UC holds a public meeting about the ice minus test.

April 21: In federal district court, plaintiffs, including FOET, agree to vacate the preliminary injunction preventing NIH from approving field tests with genetically engineered organisms and UC from proceeding with its field test on the condition that UC obtain EPA approval and that they give the plaintiffs 30 days' prior notice.

May 7: EPA gives AGS final approval for repeated tree pathogenicity tests.

May 12: EPA decision memorandum grants the EUP for UC experiment.

EPA issues document entitled "Questions and Answers on the Lindow Experiment."

May 13: UC representatives appear before the Siskiyou County Board of Supervisors to discuss the proposed ice minus field experiment.

June 2: Modoc County Board of Supervisors passes a resolution opposing the UC ice minus experiment; representatives of UC are present to provide information.

June 6: EPA and AGS reach a settlement agreement of Administrative Complaint. All charges of data falsification are removed. AGS pays a $13,000 fine for procedural violations.

June 11: Siskiyou County Board of Supervisors passes a resolution opposing the UC ice minus experiment; representatives of UC are present to provide information.

June 24: AGS submits data from repeated fruit and nut tree pathogenicity tests.

June 26: Biotechnology Science Coordinating Committee, NSF, EPA, USDA,

FDA, APHIS contribute to the Coordinated Regulatory Framework for Biotechnology in *Federal Register*.

July 23: UC issues a press release stating its intention to conduct the ice minus field test on August 6.

August 1: Californians for Responsible Toxics Management (CRTM), Concerned Citizens of Tulelake, and FOET apply to Judge Ford, California Superior Court, for a restraining order to prevent UC from planting seeds treated with ice minus bacteria. The application is denied; a second application is sent to Judge Backus.

August 4: CRTM et al. apply to Judge Backus, California Superior Court, for a restraining order to halt the UC ice minus field test.

August 6: Residents of Tulelake, CRTM, and FOET obtain a temporary restraining order from the California Superior Court blocking the ice minus field test for 19 days.

August 19: A negotiated agreement is reached between UC, Tulelake citizens, and CRTM. Under the settlement, UC will conduct a further environmental review before proceeding with the Lindow-Panopoulos experiment.

September 1: CDFA adopts revised regulations and permit procedures for biotechnology drafted by the Governor's Interagency Task Force on Biotechnology.

September 4: EPA reviews AGS's pathogenicity tests for ice minus and gives conditional approval for EUP pending site assessment.

September 18: UC representatives meet in Berkeley with attorneys for Tulelake residents for an EIR scoping session.

October 16: UC holds public meeting in Tulelake to address community concerns in response to the EIR.

October 25: UC sends notices describing the proposed field test and issues identification sheets to 2,500 post office addresses in Tulelake, California, and Merrill and Malin, Oregon.

December 17: Board of Supervisors and Agricultural Commissioners in San Benito and Contra Costa counties are notified of AGS's intent to apply for EUP at different sites.

December 24: AGS notifies EPA and CDFA of site selections for ice minus test in San Benito and Contra Costa counties.

1987

January 5: San Benito County Board of Supervisors appoints Frostban Investigation Committee.

January 7: AGS notifies EPA and CDFA of its interest in testing Frostban at two sites: one in San Benito and the other in Contra Costa County.

January 9: Jeremy Rifkin announces plans to derail AGS's test by mobilizing local opposition and possibly filing a lawsuit.

January 10: UC holds a hearing in Tulelake to hear public comments on the draft EIR. Brentwood residents in the neighborhood of test site are invited to information meeting.

January 20: The Modoc County Board of Supervisors withdraws its opposition to the UC field test of ice minus in concept but urges that alternative sites be "vigorously pursued and assessed."

January 26: San Benito citizens and some supervisors tour AGS facility in Oakland.

January 27: The Siskiyou County Board of Supervisors acknowledges the potential benefits of the UC ice minus test but urges pursuit of alternative sites. EPA, CDFA, and county agricultural commissioner conduct site inspections at Contra Costa and San Benito sites.

AGS makes public presentation to the Brentwood City Council on Frostban.

January 29: San Benito County Counsel David Pipal issues an opinion that the county is not permitted to regulate the Frostban test. A public workshop on Frostban is held by the San Benito County Board of Supervisors.

January 30: AGS makes presentation to Contra Costa Farm Bureau.

February 2: AGS VP John Bedbrook tells Monterey County supervisors that the company will drop all plans for its first test site location and concentrate on two possible sites located in a remote valley about 20 miles southeast of Hollister, Calif.

A public hearing on Frostban held at San Benito County Board of Supervisors' meeting. Sparce attendance by community prompts call for a second hearing.

February 8, 13, 16: AGS holds public workshops, presentations, and open house about Frostban and biotechnology directions and benefits.

February 9: San Benito Farm Bureau adopts resolution supporting Frostban testing in the county.

February 11: EPA approves three test sites for AGS's ice minus field experiment: (1) a San Benito County site; (2) an 86,000-acre cattle ranch in the Quien Sabe Valley; (3) a Contra Costa County location between the cities of Brentwood and Byron.

The Contra Costa Agricultural Resources Advisory Committee adopts a resolution supporting Frostban testing in Contra Costa County; the San Benito Cattleman's Association adopts resolution supporting Frostban testing in county.

February 12: The San Benito Chamber of Commerce adopts a resolution supporting Frostban testing in the county.

February 18: The San Benito county supervisors endorse the test on the Quien Sabe Valley site. Although the county supervisors' endorsement is not required, AGS's stated policy is not to test where there is community opposition.

February 19: CDFA issues tentative approval for all three sites previously approved by EPA. EPA and CDFA are the only regulatory agencies in California with jurisdiction over Frostban. Final approval awaits a 45-day public review period ending April 5, 1987.

February 20: UC issues final EIR for the ice minus field test. Public review period begins and continues until March 7, 1987.

February 26: The directors of the California Farm Bureau Federation adopt a resolution supporting Frostban testing.

March 3: A community forum on the ice minus test is held in Brentwood, Calif.

March 10: The Brentwood City Council adopts a resolution of the Agricultural Resources Advisory Committee supporting Frostban testing at a Contra Costa County site near the city limits.

March 17: The Contra Costa County Board of Supervisors adopts a resolution of the Agricultural Resources Advisory Committee supporting a Frostban test near Brentwood.

AGS notifies EPA and CDFA of the selected site for the Frostban test near Brentwood.

April 2: AGS notifies EPA and CDFA of its intent to initiate preapplication monitoring.

April 13: CDFA approves AGS Research Authorization for the field test of Frostban.

April 23: California Superior Court Judge Darrel W. Lewis denies a petition for a preliminary injunction to block AGS's field test of Frostban in Brentwood. Groups supporting the legal challenge include Earth First, the Berkeley Greens, and the Foundation on Economic Trends.

April 24: AGS sprays Frostban on strawberry plants in Brentwood, representing the first authorized release of a genetically engineered bacterium into the environment.

April 29: UC-Berkeley scientists initiate a field test of ice minus at the Tulelake Agricultural Field Station.

May 12: The Monterey County Board of Supervisors passes an ordinance regulating the location and siting of experiments involving the release of genetically engineered microorganisms into the open environment.

Acronyms

AGS	Advanced Genetic Sciences
APHIS	Animal and Plant Health Inspection Service (USDA)
BSCC	Biotechnology Science Coordinating Committee
CCM	Concerned Citizens of Monterey
CCT	Concerned Citizens of Tulelake
CDFA	California Department of Food and Agriculture
CEQT	California Environmental Quality Act
CRTM	Californians for Responsible Toxics Management
DBCP	Dibromochloropropane
DHHS	Department of Health and Human Services
DNA	deoxyribonucleic acid
EIR	environmental impact review
EPA	Environmental Protection Agency
EUP	Experimental Use Permit
FDA	Food and Drug Administration
FIFRA	Federal Insecticide, Fungicide, and Rodenticide Act
FOET	Foundation on Economic Trends
FONSI	Finding of No Significant Impact
GEM	Genetically Engineered Microorganisms
HED	Hazard Evaluation Division
INA	Ice Nucleation Active
NEPA	National Environmental Policy Act
NIH	National Institutes of Health
NSF	National Science Foundation

OPP   Office of Pesticide Programs
OSTP   Office of Science and Technology Policy
OTA   Office of Technology Assessment
RAC   Recombinant DNA Advisory Committee (NIH)
rDNA   Recombinant DNA
SAP   Science Advisory Panel (EPA)
TSCA   Toxic Substances Control Act
UC   University of California
USDA   United States Department of Agriculture

Selected Bibliography

ALEXANDER, MARTIN. "Ecological Consequences: Reducing the Uncertainties." *Issues in Science and Technology* 1(3):57–68 (Spring 1985).

BRILL, W. J. "Safety Concerns and Genetic Engineering in Agriculture." *Science* 227:381–84 (January 25, 1985).

DAVIS, B. D. "Bacterial Domestication: Underlying Assumptions." *Science* 235:1329, 1332–35 (March 13, 1987).

DEATHERAGE, SCOTT. "Scientific Uncertainty in Regulating Deliberate Releases of Genetically Engineered Organisms: Substantial Judicial Review and Institutional Alternatives." *Harvard Environmental Law Review* 11(2):203–46 (1987).

FOWLE III, JOHN R., ed. *Application of Biotechnology: Environmental and Policy Issues*. Boulder, Colorado: Westview Press, 1987.

HALVORSON, H. O., D. PRAMER, AND M. ROGUL, eds. *Engineered Organisms in the Environment: Scientific Issues*. Washington, D.C.: American Society for Microbiology, 1985.

HARLOW, RUTH E. "The EPA and Biotechnology Regulations: Coping with Scientific Uncertainty." *Yale Law Review Journal* 95: 553–76 (1986).

McGARITY, THOMAS O. "Regulating Biotechnology." *Issues in Science and Technology* 1(3):40–56 (Spring 1985).

NATIONAL ACADEMY OF SCIENCES, COMMITTEE ON THE INTRODUCTION OF GENETICALLY ENGINEERED ORGANISMS INTO THE ENVIRONMENT. *Introduction of Recombinant DNA-Engineered Organisms into the Environment: Key Issues*. Washington, D.C.: National Academy Press, 1987.

SHARPLES, F. E. "Regulation of Products from Biotechnology." *Science* 235:1329–32 (March 13, 1987).

TEICH, A. H., MORRIS A. LEVIN, AND J. H. PACE, eds. *Biotechnology and the Environment: Risk and Regulation*. Washington, D.C.: American Association for the Advancement of Science, 1985.

U.S. CONGRESS, OFFICE OF TECHNOLOGY ASSESSMENT. *New Developments in Biotechnology 2. Background Paper. Public Perceptions of Biotechnology*. Washington, D.C.: Government Printing Office, May 1987.

U.S. CONGRESS, OFFICE OF TECHNOLOGY ASSESSMENT. *New Developments in Biotechnology 3. Field Testing Engineered Organisms: Genetic and Ecological Issues*. Washington, D.C.: Government Printing Office, May 1988.

U.S. HOUSE OF REPRESENTATIVES, COMMITTEE ON SCIENCE AND TECHNOLOGY, SUBCOMMITTEE ON INVESTIGATIONS AND OVERSIGHT. *The Environmental Implication of Genetic Engineering*. Washington, D.C.: Government Printing Office, 1984.

KEY TO NEWSPAPERS

BC   = Berkeley Californian
BG   = Boston Globe
BN   = Brentwood News
BS   = Baltimore Sun
CP   = Capital Press (Salem Oregon)
CCT   = Contra Costa Times
CSM   = Christian Science Monitor
DL   = Daily Ledger (Brentwood)
F   = Fortnighter (N. Monterey County)
H&N   = Herald & News
LAT   = Los Angeles Times
LRS   = Lost River Star
MCR   = Modoc County Record
MPH   = Monterey Peninsula Herald
NYT   = New York Times
RRS   = Redding Record Searchlight
SB   = Sacramento Bee
SC   = Salinas Californian
Science = Science (AAAS)
SFC   = San Francisco Chronicle
SFE   = San Francisco Examiner
SJMN = San Jose Mercury News
USAT = USA Today
WP   = Washington Post
WSJ   = Wall Street Journal

KEY TO NAMES CITED

Djuanna Anderson, Concerned Citizens of Tulelake (CCT).
Ava Edgar, CCT.
Tino Bacchino, farmer, Tulelake.
John Bedrock, Advanced Genetic Sciences (AGS).
Frederick S. Betz, Environmental Protection Agency (EPA).
Andy Caffrey, Berkeley Greens.
Julio Calderon, AGS.
Harry Carlson, UC Agricultural Field Station at Tulelake.
Liebe Cavalieri, Sloan Kettering Institute.
Glenn Church, Concerned Citizens of Monterey (CCM).
Robert Colwell, UC-Berkeley
Perry Continente, resident, Tulelake.
Bruce Crowley, U.S. General Accounting Office.
Steven Cull, AGS.
Jeff Dearborn, crop and soils consultant, Tulelake.
Donald Durzan, UC-Davis
Norma Frey, Siskiyou County Supervisor.
Alan Goldhammer, Industrial Biotechnology Association.
Jeff Hoffman, Earth First.
Terry J. Houlihan, AGS.
Lowell Kenyon, potato grower, Tulelake.
Patti Jackson, Siskiyou County Supervisor.
Tobi Jones, California Department of Food & Agriculture (CDFA).
Georgette Kirby, resident, Tulelake.
Darrel Lewis, judge, California Superior Court.
Julianne Lindemann, AGS.
Steven Lindow, UC-Berkeley.
Lawrence Odle, executive officer, Monterey Bay Air Pollution Board.
Leon Panetta, state representative, California.
Nickolas Panopoulos, UC-Berkeley.
Marc Del Piero, Monterey County Supervisor.
Wallace Ravven, UC-Berkeley
Jeremy Rifkin, Foundation on Economic Trends (FOET).
Lee Rogers, counsel, FOET.
Ray Seidler, EPA.
Trevor Suslow, AGS.
Jeff Torlakson, Contra Costa Supervisor.
Walter Wong, Monterey County Department of Environmental Health.
Roger Zwanziger, Siskiyou County Supervisor.

Media Analysis of Ice Minus: January 1986

	SOURCE	DATE	STORY HEADLINE	#SENT CITED ON RISK	#PEOPLE CITED ON RISK	#SCIENT CITED ON RISK	NAMES CITED	DEPTH OF ANALYSIS 0–3	SYMBOLS ANALOGIES EMOTIVE EXPRESSIONS	BENEFITS CITED	RISK AND OTHER CONSIDERATIONS
1	MPH	1-7-86	Altered-bacteria test plan gets opposition.	3	3	0	Church; Rifkin; Rogers; CDFA;	0	Valley used as guinea pig		Pathogenicity.
2	MPH	1-8-86	Supervisors order review of altered-bacteria testing.	3	3	0	Church	2			Bacteria might spread; affect rainfall patterns; side effects on the environment.
3	MPH	1-9-86	County official urges relocating altered-bacteria test.	0	0	0		0			
4	SC	1-8-86	Gene test causes North County protest.	3	1	0	EPA; Church	0			
5	SJMN	1-14-86	Synthetic bacteria test set, Monterey County frets about effects.	2	2	0	Odie; Wong	0			Bacteria might spread.
6	MPH	1-14-86	Plan to test altered-bacteria in North County causes stir.	3	1	0	Wong	0	Pathogenic strains		One of the strains marginally pathogenic; high reproduction in H2O.
7	SFC	1-15-86	Genetic experiment worries Salinas.	0	0	0		0			
8	SC	1-15-86	Planned bacteria test in County hits snag.	2	1	0	Del Piero	1			One strain of P. syringae is pathogenic; bacteria multiplies rapidly in rainfall.
9	WP	1-17-86	Microbe test hits new snag.	1	1	0	Rifkin	1		Huge sums lost to frost damage.	
10	SFC	1-18-86	Court rule on genetic engineering experiment.	5	2	1	Rifkin; Suslow	0	Low probability, high risk		
11	MPH	1-19-86	Lawsuit filed to stop bacteria experiment.	5	2	1	Rogers; Cull				
12	F	1-24-86	North County residents express their concerns.	0	0	0		0			

#	Source	Date	Headline				Names				
13	MPH	1-24-86	Hearing on bacteria test draws international interest.	3	0	0	Members European & West German Parliament	0	Novel life forms		Unforeseen hazards; affect beautiful landscape of Monterey; irreversible.
14	MPH	1-28-86	Supervisors move to control release of bacteria test.	3	3	3	Suslow; Cavalleri; Lindemann	1		Lessen dependency on synthetic chemicals.	Potentially catastrophic microbes will overcome competitors.
15	SFC	1-28-86	Oakland genetics firm shifts controversial bacteria test.	5	5	4	Rifkin; Betz; Suslow; Cavalleri; Lindemann	0	Exotic organisms		Unknown effects on the environment.
16	SC	1-28-86	Supervisors try to freeze Frostban.	16	4	2	Suslow; Church; Lindemann; Rogers	3	Environmental roulette; adding a teaspoon of H2O to the ocean.		Rainfall patterns; toxicity; unpredictable effects; bacteria irritated the eyes of rabbits.
17	Science	2-14-86	Local opposition halts biotechnology test.	9	6	3	Betz; Jones; Cavalleri; Rogers; Rifkin	3			Human health; rainfall pattern altered; irritation of rabbits' eyes.
18	CSM	3-18-86	Debate over genetic antifreeze for crops.	3	1	1	Bedrock	0			No frost problem in Monterey; loss of market advantage; Ice minus can't compete
19	SJMN	1-8-86	No-Frost testing probe.	4	1	0	Church	0	Unknown hazards.		
20	F	1-20-86	Release of man-made bacteria alarms north county residents.	4	1	0	Church	1	Unknown hazards.		
21	SC	1-15-86	Public hearing ordered on plans for bacteria test.	2	1	1	Wong	0	Pathogenic bacteria.		
22	MPH	1-16-86	Smog board questions gene test.	1	1	0	Odie				Harmful bacteria released.
23	MPH	1-22-86	Panetta ask cancellation of bacteria test permit.	1	1	0	Panetta	0			
24	SJMN	1-28-86	Local protest: genetic firm holds off on spraying strawberries	5	1	1	Wong; state health official (unknown)	1	Environmental roulette.		Bacteria irritated eyes of rabbits in toxic study; bacteria might spread to other crops.
25	BS	2-11-86	Age of genetic engineering and strawberry experiment are put on hold in California.	1	0	0		0	Environmental roulette; gene experiments likened to dawn of nuclear age.		Hurt the healthful image of strawberries.

Media Analysis of Ice Minus: May 1986

	SOURCE	DATE	STORY HEADLINE	# SENT ON RISK	# PEOPLE CITED ON RISK	# SCIENT CITED ON RISK	NAMES CITED	DEPTH OF ANALYSIS 0-3	SYMBOLS ANALOGIES EMOTIVE EXPRESSIONS	BENEFITS CITED	RISK AND OTHER CONSIDERATIONS
1	H&N	5-2-86	Resistance growing to mutant bacteria plan.	6	2	1	Lindow; Church	1	Chernobyl; mutant bacteria.	Frost protection for plants	Unforeseen changes; over-production of potatoes.
2	H&N	5-?-86	Mutant bacteria questions raised.	13	3	1	Anderson;	1	Removing a single piano key does not make it a trumpet.		Markets for Tulelake potatoes could be affected.
3	MCR	5-8-86	Tulelake residents question bacteria release.	13	2	0	Lindow; Church; Edgar; Church; U.C. Scientist; Dearborn	2	Mutant bacteria; first open air release.		
4	H&N	5-6-86	Bacteria release plan questioned.	18	2	2	Lindow	3	First open air release of rDNA altered life form.		
5	SFC	5-9-86	U.C. plan to test anti-frost bacteria stirs up 2 counties.	1	1	0	Edgar	0			
6	H&N	5-9-86	Mutant bacteria talks to be Monday.	0	0	0			First open air release.		
7	USAT	5-12-86	Debate whirls around lab "creatures."	7	2	1	Crowley; Goldhammer	1	Creatures; tinkering with the natural order.		Man-made creature could throw our natural order into chaos.
8	LAT	5-13-86	Ok expected for test of engineered bacterium.	0	0	0					Liability insurance for the release.
9	LAT	5-14-86	First test of altered bacterium on potato plants ok'd by EPA.	0	0	0				$1 billion/yr losses from frost-damage.	
10	SFC	5-14-86	U.C. bacteria test approved by EPA, foe plans lawsuit.	3	1	0	Ritkin	0	Catastrophic release; like radio-activity from nuclear plant		Low probability, high risk.
11	LRS	5-14-86	Tule council hears concerns over "mutant bacteria."	7	3	2	Edgar; Lindow; Dearborn	Mutant bacteria.			

#	Src	Date	Headline				Sources cited		Notes	Concerns
12	WSJ	5-14-86	EPA clears field test on potato seeds and young plants of altered bacteria.	3	2	1	Rifkin; Panopoulos	0		Liability coverage for test
13	RRS	5-14-86	Frost bacterium test ok'd by EPA.	6	5	1	Rifkin; Lindow; Jackson; Frey; Zwanziger	0	Environmental catastrophe.	Experiment could jeopardize snowpack and increase the bug population.
14	H&N	5-14-86	EPA approves use of altered bacteria in Tulelake trail.	3	3	0	Frey; Rifkin; Rogers	0		Potato market might be affected.
15	MCR	5-15-86	Tulelake mutant experiment raises unanswered questions.	17	5	1	Kenyon; Lindow; Rogers; Church; Anderson	2	Chernobyl disaster; *Challenger* disaster; piano key analogy.	Potato market; liability; reference to Temik and watermelons.
16	LRS	5-21-86 5-27-86	University readies for Tulelake test of "genetically altered" bacteria.	5	3	1	Dearborn Rogers; Anderson	1		
17	MCR	5-22-86	Tulelake mutant hearing delayed.	0	0	0			Mutant	
18	CP	5-23-86	Controversy surrounds frost control research in Tulelake.	6	2	1	Lindow; Carlson	0	Mutant	$1.5 billion/yr in U.S. frost damage. Inexpensive antifrost spray; $5/acre.
19	SJMN	5-14-86	Genetically engineered bacteria experiment ok'd.	4	2	1	Lindow; Edgar	0		Kill potato market; bacterial spread of ice minus.
20	USAT	5-12-86	Genetic engineering: hazard or helper.	1	0	0				Risk to rainfall.
21	LAT	5-13-86	Ok on testing of engineered bacterium seen.	0	0	0				
22	BC	5-15-86	EPA allows a UCB professor to test 'ice minus.'	1	1	0	Rifkin	0		Spread of bacteria; destroy an ecosystem.
23	H&N	5-18-86	Bacteria experiment put off for now.	2	0	0		0		Effect on wildlife & sales of local potatoes.
24	BC	5-23-86	UC delays testing of 'ice minus' bacteria.	1	1	0	Ravven	0		

Media Analysis of Ice Minus: April 1987

SOURCE	DATE	STORY HEADLINE	# SENT ON RISK	# PEOPLE CITED ON RISK	# SCIENT CITED ON RISK	NAMES CITED	DEPTH OF ANALYSIS 0–3	SYMBOLS ANALOGIES, EMOTIVE EXPRESSIONS	BENEFITS CITED	RISK AND OTHER CONSIDERATIONS
1	CCT	3-19-87	Brentwood site for first Frostban test.	5	3	0	Torlakson; Bachini; Rifkin; Caffrey	0	Gulf of Tonken resolution for genetic engineering.	Less dependency on chemicals.	
2	SFC	3-19-87	Test bacteria for Contra Costa.	3	1	0	Torlakson	0			Interference with natural ecosystems.
3	SFE	3-19-87	Genetic test set for East Bay State.	1	0	0		0		Nontoxic alternative to chemical pesticides.	
4	CCT	3-22-87	New biotechnology industry raises hopes and fears.	7	2	0	Rifkin; Caffrey	0	Nuclear power and biotech.	More nutritious cheaper crops; less weed killer and water.	Unpredictable mutations.
5	DL	4-5-87	Only one local person attends protest.	3	1	0	Caffrey		Against playing God.		Unknown health risk to children & immunodeficient individuals.
6	BN	4-9-87	Frostban opponents threaten lawsuit.	6	2	0	Caffrey; Hoffman; Lindemann; Continente	0	Against playing God; against manipulating nature; nuclear industry.		Arrogant to fool with nature
7	CCT	4-12-87	Researchers transform Bentwood strawberry patch.	2	2	2	Calderon; Seidler	0			Unprecedented open air test monitoring to be conducted.
8	SFC	4-15-87	A move to block bacteria test.	0	0	0			Children as guinea pigs.		
9	SFC	4-24-87	Lab altered bacteria set for release today.	4	1	0	Lewis	0			First open air release of a genetically-engineered bacteria.

No.	Source	Date	Headline				Actors		Benefit	Risk
10	NYT	4-24-87	Judge allows field test of man-made organism.	3	2	0	Lewis; Rifkin	0		Unpredictable and dangerous ecological effects; first environmental release.
11	NYT	4-25-87	Gene-altered bacteria release outdoors in an historic experiment.	1	1	0	Rifkin	0	Frost causes hundreds of millions in damage per year.	Unforeseen consequences.
12	BG	4-25-87	Genetically altered bacterium is tested.	2	0	0		Virulence to young, older or ailing people.	Alternative to expensive and unsound methods of frost control; $1.6 billion in frost damage.	Unforeseen and unwanted side effects; effects on weather patterns.
13	SFC	4-30-87	"Ice-minus" creator has his day.	1	1	0	Kirby	0		Farmers may not need the product; market for potatoes may be hurt.
14	SFC	4-29-87	Another bacteria test on cold potatoes.	1	1	1	Suslow	0		
15	CCT	4-13-87	Brentwood residents silent on proposed bacteria test.	0	0	0		0		
16	CCT	4-24-87	Courts ok test of Frostban today near Brentwood.	4	2	0	Lewis; Houlihan	Playing God.	Frost damage to farmers is $1.6 billion annually	Atmospheric effects of releasing bacteria.
17	NYT	4-25-87	Biogenetic field test; new era, new challenge for nation.	5	3	2	Ritkin; Colwell; Durzan		Reduced need for pesticides.	
18	SJMN	4-25-87	High-tech down on the farm Frostban test opens door to release of genetically altered bacteria.	0	0	0		Creatures.	$1.6 billion in frost damage.	

Chapter 4

A NATURAL GEOLOGICAL HAZARD: THE CASE OF RADON

Introduction

"Radon, a colorless, odorless radioactive gas, is seeping into the basements of millions of American homes, and may be responsible for 20,000 lung cancer deaths annually." A statement like this has appeared in nearly every national magazine, newspaper, and television news program in the past three years. The American public is gradually becoming aware of the dangers of radon gas in the home, even outside of areas known to be "hot spots" for radon contamination.

The story of naturally occurring radon in homes is filled with paradoxes:

- The most important source of natural radon in homes is radium in the soil and rock below; yet scientists are finding elevated radon levels even where soil conditions do not appear conducive to radon formation.[1]
- Low-level radiation is one of the most thoroughly studied health hazards; yet within the scientific community widespread disagreement exists about risk estimates for radon exposure and about levels of exposure at which the government should recommend remedial action.
- The public health consequences of radon are more serious than those associated with many substances regulated by the federal government; yet radon remains a low program priority for the Environmental Protection Agency, and a great deal of responsibility for the problem has been delegated to the states, the private sector, and, ultimately, the homeowner.

130

- Radon is usually a naturally occurring hazard, without a corporate villain for affected citizens to blame; yet the radon issue has been highly politicized within the federal government.
- Government officials often go to great lengths to avoid creating panic among homeowners; yet the lack of information, concern, and action on the part of affected homeowners appears to be a more significant problem.

These and other contradictions within the radon story provide rich material for analyzing the evolution of risk communication efforts by the many actors involved: homeowners, government agencies, environmental organizations, the media, and organizations in the private sector. This case study considers the following questions surrounding risk communication about naturally occurring radon in private homes:

1. How have conflicts about the appropriate role of government in the radon issue affected the process of risk communication to the public? Radon is distinguished from many other environmental risks by the fact that it is found in private homes rather than at public sites of exposure. In addition, it is usually a naturally occurring rather than man-made hazard. There is precedent for regulation of hazards in private homes (such as banning urea formaldehyde foam insulation) as well as for regulation of natural hazards (such as building codes in earthquake-prone areas). But these features in association with the fact that radon is an invisible and long-term hazard make radon unique among risks that regulatory agencies must address.

2. What disagreements exist within the scientific community about the risks of radon, and how does scientific uncertainty about these risks affect communication? Most scientists agree that indoor radon is a significant public health hazard, but important uncertainties remain, such as the exact distribution of radon in the United States and the precise numerical risk of radon exposure at levels found in homes. Estimates of radon risk are based partly on epidemiological studies of underground miners, and scientists have used several different models to extrapolate from these data to calculate the risks faced by residents of homes with elevated radon levels. Disagreement also remains about the level of radon exposure at which homeowners should be advised to undertake remedial action.

3. How do citizen and environmental activists perceive their role in the risk communication process, and how has their activity affected its development? While most environmental organizations do not appear to have placed indoor radon high on their agendas,

several have devoted considerable attention to the issue. The Environmental Defense Fund (EDF), which campaigned against radon hazards from uranium mining and fuel production, and People Against Radon (PAR), a community-based organization in Pennsylvania, have played a significant role in bringing indoor radon to public attention, and they have brought unique perspectives to the issue of how the risk should be presented to the public.

4. How do organizations in the private sector that are involved in radon-related activities communicate radon risks to consumers? An advertisement for a home radon testing kit or a radon disclosure clause in a home purchase and sale agreement may be the first information a homeowner receives about the hazard. Radon testing and mitigation companies, homebuilders, realtors, and lenders involved in real estate transactions are potentially an important source of radon risk communication to the public.

5. What differences in the risk communication process can be discerned in a state with known radon "hot spots" (Pennsylvania) and a state thought to have average or slightly elevated radon levels (Massachusetts)? An examination of the activities in these two states reveals striking differences in the roles of state agencies and of citizen activism.

6. What role have the local and national media played in the dissemination of information about radon risks? In some instances, the local media have generated their own data about radon exposure of homeowners rather than simply reporting on the work of experts. Other features of the media role in the radon issue include their presentation to the public of scientific controversies about radon risks and the reliance by some media organizations on the Environmental Protection Agency as the main source for information about radon.

These questions will frame the analysis of this case of risk communication by government agencies, private sector organizations, the media, and citizens' and environmental organizations. Sources for the study include interviews, letters, scientific documents, congressional hearings, and public meetings.

Historical Context

The Watras Incident

The natural history of radon is part of the history of geological evolution. The effects of radon on human health were observed in the sixteenth century, when Paracelsus described "male metallorum," a disease of miners. As early as the 1950s scientists noted that radon gas could build up inside houses. The social history of indoor radon, however, did not begin until an event occurred in 1984 known as the Watras incident.

In December 1984 Philadelphia Electric Company's new Limerick Generating Station, a nuclear power facility in Pottstown, Pennsylvania, first activated its radiation monitors. Since the first day these monitors were activated, Stanley Watras, a construction engineer working at the plant, tripped every alarm he passed through, indicating the presence of high levels of radioactive contamination on his body.

Watras was not particularly concerned but became annoyed over the many hours he had to spend in decontamination facilities. On December 13, he conducted an experiment. When he arrived at work, he walked into the Technical Support Center, turned around, and walked out, without ever entering the plant's power block. When the radiation alarm went off, Watras knew that he was bringing the radiation from home.

Watras asked the Philadelphia Electric Company to test the air quality of his home. They found levels of radiation which produced a risk equivalent to that associated with smoking 135 packs of cigarettes per day. Pennsylvania's Department of Environmental Resources (DER) was then notified of this discovery; after retesting, DER hand-delivered a letter to the Watras family advising that they evacuate their home. The Philadelphia Electric Company took measures to alleviate the problem, and on July 3, 1985, DER and the Philadelphia Electric Company announced that the project was successful, and the Watras family returned to their home.[2]

Readers of national newsmagazines are probably familiar with the Watras story. News reports after the incident often implied that, before this incident, no one was aware that radioactive contamination by naturally occurring radon could be such a serious problem for U.S. homeowners, and that state and federal officials, stunned by the discovery, rushed to determine the extent of the problem in other homes and to devise a strategy to address it. It is true that the Watras home was unusual in that it contained the highest radon levels ever found in any home in the world up to that time, but it is not correct that the problem was new. Even the most cursory look at the scientific and popular literature before 1984 reveals numerous references to the hazards of radon in homes. Some scientists had been working on the issue for twenty years. Risk estimates for natural radon in homes had been devised by the Environmental Protection Agency and others in the mid-1970s. When indoor air pollution became a public concern in the late

1970s during efforts to increase the energy efficiency of homes, radon was frequently mentioned as a culprit. Finally, in Pennsylvania itself six years earlier officials had identified homes containing elevated radon levels in the Reading Prong, a uranium-rich geological formation stretching from Pennsylvania through New Jersey and New York.

It is also important to note that the Watras incident itself did not appear in the national media until May of 1985, five months after the Watras family evacuated their home. This fact plays a significant role in the analysis to be presented in this case study.

Thus, although the Watras incident became an important "hook" for public discussion of radon, the radon story is far from a simple case of risk event, media coverage, and public and government response. A deeper look reveals that as a result of complex scientific, political, and ideological dilemmas, national attention was not focused on radon until after the Watras incident. These same dilemmas, which provide the foundation for this case study, continue to shape the risk communication activities relating to radon today.

Risk Assessment

Radon, a chemically inert gas, is not itself a health hazard, but the isotope Radon-222 decays to four isotopes or "daughters" with short half-lives: polonium-218, lead-214, bismuth-214, and polonium-214. These particles are chemically active; if they are inhaled and attach to the lining of the lung, they will deliver radiation to the tissue as they continue the process of radioactive decay. Alpha particles emitted from the two polonium isotopes are thought to cause the most damage to the cells.

The following terms are commonly used by scientists to describe levels of radon gas and its decay products, and are frequently used in the communication of radon risks to the public:

- *PicoCuries per liter* (pCi/l) is a measure of the concentration of radon gas. One curie equals 37 billion radioactive decays per second, and one picocurie equals one trillionth of a curie. A level of one pCi/l means that approximately two radon atoms per minute are disintegrating in every liter of air. The average outdoor background level of radon is approximately 0.2 pCi/l.
- *Working Level* (WL) is a measure of the concentration of radon decay products, expressing how much alpha particle energy will eventually be released into the air. One WL equals the amount of radon daughters the decay of which will result in the

emission of 1.3 billion volts of electron energy; 200 pCi/l of radon gas in a home typically converts to 1 WL.
- *Working Level Month* (WLM) is an expression of cumulative exposure to radon decay products, denoting both the level of radon decay products and the duration of exposure. One WLM is equivalent to 170 hours of exposure to 1 WL.

Risk assessments for radon exposure carry a greater certainty than those for many other exposures, because extensive research has been done on the biological effects of radiation. Yet significant areas of uncertainty about radon risks remain, including the distribution of radon concentrations in the United States and the numerical value of risk per unit of exposure. Several distinct elements comprise a risk assessment for indoor exposure to radon: the source of radon in homes, the geographic distribution of radon concentrations, dosimetric models, and risk estimate models. The range of risk estimates currently used indicates that between 10 and 50 percent of nonsmoking-related lung cancers are caused by radon.[3]

Sources of Radon in Homes

There is today a consensus that the main cause of high indoor radon exposure is the ingress of radon that occurs naturally in soil gas. This conclusion was reached after investigations of such potential sources as building materials and reclaimed mining land. Department of Energy (DOE) scientists tested building materials after reports appeared from Sweden in the mid-1970s stating that concrete containing alum shale, which has a radium content hundreds of times that of ordinary rock, was producing high indoor radon concentration. The DOE researchers found that in the United States building materials accounted only for about 10 percent of radon concentration in homes.[4] In the late 1970s, EPA scientists studied radon levels in homes built on reclaimed phosphate mining lands in Florida and found that some homes *not* built on the reclaimed mines had high levels of radon.

Further study has revealed that the most significant source of radon is the natural occurrence of radium in soil and rock beneath homes. Furthermore, radon gas does not passively diffuse into a building but is actively drawn in by small pressure differentials between the inside of the lower part of the house and the outside.[5]

In a few areas such as Maine, a high concentration of radon in water from underground sources has been discovered to be an important contributor to exposure in homes. This occurs when water use, such as in showers and washing machines, releases the gas into the air.[6] The EPA estimates that 1 to 7 percent of indoor radon concentrations are attributable to radon in water.[7]

Weatherization of homes is an issue that has evolved over a decade of research on the source of the radon problem. Some scientists initially believed that energy conservation measures had a large effect on indoor concentrations of radon. Federal programs to improve home energy efficiency in the late 1970s, such as the Department of Energy's Residential Conservation Service Program, were criticized by the EPA. Agency scientists calculated that if all U.S. homes were "buttoned up," subsequent exposure to radon could lead to an increase of 10,000 to 20,000 lung cancer deaths per year. The DOE defended the programs, calling EPA's estimate of health effects "highly speculative and extremely uncertain."[8]

Today, there appears to be less emphasis on the potential of energy-efficiency of homes for creating radon problems; DOE researchers who tested the association between ventilation rates and indoor radon concentrations found "little or no correlation between the two."[9] Lowering the ventilation rate will increase indoor concentrations, but the most important factor is the strength of the radon source. Steven Page of the EPA's Radon Division states that recent data show that some old drafty homes, because of the air pressure differentials which draw radon gas out of the soil, are having just as many problems as energy-efficient homes. While energy-efficient homes may generally have elevated levels of radon, he says, "the difference is not significantly enough to send out the message that energy-efficient homes are the problem."[10]

As scientists sorted out the issue of energy conservation and increased radon exposure, the popular press was presenting the public with conflicting information. For example, *Family Safety and Health*, a publication of the National Safety Council, states in its winter 1986–1987 issue:

> *Your home could be harboring such potentially dangerous pollutants as asbestos, radon gas, benzene and many more. . . . In well-ventilated homes these indoor pollutants are found at very low levels. But as people try to save energy by adding insulation and weather-stripping their homes, they are also trapping these unwanted hazards inside.*[11]

In contrast, *Family Handyman*, in a column that advises readers on home energy problems, says:

> *In reporting the radon story, such respected publications as* Time, The New York Times, *and* The Wall Street Journal *have implied that the danger of radon is increased when you make your home more*

airtight. Allow me to dispute that. Most people who caulk and seal their homes are not endangering their families with higher radon levels, say the scientists we talked to.[12]

The unfortunate consequence of misconceptions by the media and the public on this issue is that those in nonenergy-efficient homes are less likely to consider the possibility of a potential radon problem in their homes.

Distribution of Radon Concentrations in the United States

Information on the distribution of indoor radon in the United States is incomplete. An important characteristic of the radon hazard is its unpredictability: one home in a community may be heavily contaminated with radon while an adjacent home has negligible levels. Past efforts to characterize radon distribution have had different scientific objectives and have employed divergent techniques and procedures for selecting homes, so that results have varied significantly.

The EPA has developed a map indicating potential high indoor radon areas based on geological reports and on National Uranium Resource Evaluation data collected by DOE in the early 1970s to locate sources of radioactive fuel for nuclear power plants. In presenting this map, EPA cautions that as yet "there is no completely reliable method for predicting where high indoor radon levels will occur."[13] The agency has estimated that 12 percent of the nation's 75 million homes may have radon levels above 4 picocuries per liter. Preliminary results from a new ten-state EPA survey released in August 1987 indicate that one in five U.S. homes may have elevated levels of radon.

Scientists at the Lawrence Berkeley Laboratories calculate that the average indoor radon concentration is 1.5 picocuries per liter, and that 1 to 3 percent of U.S. homes have concentrations of 8 picocuries per liter or more. Scientists generally agree that high concentrations of radon may exist anywhere in the country and that most of these high-risk areas have not yet been identified.[14]

Dosimetric Modeling

Several different models are used by scientists to determine the dose of alpha radiation that is delivered by radon daughters. Data concerning both atmospheric parameters (the free ion fraction and particle size for radon daughters attached to aerosols) and biological parameters (breathing rate and target cell depth) must be used to calculate the dose.

The International Commission on Radiation Protection (ICRP), the Organization for Economic Cooperation and Development/ Nuclear Energy Agency (OECD/NEA), and the National Council on Radiation Protection (NCRP) have developed three divergent sets of input parameters to calculate dose. The numerical results of the calculations produced by each model, however, are similar.[15]

Risk Projection Modeling

Several major epidemiological studies of underground miners form the basis for estimating the risk of indoor radon exposure. A 1980 report by the National Research Council's Committee on the Biological Effects of Ionizing Radiation, *Effects on Populations of Exposure to Low Levels of Ionizing Radiation (1980)*, also known as BEIR III, summarized the major studies, which showed a correlation between lung cancer and occupational exposure to radon. Subsequent studies have improved on the risk estimates by incorporating longer follow-up of the subjects and by using better controls for cigarette smoking.

Several problems in the use of the miner epidemiological data for estimating population risk are noteworthy:

- Precise radon exposure data for miners are lacking, especially for earlier years.
- Most of the studies are not yet complete—all subjects have not died from cancer or other causes.
- The number of miners in these studies is relatively small.
- Earlier studies had inadequate controls for cigarette smoking.
- The particular susceptibility of women, children, and fetuses to radon is unknown.
- The shape of the dose-response curve at the lower levels of exposure to which the general population is likely to be subjected is uncertain.

A number of different scientific organizations have published risk projection models that allow calculation of the lifetime risk of radon exposure. This modeling is necessary because none of the epidemiological studies of miners can yet provide information on lifetime risk. The major differences in the models lie in their assumptions about how to project the risk beyond the period of direct observation. Some use an absolute risk model, which assumes that the number of excess cancer cases per unit dose of radiation remains

Table 4–1   Models and Risk Estimates for Radon Progeny in the General Population

Model	Type	Exposure Period	Duration Expression	Fatalities per 10^6 Person WLM
BEIR-III (Committee on the Biological Effects of Ionizing Radiation)	Absolute[a]	Lifetime	Lifetime[b]	730
AECB (Atomic Energy Control Board)	Relative	Lifetime	Lifetime	600[c]
UNSCEAR (United Nations Scientific Committee on the Effects of Atomic Radiation)	Absolute	Lifetime	40 years	200–450
ICRP (International Commission on Radiation Protection)	Absolute	Working Lifetime	30 years	150–450
NCRP (National Council on Radiation Protection)	Absolute	Lifetime	Lifetime[d]	130

[a]Age-dependent.
[b]No expression before age 35.
[c]Adjusted to reflect sex ratio of the 1970 U.S. population.
[d]No expression before age 40.
SOURCE: William Ellett, "Epidemiology and Risk Assessment: Testing Models for Radon-Induced Lung Cancer," in Richard Gammage and Stephen Kaye (eds.), *Indoor Air and Human Health*, "Part I: Radon" (Chelsea, Mich.: Lewis Publishers, 1985), p. 82.

constant over each year of an individual's lifetime. A relative risk model, on the other hand, assumes that the incidence of cancer at any given time is raised by some constant factor for any given dose of radiation.[16] Table 4–1 summarizes results obtained from risk projection models of five different scientific organizations.

The EPA has used a relative risk model, which projects 860 excess lung cancers per 10^6 persons per working level month—that is, out of every million persons exposed to approximately 200 pCi/l for 170 hours, 860 would be expected to develop lung cancer. In an effort to help resolve the differences in the risk estimates among the various scientific organizations, the EPA asked the National Academy of Sciences to review animal and human data on radon exposure in order to develop a current estimate of quantitative risks. That study (BEIR IV), which analyzed original data from four of the major epidemiological studies of underground miners, estimates a lifetime risk of 350 lung cancer deaths per 10^6 persons per WLM.[17] In addition, epidemiological studies in high-radon areas are currently being carried out in three states by state and federal agencies.

Scientific Controversy and the Public

Disagreements among scientists about risk estimates for radon have surfaced in the media, particularly after EPA developed guideline levels for remedial action by homeowners. The NCRP, for example, recommends that action be taken at 8 picocuries per liter rather than the 4 picocuries per liter recommended by the EPA. Naomi Harley, a radiation scientist at New York University who has developed radon risk estimates for NCRP, maintains that EPA's risk estimates "border on fiction."[18] Her comments are frequently quoted in newspaper stories as a counterpoint to EPA's evaluations of radon risks; a *New York Times* article of September 2, 1986, "Radon: Threat Is Real But Scientists Argue Over Its Severity," quotes Harley as characterizing EPA's risk estimates as "outlandishly high."[19]

Scientists working for the Department of Energy are also frequently portrayed as disputing EPA guidelines. Anthony Nero is quoted by the *New York Times* as saying that the risk of exposures to less than 10 picocuries per liter, while not to be ignored, is "comparable to the risks people normally accept in their lives, and less than the risk of driving a car." Media accounts have even appeared about scientists who posit that radon at low levels is actually beneficial to human health.[20]

The EPA's Richard Guimond believes that scientists have not served the public well by airing the controversy about radon risk assessment in the media. While differences remain about the numerical risks associated with radon and where to set action guidelines, most scientists believe that radon is a serious health threat to the population. Public interest may be generated by media portrayal of a debate among scientists, but when people are aware of a scientific controversy surrounding a risk, says Guimond, they may believe that the debate is about whether there is a risk at all and therefore be less likely to take the problem seriously.

EPA Guidelines and "Safety"

It is not difficult to find media reports referring to EPA's action guidelines for homeowners as "the government's safety level." The EPA, however, does not claim that radon exposure below its guideline level of 4 picocuries per liter is safe. (For example, it estimates 13 to 50 lung cancer deaths per 1,000 people exposed to 4 pCi/l for a lifetime, and 7 to 30 lung cancer deaths per 1,000 people exposed to 2 pCi/l for a lifetime.)[21] It says, rather, that it has determined this to be an appropriate action level "as a result of balancing of risk and remediation costs."[22] Describing his disagreement with DOE scientist Anthony Nero on this issue, Richard Guimond says that Nero argued strongly for a guideline level of 20 picocuries per liter in order to target the worst homes, but Guimond stated that the public would then begin to think of 20 picocuries per liter as a safety level, and that, in addition, such a guideline would only reduce the total public health impact of radon exposure by 1 percent.

Recognizing the likelihood that many members of the public will perceive government guidelines as safe levels of exposure, the Environmental Defense Fund (EDF) has been a prominent critic of the government on this issue. The EDF's Robert Yuhnke wrote to members of Congress in 1985,

> *According to the risk assessment EPA performed . . . this standard continues to expose residents of these homes to a 1 in 65 chance of fatal lung cancer. . . . Not even EPA would go so far as to describe such a high risk as "safe". . . . If Congress now places a stamp of approval on EPA's residential radon standard as "safe," it would confirm the radical reversal in national health policy that has been introduced by EPA and OMB in the last three years. . . . Please don't contribute to the grand illusion that will mislead most Americans into thinking that homes meeting EPA's standards are "safe."*

The EDF takes its criticisms of EPA on the issue further in its publication "Radon: The Citizens' Guide":

> *The EPA advises the public that levels of radon as high as 4 pCi/l are safe. The Environmental Defense Fund believes that the cancer threat of 4 pCi/l is much too high, and that EPA is misleading the public into believing that dangerously high levels of radon exposure are safe.*[23]

This controversy surrounding guidelines and safety contains key questions about risk assessment, policy, and the public perception of risk. Homeowner interpretation of EPA guidelines is a potentially important element in the evaluation of radon risk communication efforts.

Legislative and Regulatory Framework

A myriad of federal agencies have been involved in the issue of indoor radon, including the Environmental Protection Agency, the Department of Energy (including the federal utilities such as the Bonneville Power Administration and the Tennessee Valley Authority), the Department of Housing and Urban Development, the National Bureau of Standards, the U.S. Geological Survey, the

Centers for Disease Control, and two interagency groups, the Committee on Indoor Air Quality and the White House Committee on Interagency Radiation Research and Policy Coordination.

The activities of all these agencies, however, have not coalesced into a clear, unified approach at the federal level to deal with naturally occurring indoor radon. The legislative and regulatory framework for radon has been characterized by ambiguity and controversy about whether particular agencies have the responsibility and authority to address the problem.

Outstanding issues in establishing a clear legislative/regulatory framework for radon include:

1. Should government regulate air quality in private homes? This question arose with regard to the use of the Clean Air Act to deal with radon. In 1980 the General Accounting Office (GAO), the research arm of Congress, urged Congress to amend the Clean Air Act to give EPA responsibility for overseeing a program to address radon and other indoor air pollutants. In 1985, however, the GAO concluded that EPA responsibility under the act "does not appear to extend to pollution in the indoor environment."

2. Should the federal government intervene to remediate naturally occurring as opposed to man-made pollution? An EPA official declared in a national news broadcast in April of 1985, "All kinds of things happen naturally. The question is, do you want the federal government involved? I don't think you do." This question has been debated with regard to use of the Superfund law to mitigate radon-contaminated homes. The EPA has occasionally used Superfund money for indoor radon work but only when man-made sources, such as uranium mill tailings and radium watch dial painting residues, produced the contamination. Congress, as well as EPA and the Reagan administration, have been reluctant to broaden the interpretation of Superfund to include "natural" pollution. [24]

3. Should government bear the cost of remediating the radon problem? Given that at least 8 million homes in the United States are believed to have elevated levels of radon, with the cost of mitigation ranging from $100 to $5,000, the potential overall cost of remediation is enormous. Here it should be noted, however, that some scientists have calculated that one life could be saved for every $10,000 spent on radon reduction, a figure far lower than those for many other risks regulated by the federal government. [25]

Environmental Protection Agency

Throughout the 1970s the Environmental Protection Agency assisted states in assessing the impact of indoor radon from man-made sources, such as uranium mill tailings and phosphate slag in building materials. Controversy about EPA's role in protecting public health from radon risk emerged in the early 1980s with the agency's handling of standard-setting for radon emissions from uranium mill tailings. In 1978 Congress passed the Uranium Mill Tailings Radiation Control Act in response to public concern about radiation accumulated in the production of uranium for nuclear reactors. The EPA was directed to issue standards by 1980. When the agency had not done so by the end of 1982, Congress required that it issue standards by September 30, 1983, or lose its jurisdiction to the Nuclear Regulatory Commission.

The EPA standard that ensued, which allowed a one in one thousand risk of lung cancer from radon released after a uranium mill was closed, was assailed by environmentalists as well as by some EPA career officials. They argued that the new EPA administrator, William Ruckleshaus, was following a pattern of standard-setting at far higher risk levels than the agency had allowed prior to the Reagan administration.[26]

On the other hand, the new standards were strongly criticized as too restrictive by a number of scientific organizations and by organizations with an interest in the development of nuclear energy. The Nuclear Regulatory Commission, the American College of Nuclear Physicians, the National Council on Radiation Protection, the Health Physics Society, the Department of Defense, and the American Mining Congress presented testimony opposing the EPA standards in October 1983 hearings before the U.S. House Armed Services Committee.

It is noteworthy that, throughout these hearings, EPA was criticized for giving priority to the uranium mill tailings problem when naturally occurring radon contributes most of the radon dose to the U.S. population. Senator Samuel Stratton, chairman of the House Armed Services Committee, wrote to William Ruckleshaus, as follows:

> *Please understand that there are some 100,000 to 125,000 lung cancer deaths per year in the United States, 10,000 to 25,000 of which are attributable to the emission of radon from sources in no way related to mill tailing piles.*[27]

The EPA itself estimated at this time that "there may be many thousands of homes," completely removed from exposure to mill tailings, whose radon levels exceed the EPA standards for radon release from inactive mills.[28]

The EPA's interest in naturally occurring radon began in 1975, when the agency studied homes built on spent mines in Florida and found that high indoor radon levels were occurring naturally as well. The Florida study produced risk estimates used by EPA today, projecting that 10 to 20 percent of lung cancer morbidity in the United States is caused by indoor radon. Despite this knowledge, top EPA officials argued before 1985 that they had no legislative mandate to remediate indoor air pollution and did not wish to intrude into people's homes.[29] This approach to indoor air pollution in general, and radon in particular, flourished in the antiregulatory climate of the Reagan administration, which deleted EPA's entire proposed fiscal 1982 budget request for indoor air pollution research. Between 1981 and 1984, EPA's Office of Research and Development received no money for radon research. Anne Burford, EPA administrator between 1981 and 1983, is quoted as saying that radon did not become a public issue during this time "because the media did not pick it up."[30] Meanwhile, EPA issued no information about radon during her tenure and requested zero funding for indoor air research in 1984.

Agency officials testified during congressional hearings in August 1983 that EPA was "getting more aggressive in the problem definition part of the research effort" on radon and indoor air pollution, and that this was the result of a different perspective brought by recently appointed administrator William Ruckleshaus. It appears, however, that it took widespread media attention to the radon problem after the Watras incident in 1984, coupled with congressional initiatives, to produce substantially greater effort by EPA to address the problem.

In July of 1985, EPA developed, at the request of Administrator Lee Thomas, a report entitled "Health Risks Due to Radon in Structures: A Strategy and Management Plan for Assessment and Mitigation," which outlined a federal strategy for addressing radon. Several different options for federal involvement were considered (see Appendix A at the end of the Chapter). The strategy chosen included assessing the national distribution of radon, identifying high-risk lands, and developing state and private-sector capabilities for mitigation and prevention. It was projected to prevent several thousand lung cancer deaths per year and would cost $10 million over five years.

This report was made public during congressional hearings on radon and indoor air pollution in October 1985. Senator Frank Lautenberg of New Jersey reported that it had been replaced by a "truncated, sanitized version," in which information about options considered, budget data, and the potential number of lives saved was omitted.[31] These changes, according to Lautenberg, were the result of pressure from the Office of Management and Budget

(OMB), which, he stated, "would rather save money than save lives."[32]

Despite opposition within the Reagan administration, EPA is now pursuing a Radon Action Program with a focus on developing and disseminating technical knowledge and on working with states and the private sector to develop assessment and mitigation capabilities. However, the radon effort remains a low-priority program within the agency. A 1987 EPA report, entitled *Unfinished Business: A Comparative Assessment of Environmental Problems*, describes radon as an area of "relatively high risk but low EPA effort."[33]

In the summer of 1986, EPA proposed its "action guidelines," the levels of indoor radon at which remedial action would be recommended to homeowners. The EPA stressed that its figures were guidelines and not standards, although several media accounts reported at the time that standards had been set for indoor exposure to radon.

Although the guidelines were set at levels at which approximately one to three individuals in one hundred would be expected to die of lung cancer, several other federal agencies, including the Department of Energy, Department of Housing and Urban Development, and Office of Management and Budget, strongly objected to the guidelines when they were proposed, arguing that they were too stringent. However, the Centers for Disease Control, considered "the other public health agency" within the federal government, backed the EPA proposal. The guidelines were issued in August of 1986.

Department of Energy

The Department of Energy has conducted research on indoor radon since 1977. Its research has focused on four areas: (1) mechanisms and rates at which radon enters buildings; (2) techniques to identify geographical areas and building characteristics that might result in high indoor radon concentrations; (3) health effects of indoor radon exposure; and (4) methods to measure and reduce indoor radon concentrations.

The agendas of DOE and EPA are quite different, and this is reflected in their approaches to the radon issue. While DOE has concentrated on basic research, EPA's philosophy is that "we now know enough to get something going." The Committee on Indoor Air Quality, an interagency group established in 1979 to develop a coordinated federal response to radon, has reportedly suffered over the years from conflicts between DOE and EPA.

The EPA's Richard Guimond believes that deep-rooted philosophical differences exist between those who work for the two agencies. He says:

> *DOE's entire existence has been based on promoting nuclear energy and using radiation. It is not a public health agency, and it is misguided to spend so much at DOE for health research. If EPA spent fifty million dollars on health research, we would elucidate the problem by no more than a factor of two. But if we spent that on education and assistance to the states, we could fix most of the problem.*

The DOE, however, has maintained that before work to control radon can be justified, accurate distribution, exposure, and risk assessments should be completed.[34]

Congressional Initiatives

A coalition of legislators from New Jersey, New York, Pennsylvania, Maine, and Rhode Island, sensitive to the growing concerns of constituents about radon contamination, has been working to provide EPA with the funding and authority it needs to pursue the radon issue. In 1984 and 1985, Congress appropriated funds for radon research at EPA for mitigation methods and measurement standards.

Several bills addressing radon were attached as riders to legislation renewing the Comprehensive Environmental Response, Compensation, and Liability Act (CERCLA) of 1980, commonly known as Superfund. In 1986, Congress passed an amendment to CERCLA which directs EPA to assess the extent of the indoor radon problem in the United States and to conduct a demonstration program to test methods of reducing the public health threat of radon. Thus, Congress established EPA as the lead agency on radon, although the amendment does not direct Superfund dollars to these activities.

Recent actions by the Office of Management and Budget, however, may undermine the intent of Congress. The EPA requested $7.5 million for its radon program in 1988, but the Reagan administration proposed that this request be cut to $5.7 million. An OMB spokesman, in a prepared statement to a Pennsylvania newspaper, explained the budget cut by saying,

> *The position of the president, and we are an arm of the president, has long been that the federal government should not get into the business of regulating radon. . . . The EPA's program was leading down the road to regulation. That's why they were cut back.*[35]

The newspaper reports that, in an unusual move, OMB asked to retract part of this statement the next day, stating,

We don't want a regulatory program, that's true, but the EPA's program is not leading down the road to regulation. . . . We're squeezing everybody's budget.[36]

The Department of Energy's budget for radon, however, is not being squeezed: it requested $4.5 million for basic radon research and is scheduled to receive an additional $10 million. Congressional leaders on the radon issue complain that, through the budget process, the administration is attempting to subvert the new law designating EPA as the lead agency on radon.[37]

Alternative Approaches

Several other avenues for responding to radon are being proposed by affected citizens, public interest groups, and legislators. A common theme among many who have advocated greater attention to the radon problem is that radon contamination constitutes a natural disaster. This position, whose adherents include the Environmental Defense Fund, Ralph Nader's Public Citizen, and People Against Radon, moves the issue away from the individual decisions and actions and into a more collective framework. It implies that greater responsibility for the problem should be borne by the federal government.

The Environmental Defense Fund proposes, among other things, a publicly funded radon testing service and a federally supported loan fund for remedial work by homeowners. Public Citizen advocates a public information program at the national level, emergency funds for disaster relief in high-radon areas, and national building codes for new homes. People Against Radon also urges emergency federal response to high radon levels, stating that "in *no* time in U.S. history has the Government acted so slowly in the response to a natural disaster."[38]

Some of the radon-related bills currently being considered by Congress include a proposal to give homeowners federal tax credits for radon mitigation work and a proposal to amend the 1974 Disaster Relief Act to include disasters from indoor radon contamination.

It is interesting to note that, despite fears repeatedly voiced within the Reagan administration that efforts to control radon would lead to regulation, it appears that no environmental or citizens group has proposed federal regulation of indoor air quality as a solution to radon contamination of private homes.

There is precedent, however, for state standards for exposure to natural radon. Florida's Department of Health and Rehabilitation Services (HRS) has promulgated rules that specify construction standards and testing provisions for new homes. Newly constructed homes of certain designs must be tested for radon and remediated if necessary to meet a 4-picocuries-per-liter standard before the buildings can be occupied. This standard also applies to new schools and commercial buildings. It will be applied in all areas of the state with elevated radon levels; HRS is currently working to identify these areas.[39]

Risk Communication

Environmental Protection Agency

Because EPA's radon strategy relies on decisions by individual homeowners to test and remediate their homes, risk communication might be seen as the key link in federal efforts to address the radon problem. At the center of EPA's radon risk communication are two booklets, "A Citizen's Guide to Radon: What It Is and What to Do About It" and "Radon Reduction Methods: A Homeowner's Guide." These booklets describe the risks from indoor radon and explain what actions homeowners can take to reduce radon exposure. They were issued in August 1986, at the same time that EPA issued its radon action guidelines. The EPA provides camera-ready copy of these booklets to states, which reprint and distribute them to residents.

The "Citizen's Guide" did not escape the controversy that has surrounded every attempt to tackle the radon problem. One chart in the booklet, known as the "head chart," depicts lung cancer deaths at various radon levels using black boxes among human heads (see Appendix B at the end of the chapter). The chart was reportedly criticized, both within and outside the EPA, as "inflammatory." The OMB, for example, was said to have feared the chart would "scare people" and "create panic."[40] Defenders of the chart within EPA responded:

> *Providing citizens with risk information in the complicated, highly qualified language of analysts and scientists would be "worse than nothing."... Only those in the federal establishment dislike the chart.... People in the outside world think it is effective and appropriate.[41]*

The EPA's Richard Guimond underscores the importance of clear, easily understood information about risks and available remedial actions. He says that he tries to create "controlled concern" among homeowners, noting that "sometimes it's easier to get a homeowner with a level of four or five picocuries per liter to understand the risks in a logical, rational way, than it is to convince someone with a very high level to do something."[42] He points out that, because radon is invisible, because health effects occur over years of exposure, and because no *individual* case of lung cancer can be definitively linked to radon exposure, it can be very difficult to motivate homeowners to test and mitigate their homes.

The EPA has worked closely with reporters from various media to develop radon stories. Popular magazines have the potential to exert a much wider impact on public awareness than government brochures. Most of these magazines have carried at least one story about radon in the past three years. "Overall, we are pleased with the results," says EPA's Steven Page. "Occasionally an article appears which will attribute the problem to radon in water or in building materials, which contribute less than one percent of the problem. But generally, reporters are very conscientious in working with us and getting the facts right."[43]

The EPA has also undertaken a number of experiments to measure the effectiveness of various risk communication strategies and techniques. One study of Maine households examined perceived risk and mitigation actions by homeowners who received new information about the risks of radon exposure; it identified no statistically significant relationship between mitigating behavior and "objective risks."[44]

Another study in New York State is evaluating the comparative effectiveness of risk information provided in four experimental brochures, the EPA's "Citizen's Guide," and a brief fact sheet.[45] This study is concerned with developing a model for effective risk communication that can be used by the states. Two central issues addressed in the study are the relative effectiveness of quantitative versus qualitative risk information and of the use of a "command" tone, which tells people what they should do at a given radon level, versus a "cajole" tone, which suggests what they might consider in deciding what to do. A highly complex research design has been incorporated to determine the optimal packaging of radon risk information.

While this approach cannot be faulted as a rigorous methodological exercise, psychometric and decision theory parameters frame the definition of "effectiveness." The absence of a cultural or sociological element in the design of this experiment reflects the traditional risk perception emphasis of the EPA research perspective. The search is for an effective brochure to communicate expert opinion on the risk of radon; no attention is given to generating a citizen-derived definition of risk to better inform experts.

Pennsylvania

Risk communication activities in Pennsylvania have taken place in the context of a public health crisis, which came to light when the

Watras incident triggered the discovery of extremely high radon levels in some homes. The state's environmental agency became very active in responding to the radon issue, working closely with EPA to develop testing programs, mitigation efforts, and risk information for the public. Some affected homeowners responded by organizing a citizens' group, which has played an important role in spurring government action and in disseminating information. Extensive local newspaper coverage has also been important in the Pennsylvania radon story.

Department of Environmental Resources. In one of the many ironies of the radon story, the same state that was forced to deal with the accident that released radioactivity from the Three Mile Island nuclear power plant was then confronted with radiation from *natural* sources, of potentially far greater public health consequences, after the Watras incident in 1984. Actually, the state's Department of Environmental Resources (DER) had in 1979 discovered high levels of natural radiation from uranium deposits in the Reading Prong area. It considered conducting a radon survey, but the idea was preempted by activities following the Three Mile Island incident.[46]

After the Watras incident, DER's Bureau of Radiation Protection began a survey in Colebrookdale Township, in which the Watras home is located. Many other homes were found to have elevated radon levels, some higher than one working level, a level at which EPA recommends that remedial action be taken within one week.

The DER decided to continue its survey by placing advertisements for free radon monitors in local newspapers. (This decision was made after disappointing results in a door-to-door campaign to distribute free test kits, in which less than 50 percent of homeowners approached in person agreed to test their homes.) The DER received 25,000 requests for monitors following the newspaper advertisements and sent out 21,000 monitors. Each homeowner who participated in this survey was given information on the home's radon level along with fact sheets developed with EPA assistance. Nicholas DeBenedictus, Secretary of DER, described the efforts of his agency to develop appropriate risk information for homeowners:

> *One of the first things we discovered is what is the right level, and I think we are still groping for that. . . . [The information] is pretty detailed. We didn't want to make it like a first grade primer, so what we did was actually give some scientific data in layman's language, hopefully, so people could understand what radon is, get rid of some of the unknown, but it fell short of saying "At this level this is what you should definitely do."[47]*

This experience foreshadows decisions about radon policy that would be made later at the federal level. To date, 17,000 homes in several counties have been screened, and 50 to 60 percent have been found to have radon levels higher than 4 picocuries per liter. Homes that exceeded that level were given another monitor for long-term testing. Homes with 20 picocuries per liter or higher were offered a visit by DER staff, who would make a more detailed evaluation and recommend remedial action. A Colorado firm developed for DER a guide to remediation, which was provided to 10,000 homeowners.

The DER coordinated its efforts with other state agencies, including the Health Department, the Department of Education, the Department of Community Affairs, the Governor's Energy Council, and the Council on the Hispanic Community. The DER's Bureau of Radiation Protection added twenty-one people to its staff. Because DER determined public information and community relations to be an essential part of its radon program, it established a Radon Information Center, hired a community relations specialist, and set up a toll-free radon information telephone number.

State spending on radon activities has been significant. Initially, $9 million was allocated to this work from DER's existing budget. The radon monitoring budget for 1986 and 1987 was $1.3 million, and the state legislature, partly in response to lobbying by affected citizens, authorized an additional $1 million for research and demonstration of remedial measures.

People Against Radon. As the radon picture in Pennsylvania began to come into focus, citizens from Colebrookdale Township became active in publicizing the dangers of radon and demanding relief from the state and federal government. Stanley Watras initially shunned publicity about his family's plight but soon began giving frequent media interviews, testified before Congress, and even negotiated with a Hollywood producer about a proposed television movie.[48] He has become an important source of information, as well as a symbol of the radon problem, for news organizations throughout the country.

Two of Watras's neighbors, Kay Jones and Kathy Varady, formed the organization People Against Radon (PAR). Both Jones and Varady had been found to be living with high radon levels during DER's survey in early 1985. Their homes were among eighteen in the area that were remediated through an EPA-funded demonstration project.

A short time after they learned about radon in their homes, the two women say, "lack of knowledge turned to outrage." They have made it their business to publicize the need for government involvement, holding community meetings and speaking to politicians, federal and state officials, scientists, and reporters. Thus Jones and Varady, both housewives with high-school educations, join the new tradition of women environmental/community activists who became self-educated experts in order to organize their communities. In fact, the two initially contacted Lois Gibbs, known for her organizing work at Love Canal, for advice about their fledgling efforts.

Another important connection to the environmental movement was made when Kay Jones contacted the Environmental Defense Fund's Robert Yuhnke, who had previously campaigned against radon hazards from uranium mining and uranium fuel production, to obtain information about radon. Jones relates that Yuhnke thought her report of a 2.1 working-level reading in her home was "exaggerated."[49] Yet once he was convinced, both Jones and Varady report that he became a source of strength for them. He provided information and support at a time when they believed that they were being told nothing by Pennsylvania officials.

Yuhnke also gave the Pennsylvania story to *New York Times* reporter Philip Shabecoff. On May 19, 1985, five months after the Watras incident, the *New York Times* ran the story on page one. Shortly after this story appeared, radon became news on television stations and newspapers across the country. Thus, Jones and Yuhnke are directly responsible for catapulting the radon issue into the national spotlight.

Perhaps as a consequence of their collaboration on this issue, PAR and EDF use very similar language in referring to the radon risk. For example, the phrase "cancer time bomb" to characterize the indoor radon hazard appears in the literature and other statements of both organizations.

Kathy Varady's involvement in the radon issue began when Kay Jones asked her to sign a petition seeking government help in radon remediation and education. Their collaboration helped them through discouraging initial efforts to organize their neighbors, efforts that, Varady says, were met with apathy or "bad feedback" resulting from fear for property values.[50]

Varady said that she knew from the outset that radon was a national issue. Her focus is on the dissemination of information that will make homeowners aware of radon and information about initial testing. She is critical of publicity sensationalizing the situation of Stanley Watras, in which he was portrayed as "stripped to his underwear and forced out of his home."

Varady and Jones characterize their organization partly as a support group for victims of radon:

First of all we were there for anyone who needed a sympathetic ear.
Even though we were over the initial shock of learning about our
radon problem, we knew exactly how the homeowners were feeling.
The anger, the fear, and the helplessness were all emotions that are
characteristic of the radon victim.[51]

Jones and Varady frequently stress the mental health consequences
of learning about radon in one's home and decry the lack of support
systems which usually develop in the aftermath of natural disasters.
Varady wrote in 1986:

In every catastrophic situation teams of mental health people come
into an area to help relieve the stress brought on by a tragedy. It has
been 18 months since extreme levels of radon have been found in
Pennsylvania homes. No agency, state or federal, has focused on the
stress factor of dealing with this frightening situation. To avoid
dealing with this problem many have chosen to ignore it completely.
This apathetic attitude will have tragic results.[52]

To address this need, PAR advocates "mental health hotlines" to
help alleviate the fears and concerns of homeowners.

In considering responses to radon-related stress, it is intriguing
to look at the comparison made by Kasperson and Pijawka[53] of the
communities that had experienced natural and technological disasters. After most natural disasters,

loss is directly observable, and thus therapeutic response can be
focused, defined, and targeted. . . . [T]he onset of most natural
disasters occurs relatively quickly and tends to be short-lived. There-
fore energies are focused on recovery; mutual aid systems enhance
the rebuilding process.[54]

By contrast, technological disasters, such as chemical spills,

are a new societal phenomenon, and because of the technical nature
of the threat, substantial dependence rests on scientific and regula-
tory institutions rather than on individual family and community
efforts. . . . If exposure poses risk to only part of a community and
there are scientific uncertainties about the nature of the hazard,
then the potential adverse impacts on the economy of the area, pro-
longed debate, political activity among victims, and intense media
debate may induce substantial community conflict.[55]

Spontaneous altruistic activities of institutions and individuals in
a community, then, are much less likely to emerge after a technological disaster. The events after the Watras incident contained
many of the characteristics common to a technological rather than a
natural disaster: scientific uncertainties about the risk, dependence
by victims on scientific and regulatory agencies, and intense politi-

cal activity by affected citizens. Radon contamination of homes is an unusual natural disaster; these characteristics may help to explain the apathetic community response which PAR seeks to transform.

Risk communication by citizens and citizens' groups can rarely be separated from citizen evaluation of government performance in the face of a risk. PAR members took both state and federal officials to task for their handling of the radon crisis. Varady, for example, wrote that Pennsylvania Governor Richard Thornburgh and the state's health department were oblivious to requests for help in initiating federal emergency assistance for disaster relief. She also criticized EPA for its slowness in issuing action guidelines.

At the same time, PAR has developed a cordial relationship with EPA and DER, a situation that, as noted elsewhere,[56] is uncharacteristic for most environmental organizations. Kathy Varady writes of DER officials:

> *In all fairness to Mr. Tom Gerusky and Maggie Reilly of DER, we have found them always to be there for us whether we call them with a question or just call to let off some steam. We will always be grateful for these people who have been the glue that held us together many times. Maggie and Tom know that Kay and I have always been ones to speak our minds but we have always been totally honest with each other and that makes for a very good working relationship.*[57]

The DER, for its part, has praised both PAR and the Watras family "for their active citizen input and the fact that they have allowed us to work with them."[58]

Local News Media. Citizen activists in Pennsylvania have worked extensively with the local news media to publicize the radon problem, and the local media have taken the issue seriously. A media analysis conducted at Lehigh University in Bethlehem, Pennsylvania, found that three local newspapers ran a total of 138 articles on radon in the nine-month period following the Watras incident. In contrast to the aggressive role of local newspapers, two national newspapers published in the region, the *Philadelphia Inquirer* and the *New York Times*, printed 17 and 6 articles, respectively, in the same nine-month period. They apparently viewed radon as a local problem that would not interest their readers.[59] The important role of local newspapers was underscored by a public opinion poll in the area, in which 61 percent of respondents said that newspapers were the main source that alerted them to the radon problem.[60]

Massachusetts

Radon risk communication activities in Massachusetts have taken place in a context where the magnitude and severity of the radon problem are essentially unknown. In contrast to the situation in Pennsylvania, the state agency in Massachusetts responsible for indoor radon has taken a low-key, nonaggressive approach to the problem. No citizens' group has appeared to prod government officials to act. In the absence of strong governmental initiatives, local television has played a central role in beginning to assess the problem and in bringing the radon issue to public attention.

Radiation Control Program. The Massachusetts Department of Public Health's Radiation Control Program, which is responsible primarily for testing x-ray equipment in the state, handles requests for radon information by Massachusetts residents. According to the program's director Robert Hallissey, between 1980 and 1986 three to four homeowners per year requested information about radon, usually after learning about the problem from magazine articles.

The number of requests swelled in the summer of 1986 with newspaper coverage of EPA initiatives in issuing guidelines. The Radiation Control Program continued to respond to these homeowner requests, although no particular segment of its budget was alloted for radon. The EPA's "Citizen's Guide" was reprinted as a Massachusetts booklet and provided to homeowners who contacted the state. A list of organizations that conduct testing was also developed for homeowners. Officials of the Radiation Control Program began to give talks and seminars on request to interested groups such as realtors and community organizations.

The state has monitored a small number of homes in the state, revealing some elevated levels, and the Radiation Control Program requests that homeowners report results of privately obtained screening measurements. Hallissey emphasizes that these data in no way represent a statewide assessment, since the measurements are "generally from homeowners who perceive that they have a problem and who may or may not be representative of the radon problem."[61]

A state budget request of $150,000 for 1988 has been submitted for a study of the radon issue and for a full-time radon position in the Radiation Control Program. A statewide assessment study proposal was also submitted to EPA in 1987. But Massachusetts was not among the states initially chosen for these studies, but subsequently it was funded to undertake statewide monitoring.

The position of the Radiation Control Program to date is that the radon problem in Massachusetts is probably not severe. "None of the data indicates that we have anywhere close to the radon problems of Pennsylvania and New Jersey," says Hallissey.[62] He also questions some of the assumptions in EPA's risk assessment for radon:

*We don't entirely buy into EPA's risk estimate. . . . While elevated
levels for an extended period of time can have serious health effects,
we are not sure whether the risk of lung cancer is linear at low doses
such as 4 picocuries per liter. In addition, EPA's risk estimate is
based on spending 75 percent of your time in your home over seventy
years, which we consider almost an impossibility.*

In explaining the state's activities on the radon issue, Hallissey
states that "we feel strongly, as do other states, that initial screening is up to the homeowner." He expresses concern about the
reluctance of homeowners to address the radon issue for fear of a
negative effect on real estate values. "Our response," he says, "is
that radon is an easily controlled problem—don't be frightened that
a home with elevated radon levels is no longer marketable."[63]

Local News Media. In contrast to the low-key, reactive approach to the state's Radiation Control Program is the aggressive
radon series produced by television station WNEV in Boston during November 1986 and February 1987. An assistant news director
at the station had produced a radon story at a television station in
Chicago and suggested a similar story be developed in Boston.
Special projects producer Susan Walker called state agencies to
find out if a radon survey had been conducted in Massachusetts.
She discovered that no survey had been done and that "a kind of
obscure program," the Radiation Control Program, was responsible
for dealing with radon in the state. The station decided to conduct
its own survey on radon in Massachusetts homes.

Walker obtained the assistance of Tufts University's Center for
Environmental Management and Environmental Studies program
in designing and carrying out the survey. She learned that the
station could not carry out a "truly definitive" survey, but, as a way
of teaching people about radon, it could get an initial indication of
how many homes in Massachusetts have elevated radon levels.
Two hundred homes in 42 communities were chosen, including 10
homes in each of 5 communities with "suspicious rockbed." Results
from 189 of the survey homes were obtained; 52 of these homes, or
more than one in four, had a level greater than the EPA guideline
of 4 picocuries per liter.

The initial television series based on these results, presented by
consumer specialist Phyllis Eliasberg, aired during the week of
November 11, 1986. The series producers attempted to convey
three basic messages: (1) what radon is and what its dangers are; (2)
that no one in Massachusetts is doing much about the radon problem; and (3) that radon is easy to detect and easy to take care of. "It
was a real challenge to make a visual series out of a gas you can't
see, to explain something that's pretty complicated and still be

accurate yet not sensational," says Walker. "Viewers want to be shown a 'victim of radon,' but we can't point to one person who died of lung cancer and say 'this is because of radon.' It's a tough thing to communicate."[64]

The series employed several comparisons to communicate the risk of radon. The analogy to cigarette smoking was used frequently, because it was considered to be a "common denominator" for interpreting a risk and well understood by the public. The series also stated that one home had more radioactivity than was emitted in the 1979 accident at Three Mile Island. "We received criticism for this statement as sensationalist," says Walker, "because the public perceives that a lot more radiation was released in that accident than really was."[65]

Some homes were described in the broadcast as having levels of radon at which, if they had been uranium mines, they would be shut down by the government. This comparison was used, explains Walker, to "point out some strange policies the government has, where it would close mines but would not consider a national survey or provide funds to the states to address radon in homes."

Several other images and techniques were used in the WNEV survey to communicate radon risks to viewers. One out of four pictorial representations of homes "glowed" to portray elevated radon levels. Experts were interviewed, including federal and state agency officials, university scientists, and physicians. Stanley Watras was interviewed, and his home was shown in one segment. A Massachusetts homeowner whose home initially measured 18 picocuries per liter reported that she gave up smoking after this finding and that her young son asked her if it was "okay to take a deep breath" in their home.

The station was stunned by viewer response to the radon series. In the broadcasts interested viewers were invited to write for more information about radon and how to test for it; 13,000 people requested this information, comprising the largest response to any story the station had broadcast.

Station ratings rose during the week of the series. The station ran a half-hour audience participation special on radon the following week, and in February 1987 it broadcast another series, which reported the results of a follow-up study of modern energy-efficient homes.

The November series was nominated for a local Emmy award. Several other stations subsequently ran "copycat" series reporting on environmental dangers in the home. WBZ-TV ran a report called "Something in the Air" in the same week as WNEV's February series. That report discussed the "invisible enemies" in the home, "from tiny organisms that feed on your skin to fumes from office products."

The radon series had effects beyond the medium of television. A number of moderate-size newspapers in Massachusetts (some of which are members of a "news exchange" with WNEV) published stories based on the station's survey. The *Boston Globe* did not run the story, possibly because it did not want to give publicity to a survey done by a television station.

Smaller local papers did pick up the story, sometimes generating further local activity. For example, the *Good Neighbor,* a small advertising-supported newspaper in Arlington, Massachusetts, ran a radon story and ads for testing kits; 150 local residents bought kits to test their homes.

The series provided a boost for a bill in the Massachusetts legislature to establish an Indoor Air Pollution Commission. The bill, which had been introduced by Representative Patricia Walrath of Stow, was passed three weeks after the series aired, in large part because of publicity about the radon issue generated by the series. The commission is currently holding hearings to determine the extent of the radon problems in the state and to develop recommendations about who should address the issue.

WNEV producers believe that the story also had some effect on radon policy within the state's Department of Public Health. Frustrated by the Radiation Control Program's assertions that Massachusetts's radon problem was not as bad as New Jersey's or Pennsylvania's when no state survey had been conducted, the WNEV people thought that DPH commissioner Bailus Walker took the issue more seriously after the series. Walker initiated several (albeit poorly attended) public information meetings on radon and pushed behind the scenes for more attention to radon.

A conflict with the Radiation Control Program also arose after the series, according to Susan Walker: officials were upset that the station would not release the names of people who had been found to have homes with elevated radon levels. Walker says:

> *Producers of radon series in other cities had told me that public agencies would want to get names, but we had promised homeowners that the testing would be confidential. It is commendable that the Radiation Control Program wants to do follow-up testing and help people with high readings—but that is after they have said they don't think Massachusetts has a radon problem, and we then do all the legwork to find there is a problem.*[66]

Walker also expressed concern about the way the series was used by a few unscrupulous entrepreneurs who used the publicity about radon generated by the series to sell home testing kits at exorbitant prices. One firm ran ads in the *Boston Globe* for testing devices during the week of WNEV's second radon series: "Can you afford not to know? . . . Send $50 for your radon sampling kit and analysis." (Similar tests are available elsewhere for less than $20.)

Local television reporting on radon such as the WNEV series almost always generates enormous viewer response. A station in Buffalo, New York, distributed 13,000 test kits; a Chicago station conducted a survey of 260 homes and received 10,000 requests for information. A trade newsletter for local stations, *The Rundown*, ran a story in its December 1986 issue describing the success of local radon reporting, "Radon Gas: Series Draws Responses." The EPA received eighty requests for information from television stations as a result.

The success of these stories raises two questions about local television reporting on radon. First, what accounts for the large amount of viewer interest in these stories, while others have reported apathy as a major problem in risk communication about radon? Susan Walker of WNEV in Boston says, "People had heard a little about radon and wanted to know more. And we weren't just saying here's another risk that's out there; we were saying, this is what we've actually found *here*."[67] A WJW-TV (Cleveland) reporter says, "Everybody is concerned about his health and this affects just about everybody."[68]

Viewer response to local series can probably be attributed to several other factors. Very concrete and specific activities were presented for viewers to perform: "send for information" or "send for a test kit—here is the address" rather than "contact your local state agency." All the series stressed that radon is easy to detect and usually inexpensive to correct, leaving viewers with the feeling that they could do something about the problem. Finally, the television series depicted local residents who had confronted a radon problem and did not dwell on abstract comparisons.

A second question one may ask is, What are television stations doing conducting quasi-scientific surveys? Generally, they conducted such studies because no one else at the state level had the information they deemed important, and because the stations can enhance their public image by performing this community service. Television producers also perceive that the proper role of the media on this issue is both to conduct studies and to report on them: "The government is expecting us to do these stories. This is how the word is spreading around the country on this particular environmental hazard." "This is the kind of thing the press has to do. Who else is going to do it?"[69]

The phenomenon of television carrying out studies has at times produced an interesting twist on media use of scientific experts in reporting. WJLA-TV in Washington, D.C., ran a series in February 1986, which reported on its survey of forty homes in the area; it found that one-half of the homes had radon levels greater than 4 picocuries per liter. The station interviewed Naomi Harley (a radiation scientist at New York University) about these results; she stated, "I'm quite surprised—I had no reason to think there was anything unusual in the area." Here the media, rather than the scientist, originated useful risk information.

The Private Sector

In most instances, the private sector becomes involved in communication about risk in response to allegations of a hazardous product or process. No firm or industry need defend naturally occurring radon in homes, but organizations in the private sector have become an integral part of radon risk communication, as they carry out activities in the areas of radon testing and mitigation, home building, real estate, and banking.

The EPA has made collaboration with the private sector an important element in its strategy for addressing radon. The agency works with a number of private-sector groups, including regional and national organizations of homebuilders, realtors, and mortgage lenders, to develop their members' capabilities in radon assessment and mitigation. In EPA's view, the states and the private sector should assume the primary responsibility for addressing the radon problem, with the federal government playing a supporting role.[70] It follows that organizations in radon-related fields bear some responsibility for timely and accurate communication about risk to homeowners. Given motivations ranging from avoidance of lawsuits to profits to civic duty, the degree of fulfillment of this responsibility by private-sector organizations varies widely.

Testing and Mitigation. The increased demand for radon testing by homeowners has made this a growth industry in recent years. In order to give states and homeowners confidence in radon measurements, to encourage firms to go into the testing business, and to weed out the inevitable unscrupulous operators who test homes with Geiger counters or mayonnaise jars, EPA has developed a voluntary Radon/Radon Decay Products Measurement Proficiency Program. A list of testing firms whose work is certified by EPA as accurate is sent to state agencies, which provide the names of these companies to interested homeowners.

Because the testing firm is often the first organization to contact the homeowner with radon test results, the quality of risk information it provides is important. There appears to be room for improvement in this area. For example, one major firm, which manufactures and performs laboratory analysis of short-term test kits, sends a letter with laboratory results, along with interpretive information, to homeowners. The information provided is a condensation of radon testing protocols from EPA, presented in the form of dense text, without graphs or charts that might help homeowners to conceptualize the risks to which they are exposed (see Appendix C at the end of this chapter). For example, the reader may wonder whether the guideline levels of 4 and 20 pCi/l are based on extensive exposure, or how a basement reading of 4 pCi/l would translate into exposure in living areas. The letter provides little assistance in translating different levels of exposure.

Mitigation firms have not experienced the same rapid growth as testing companies, perhaps because the task of remediation is much more complex than testing for the presence of radon. The EPA provides training courses for contractors and homeowners interested in performing mitigation work; to date, training sessions have been conducted in New Jersey, New York, and Pennsylvania, with participation by more than a thousand people representing forty states.

The EPA has also issued an extensive manual entitled "Reducing Radon in Structures," which not only provides technical information but also gives advice to the contractor about communicating with homeowners about radon risks. It includes an article from a popular science magazine about risk perception,[71] and urges the contractor to "acknowledge the homeowner's feelings," "communicate facts in understandable terms," and "actively involve the homeowner" in decision-making about the mitigation process.[72]

Advertising by testing and mitigation companies is a potentially important form of radon risk communication to the public. Several examples of clearly exploitive advertising have appeared in areas where high radon levels have been discovered. For example, Radiation Testing Services of Little Ferry, New Jersey, mailed a highly sensationalized five-page brochure to homeowners, warning that radon may be "the greatest immediate threat to your life" and that "every breath you take in the 'safety' of your home may be deadly." The firm, which has since gone out of business, was reprimanded by New Jersey's Department of Environmental Protection for "promoting panic."[73]

Several radon firms have formed a new professional association, the American Association of Radon Scientists and Technologists, stating that they wanted to counter "a growing impression that companies are put together by opportunists seeking a fast profit from exaggerated public hysteria." Ironically, Radiation Testing Services was a charter member, but the association says it dropped the firm after it was reprimanded by state officials.

Association president David Shutz says that the group relies on peer pressure to curb unscrupulous practices; it has drafted an ethics code to guide customer relations, marketing, and use of the media. State officials encouraged formation of the association and believe that it has had "a good leveling influence" on the industry.[74]

Homebuilders. Homebuilders will have an increasingly important role in the prevention of radon exposure in coming years, since much of the problem can be prevented in new homes by relatively inexpensive construction methods. For example, pipes can be installed around the interior of a basement under floor slabs; if elevated radon is detected, fans can then be installed to pull radon gas out of the home.

Techniques such as this seem to hold promise for reducing radon risks to homeowners. However, in the homebuilding industry, it reportedly takes ten to fifteen years for half of all builders to adopt a new building technique, even if that technique has been proven to work.

The National Association of Home Builders (NAHB) has been active on the radon issue. In 1984 it gave $12,000 to its research foundation to study radon and indoor air pollution. The NAHB works with EPA to provide information about radon to homebuilders around the country.

Lawsuits against building contractors for radon contamination are expected to loom large in the future. Joel Nobel, a Pennsylvania resident who discovered very high levels of radon in his home several years before the Watras incident, brought a lawsuit against the contractor who installed the home's ventilation system. "You can't sue God," states a counsel for the National Association of Home Builders. "Contractors have the next deepest pockets."[75] Fear of such lawsuits may provide an impetus for builders to incorporate radon-resistant building techniques in the future.

The Real Estate Industry. Fear of a decline in real estate values is often cited as a reason why homeowners may be unreceptive to information about radon contamination. It appears, however, that this fear is not being borne out, even in an area such as Colebrookdale Township in Pennsylvania where the risk is known to be high. David Weiss of the National Association of Realtors (NAR) remarks:

> *We have not seen any decline in values in the Reading Prong. I'm not saying that no house anywhere has been affected, and these are good*

*real estate years, so the effect might be hidden. But people are
getting the message that the problem can be cured.*

The NAR disseminates information about radon to local boards
and state associations. It once provided its own collection of articles; it now distributes "many hundreds" of EPA's citizens' booklets. It has set no national policy about measures being used in
some markets, such as radon clauses in purchase and sale agree-
ments. "It's too early for a national effort," says Weiss. "We urge
caution until there is a better understanding of the radon problem."

As the radon story has unfolded differently in Massachusetts and
Pennsylvania, so do the attitudes of the real estate community
toward the radon problem in the two states differ. In Massachusetts, realtors are breathing a perhaps premature sigh of relief
about radon in the state. The Massachusetts Association of Realtors
has not developed a policy on educating its members because it
hasn't "had any complaints about it or heard of any problems in this
state." A trade magazine reports that fears of radon gas are "ground-
less" and that in the Boston area only a few "radon positive sites"
have been located.[76]

In Pennsylvania, events have forced realtors to adopt much more
aggressive measures. Because real estate companies have been
held legally responsible in the past for problems in homes, such as
contamination by urea formaldehyde foam insulation, the Pennsylvania Association of Realtors advises its members to utilize disclosure forms. These forms discuss radon, direct the seller to test the
home, and require disclosure of test results.

The accuracy of testing is an issue of concern for both realtors
and homebuyers, who point to an incentive for sellers to minimize
or even falsify readings. In 1986 Pennsylvania realtors came under
fire for suggesting use of the Kusnetz testing method in homes.
This method, originally used in uranium mines, draws a 200-liter
sample of air into a container in a five-minute period; it is far less
reliable for predicting long-term exposure than other methods,
which test over a period of days or months. Critics charged that
realtors had recommended use of the Kusnetz method to "minimize the disruption of the selling process and protect the interests
of realtors as far as possible." Realtors responded that they wished
to take into account both testing needs and transaction deadlines,
and that the industry was still learning about the problem.[77]

One proposed solution to the dilemma of testing accuracy is to
set up an escrow account for radon remediation. These funds would
then be available if the buyer subsequently discovers unaccount-
able levels of radon in the home.

Banks and Other Lending Institutions.   Some banks in high-radon areas have expressed concern about losing money on foreclosures of radon-contaminated homes. A number of Pennsylvania banks are discussing avenues such as mandatory radon testing to protect lenders and homebuyers, but most appear to have made no decision on the issue. An EPA official speculates that "eventually testing may be controlled by the lender, handled more like a termite inspection as a standard part of the transaction."[78]

Thus, organizations within the private sector are already carrying out a wide variety of risk communication activities about radon in the course of their business. These activities vary greatly in quality and sophistication and also differ widely from state to state. Many private-sector organizations are taking a wait-and-see attitude before setting uniform policies about radon.

Conclusion

Several preliminary conclusions can be drawn from the material on radon risk communication activities gathered for this case:

1. The manner in which the radon issue has unfolded clearly demonstrates that, for a risk to emerge as a public concern, it is not sufficient for the scientific community to be aware of that risk: at least six years before radon came to national attention, EPA researchers as well as other scientists had projected significant mortality estimates for lung cancer due to naturally occurring radon. Nor is it sufficient, for an issue to be validated in the societal context, that a dramatic event highlighting the risk take place: the national media did not focus on radon until five months after the now-famous Watras incident. In the case of radon, it took the combined efforts of active citizens in a severely affected area and an environmental activist with experience in the problems of radioactive contamination to push the issue into the national spotlight.

2. Scientific uncertainty about radon risks has led to the dissemination of conflicting and ambiguous information. The EPA, which presents information to the public in the form of action guidelines (among other things), is criticized from two sides: some scientists maintain that the guidelines are too conservative, while some environmentalists argue that the guidelines are too high and give homeowners a false sense of security.

3. An increasingly common theme in public-health-oriented fields is that educational programs to change individual behaviors are the preferred approach to protecting health and safety; of all

environmental risks, radon is among them most amenable to this approach. Because the risk is located in private homes, it can be framed as an *individual* risk, and testing and mitigation can be viewed as individual risk-related behaviors. This approach has dovetailed neatly with the antiregulatory climate that has prevailed in Washington as radon has come into public view. Evaluation of behavioral-change programs, however, reveals that "success is claimed more often than demonstrated, and failure is experienced more often than admitted."[79] One of the common assumptions of behavioral-change programs is that when individuals are informed of a risk, they will take appropriate action to reduce that risk. It remains to be seen whether this assumption is valid in the case of radon in homes.

4. Citizen and public interest groups tend to advocate greater federal involvement in the form of public information campaigns and funding for radon testing and remediation, but regulation of private homes is rarely mentioned as a solution. Perhaps the notion of regulating in private homes—like laws requiring automobile seat belt use—violates deeply held cultural values about privacy and individual choice.

5. The distinction between radon as a natural hazard and man-made hazards is ultimately useful only for those who oppose government involvement in the issue for ideological or other reasons. For homeowners faced with radioactive contamination and for those concerned with the public health consequences of radon in homes, it makes little difference whether the source is an inactive uranium mine or the rock and soil occurring naturally beneath the home.

6. The Environmental Protection Agency has been a primary source of information about radon for many news organizations, state agencies, and public interest groups; it therefore has greater control over public information about radon than about a number of other risks. It is possible that EPA achieved this position of high credibility on the radon issue because the agency developed risk estimates and policies that are more conservative with respect to public health than those of other agencies and scientific organizations.

7. In a state where little is known about the extent of the radon problem and where government activity is minimal (Massachusetts), a local television station addressed the problem and generated an enormous viewer response. A quasi-scientific study carried out by the station, an emphasis on the local impact of radon, and specific suggestions to viewers led many homeowners to take a first step to learn more about radon. The full impact of this risk communication event cannot be evaluated, however, since it is not known

how many of those who received additional information tested their homes or how many of those who had unacceptable radon levels took remedial action.

8. The environmental agency of a state with homes found to have extremely high levels of radon (Pennsylvania) developed an aggressive radon risk communication program after the Watras incident. Events in that state also spawned an active citizens' group, which became key in disseminating information about the radon problem at both the state and national levels. But it is unlikely that citizens' organizations would play the same role in any state where radon was found at only moderately elevated levels, since the problem is then not perceived as so immediate and disastrous.

9. The private sector provides a significant part of the radon information received by the public. Its importance is likely to grow in coming years, as the federal government promotes a partnership with the private sector, as more homeowners seek out testing and mitigation firms, as more homebuilders incorporate radon-proofing techniques, as radon testing becomes a routine part of real estate transactions, and as radon-related lawsuits become more common. The quality of risk communication activities by the private sector is highly variable. It might be noted that fear of liability has never inspired laudable risk communication activities in U.S. business history; other positive incentives for businesses to develop useful risk information for the public might be considered.

Endnotes

1. Cass, Peterson, "One in Five Homes Tested Has Too Much Radon," *Boston Globe*, August 4, 1987.
2. U.S. Congress, House of Representatives, Subcommittee on Natural Resources, Agriculture Research, and Environment, Committee on Science and Technology, "Testimony by Stanley J. Watras," *Radon and Indoor Air Pollution* [Hearing], October 10, 1985. Washington, D.C.: U.S. Government Printing Office, 1986, pp. 89–97. (Hereafter, Hearings on Radon and Indoor Air Pollution.)
3. Wayne Lowder, "Part One [Radon]: Overview," in Richard Gammage and Stephen Kaye (eds.), *Indoor Air and Human Health* (Chelsea, Mich.: Lewis Publishers, 1985), p. 41.
4. Anthony Nero, "The Indoor Radon Story," *Technology Review* 89(1):37 (January 1986).
5. Anthony Nero, "Indoor Concentrations of Radon-222 and Its Daughters: Sources, Range, and Environmental Influences," in Gammage and Kaye, *Indoor Air and Human Health*, pp. 46–51.
6. Ibid., p. 50.

7. "Radon in Water: Fact Sheet," EPA, August 1986.

8. GAO Report to Congress, "Indoor Air Pollution: An Emerging Health Problem," Washington, D.C., 1980, p. 32.

9. Nero, "The Indoor Radon Story," p. 37.

10. Steven Page, interview, April 20, 1987.

11. Mitch Coleman, "Air Pollution: The Inside Story," *Family Safety and Health*, Winter 1986–1987, p. 5.

12. Larry Stains, "Stopping the Radon Scare," *Family Handyman*, February 1986, p. 20.

13. "Potential Areas with High Radon Levels: Fact Sheet," EPA, August 1986.

14. Nero, "The Indoor Radon Story," p. 31.

15. Naomi Harley, "Comparing Radon Daughter Dosimetric and Risk Models," in Gammage and Kaye, *Indoor Air and Human Health*, pp. 69–73.

16. Julie Overbaugh, "Discussion of the Lung Cancer Risk Resulting from Radon Exposure to the General Population," Environmental Defense Fund Report, September 1984, p. 4.

17. National Research Council, *Health Risks of Radon and Other Internally Deposited Alpha Emitters (BEIR IV)*, National Academy Press, Washington, D.C., 1988, p. 76.

18. Harley, "Comparing Radon Daughter Dosimetric and Risk Models," p. 75.

19. Erik Eckholm, "Radon: Threat Is Real, But Scientists Argue over Its Severity," *New York Times*, September 2, 1986, p. 10.

20. "Some Scientists Claiming Lower Levels of Radon May Be Beneficial," *Reading (Penn.) Eagle*, January 19, 1986.

21. U.S. Environmental Protection Agency, "A Citizen's Guide to Radon: What It Is and What to Do About It," August 1986.

22. "Additional Questions for the Record Submitted by the EPA," Hearing on Radon and Indoor Air Pollution, House Committee on Science and Technology, August 10, 1985, p. 275.

23. Environmental Defense Fund, "Radon: The Citizens' Guide," 1987, p. 6.

24. Marcia Coyle, "Superfund Can't Finance Most Cleanups," *Morning Call* (Allentown, Penn.), Ocotber 28, 1986, p. 4.

25. Bernard Cohen, "Radon: Our Worst Radiation Hazard," *Consumer's Research*, April 1986, p. 13.

26. "Higher Risks Seen in New EPA Rules—Agency Aides Say Two Proposals Exceed Past Acceptability," *New York Times*, September 18, 1983.

27. U.S. Congress, House of Representatives, Procurement and Military Nuclear Systems Subcommittee, Committee on Armed Services, *EPA Radon and Radionuclide Emission Standards* [Hearing], October 6, 1983. Washington, D.C.: U.S. Government Printing Office, 1983, p. 422.

28. "EPA's Responses to Questions on Standards for Remedial Actions at Inactive Uranium Processing Sites," ibid., p. 522.

29. Michele Galen, "Nowhere to Run from Radon," *The Nation*, February 14, 1987, p. 181.

30. Jason Adkins and Daniel Pink, "Radon: What You Don't Know Can Hurt You," *Public Citizen*, April 1987, p. 13.

31. Hearing on Radon and Indoor Air Pollution, p. 14.

32. Ibid., p. 15.

33. U.S. Environmental Protection Agency, *Unfinished Business: A Comparative Assessment of Enviromental Problems*, February 1987, p. xv.
34. Statement of John P. Millhone, Hearing on Radon and Indoor Air Pollution, House Committee on Science and Technology, August 10, 1985, p. 142.
35. Scott Higham, "Radon Budget Request Cut," *Morning Call* (Allentown, Penn.), April 5, 1987, p. 1.
36. Ibid., p. A22.
37. Ibid.
38. Kathleen Varady, personal communication, July 9, 1987.
39. J. Daniel Nash, "Working with the Phosphate Factor," *Environment* 29(2):15.
40. "Despite Criticisms EPA Staff Stands Ground on Radon Action Guidelines," *Inside EPA* 7(23):2.
41. Ibid.
42. Richard Guimond, interview, June 29, 1987.
43. Steven Page, interview, April 20, 1987.
44. F. Reed Johnson and Ralph A. Luken, "Radon Risk Information and Voluntary Protection: A Natural Experiment," *Risk Analysis* 7(1):97–107.
45. V. Kerry Smith, William H. Desvousges, Ann Fisher, and F. Reed Johnson, *Communicating Radon Risk Effectively: A Mid-Course Evaluation*, EPA-230-07-87-029, July 1987.
46. Allan Mazur, "Putting Radon on the Public's Risk Agenda," *Science, Technology & Human Values* 12(3 & 4):86–93 (Summer/Fall 1987).
47. Hearing on Radon and Indoor Air Pollution, House Committee on Science and Technology, August 10, 1985, p. 147.
48. "Hollywood Comes Calling Radon Victims," *Pottsdown (Penn.) Mercury*, September 15, 1986.
49. Kay Jones, interview, June 1987.
50. Kathy Varady, interview, June 1987.
51. Kathy Varady, "The Impacts of Living in High Level Radon Homes," presented to the Health Physics Society, July 2, 1986.
52. Ibid.
53. Roger Kasperson and K. David Pijawka, "Societal Response to Hazards and Major Hazard Events: Comparing Natural and Technological Hazards," *Public Administration Review* 45:7–17 (January 1985).
54. Ibid., p. 16.
55. Ibid., p. 16.
56. Mazur, "Putting Radon on the Public's Risk Agenda."
57. Varady, "The Impacts of Living in High Level Radon Homes."
58. Hearing on Radon and Indoor Air Pollution, p. 146.
59. Sharon Friedman et al., "Reporting on Radon: The Role of Local Newspapers," *Environment* 29(2):4 (March 1987).
60. Ibid.
61. Robert Hallissey, interview, February 27, 1987.
62. Ibid.
63. Ibid.
64. Susan Walker, interview, May 18, 1987.
65. Ibid.
66. Ibid.

67. Ibid.
68. "Radon Gas: Series Draw Record Responses," *The Rundown* 6(49):361–365 (December 1986).
69. Ibid.
70. "EPA's Radon Action Program," EPA Fact Sheet, August 1986.
71. William Allman, "Staying Alive in the 20th Century," *Science* 85 (October 1985).
72. "Reducing Radon in Structures," EPA, Washington, D.C., pp. I-20–21.
73. Scott Higham, "Remediation Industry Has Growing Pains," *Morning Call*, October 28, 1987, p. 7.
74. Ibid.
75. William Stevens, "Big Increase Expected in Radon Pollution Suits," *New York Times*, September 28, 1986, p. 54.
75. Lauren McCarthy, "Radon Gas," *Boston Homes and Condominiums*, July-August 1987.
77. "Gas Test Assailed in Pennsylvania," *New York Times*, November 2, 1986.
78. Mike McClintock, "Residential Radon Testing," *Washington Home*, June 4, 1987; p. 6.
79. Leon Robertson, *Injuries* (Lexington, Mass.: Lexington Books, 1983), p. 91.

Sources

Interviews

MARY SUE BARRETT, Energy Conservation Coordinator, Buyers Up, May 26, 1987.
JANICE CASWELL, Public Information, Environmental Defense Fund, April 27, 1987.
PHYLLIS ELIASBERG, Consumer Specialist, WNEV-TV Boston, May 20, 1987.
ANN FISHER, Environmental Protection Agency, February 13, 1987.
RICHARD GUIMOND, Director, Radon Division, Environmental Protection Agency, June 29, 1987.
ROBERT HALLISSEY, Director, Radiation Control Program, Massachusetts Department of Public Health, February 27, 1987.
KAY JONES, People Against Radon, June 1987.
MICHAEL LeFAVORE, Author, *Radon: The Invisible Threat*, June 1, 1987.
ROBERT LEWIS, Radiation Health Physicist, Pennsylvania Department of Environmental Resources, June 1987.
STEVEN PAGE, Public Information and Policy, Radon Division, Environmental Protection Agency, April 20, 1987.
JEFFREY PETERSON, Legislative Aide to Senator George Mitchell (Maine), May 1, 1987.
ROGER STIX, Editor, *Good Neighbor* (Arlington, Massachusetts), June 16, 1987.
KATHY VARADY, People Against Radon, June 1987.
SUSAN WALKER, Special Projects Producer, WNEV-TV Boston, May 18, 1987.
DAVID WEISS, National Association of Realtors, June 17, 1987.

Meeting

MASSACHUSETTS INDOOR AIR POLLUTION COMMISSION, June 3, 1987.

Chronology

1531

Paracelsus describes "male mettalorum," a disease of miners.

"Mountain disease" of European miners is identified as lung cancer.

1950

Beginning of epidemiological studies of U.S. uranium miners exposed to high radon levels (these and later epidemiological studies become the primary basis of risk assessment for indoor air pollution from radon).

1953

Andrew Gabrysh at Oak Ridge National Laboratory finds that radon can build up indoors, especially if there is radioactivity in building materials.

1969

Interest in indoor radon is aroused when exposures from the use of uranium mill tailings in structures in Grand Junction, Colorado, come to national attention. The Colorado Department of Public Health carried out measurements of hundreds of such homes through a federally sponsored program beginning in 1970.

1970–1975

U.S. Department of Energy collects geological reports to locate sources of radioactive fuel for nuclear power plants. A uranium belt in the Reading Prong area is discovered.

1971

Under authority inherited from the Federal Radiation Council, EPA finalizes a new radon occupational exposure limit for uranium miners: the exposure limit is reduced from 12 working level months to 4.

1975

EPA begins a study in conjunction with Florida state and county agencies to assess the radiation impact on people living in structures built on phosphate land.

1976

Data from epidemiological study of Czechoslovakian miners associate lung cancer with radon exposure.

1978

Environmental Measurements Laboratory of DOE conducts a program to measure indoor radon levels in homes in the New York City area to determine normal environmental levels of indoor radon.

Tennessee Valley Authority measures radon levels in homes in Alabama and neighboring states where phosphate slag was used in house construction.

EPA and Montana Department of Health and Environmental Sciences begin taking measurements in homes because of intensive local use of phosphate slag in concrete block. Very high radon levels are found in homes with and without phosphate block.

Uranium Mill Tailings Radiation Control Act is passed. Congress agrees to commit federal funds to remediate homes built with uranium mill tailings in the 1950s and 1960s.

1979

April: The Committee on Indoor Air Quality, a workgroup of 16 federal agencies, is formed to develop a coordinated federal response to indoor radon.

July: EPA recommends that remedial action be taken in existing residences where radon levels exceed .02 working levels, and to average background levels for new residences. EPA estimates that 10 to 20 percent of U.S. lung cancer morbidity is caused by indoor radon.

Bonneville Power Agency's home energy conservation program in the Pacific Northwest discovers radon contamination in its customers' homes in Oregon, Washington, Idaho, and Montana.

DOE proposes nationwide weatherization program; EPA concludes that this program will result in 10,000 to 20,000 lung cancer deaths per year due to indoor radon exposure.

Department of Housing and Urban Development requires homeowners in Montana and Florida to test for radon before seeking federally guaranteed mortgages.

Lawrence Berkeley Laboratory begins measuring indoor radon levels and infiltration rates of houses in the San Francisco area.

1980

August: U.S. Radiation Policy Council issues "Report of the Task Force on Radon in Structures," the first governmental report appraising the indoor radon problem as a whole. It concludes that although attention to the problem is warranted, wide-ranging national programs should not be undertaken until more is known about the prevalence of high exposures and ways of controlling them.

September 24: General Accounting Office issues report identifying radon in homes as a health concern, recommending that Congress amend the Clean Air Act to provide EPA with authority and responsibility for air quality in the nonworkplace.

1981

National Academy of Sciences reports to EPA that indoor air pollution is "a serious and growing problem that can cause discomfort, illness, and even death."

Pennsylvania Power and Light Company begins a home energy conservation project; it reports to the state's Department of Environmental Resources that it has discovered elevated radon levels in some homes.

EPA begins studies of measurement methods for radon decay products in air, effects of air circulation systems on radon progeny levels, and health effects of radon daughters in water.

August: The Reagan administration deletes EPA's entire proposed fiscal 1982 budget for indoor air pollution, arguing that EPA has no statutory authority for this problem.

September: National Bureau of Standards issues a report requested by EPA on radon transport through and exhalation from building materials. The report calls for an assessment of radon measurement methodologies and development of measurement standards, calibration facilities, and measurement assurance mechanisms.

1982

EPA requests $2 million for research on indoor air pollution, including radon; entire request is deleted by the Office of Management and Budget.

1983

January: EPA provides standard under UMTRCA mandating levels to which structures contaminated with uranium mill tailings must be remediated.

August: U.S. House Committee on Science and Technology conducts hearings on indoor air quality research.

October: U.S. House Committee on Armed Services conducts hearings on EPA radon and radionuclide emission standards.

1984

Florida Department of Health and Rehabilitative Services promulgates standards for radon levels in new homes.

Congress appropriates $300,000 for radon in EPA budget.

December: Stanley Watras, a Pennsylvania nuclear power plant worker, sets off radiation alarms on his way in to work. His home is discovered to be the most radioactive home found anywhere in the world.

EPA begins tests of thousands of homes throughout the United States.

1985

Pennsylvania Department of Environmental Resources tests 17,000 homes in the Reading Prong area, discovering that 60 percent have elevated radon levels.

The Radon Subpanel of the White House Committee on Interagency Radiation Research and Policy Coordination is chartered to develop a federal consensus on scientific issues regarding radon exposure.

April: Pennsylvanians Against Radon (later People Against Radon) is formed by Kay Jones and Kathy Varady.

April 17: EPA Acting Deputy Administrator A. James Barnes announces at his confirmation hearing that indoor air pollution and radon remain a low agency priority.

May 19: First national news media coverage of the Watras incident; story by environmental reporter Philip Shabecoff appears on page one of the *New York Times*.

August: Senate Committee on Environment and Public Works conducts hearings on S.1198, a bill to establish within EPA a research program on indoor air quality.

September: EPA Administrator Lee Thomas creates the Radon Action Program to assist states in dealing with indoor radon.

October: Senate Committee on Science and Technology conducts hearings on radon and indoor air pollution.

1986

May: EPA releases its first list of home radon testing companies that meet its measurement proficiency standards.

June: EPA's Office of Research and Development issues the document, "Radon Reduction Techniques for Detached Houses: Technical Guidance" to supply state radiological health officials, state environmental officials, building contractors, and homeowners with information on mitigation techniques.

July: GAO proposes reorganizing and expanding federal efforts on radon.

August: EPA sets homeowner "action guidelines" for radon. It releases preliminary findings that one in eight U.S. homes contain levels above 4 picocuries per liter. EPA also releases two booklets: "A Citizen's Guide to Radon" and "Radon Reduction Techniques: A Homeowner's Guide."

September: EPA proposes limits on radon in drinking water.

American Association of Radon Scientists and Technologists is formed.

October: Congress amends Superfund law to direct EPA to assess the U.S. radon problem and to conduct a demonstration project on remediation methods.

November: First Massachusetts television series on radon by WNEV-TV; 13,000 viewers request additional information.

December: Massachusetts legislature establishes an Indoor Air Pollution Commission.

1987

EPA conducts state radon surveys in conjunction with ten states.

February: EPA issues policy report ranking radon as a relatively high health risk but a low agency priority.

April: Senate Committee on Environment and Public Works conducts hearings on S.744, a bill to assist states in responding to the threat to human health posed by radon.

August: Preliminary results from a new national EPA survey indicate that one in five U.S. homes may have elevated levels of radon.

Acronyms

AECB	Atomic Energy Control Board
BEIR	Biological Effects of Ionizing Radiation
BPA	Bonneville Power Administration
CERCLA	Comprehensive Environmental Response, Compensation, and Liability Act (Superfund)
DER	Department of Environmental Resources (Pennsylvania)
DOE	Department of Energy
DPH	Department of Public Health (Massachusetts)
EDF	Environmental Defense Fund
EPA	Environmental Protection Agency

GAO   General Accounting Office
HRS   Health and Rehabilitation Services (Florida)
ICRP   International Commission on Radiation Protection
NAHB   National Association of Home Builders
NAR   National Association of Realtors
NCRP   National Council on Radiation Protection
NRC   Nuclear Regulatory Commission
OECD/NEA   Organization for Economic Cooperation and Development/
               Nuclear Energy Agency
OMB   Office of Management and Budget (U.S.)
PAR   People Against Radon (formerly Pennsylvanians Against Ra-
               don)
pCi/l   picoCuries per liter
UMTRCA   Uranium Mill Tailings Radiation Control Act
UNSCEAR   United Nations Scientific Committee on the Effects of Atomic
               Radiation
WL   Working Level
WLM   Working Level Month

Selected Bibliography

ENVIRONMENTAL DEFENSE FUND. *Radon: The Citizen's Guide*. New York: Environmental Defense Fund, 1987.

FRIEDMAN, SHARON, et al. "Reporting on Radon: The Role of Local Newspapers." *Environment* 29(2):4–5; 45 (March 1987).

GAMMAGE, RICHARD, AND STEPHEN, KAYE, eds. *Indoor Air and Human Health*, Part 1: *Radon*. Chelsea, Mich.: Lewis Publishers, 1985.

LAFAVORE, MICHAEL. *Radon: The Invisible Threat*. Emmaus, Penn.: Rodale Press, 1987.

MAZUR, ALLAN. "Putting Radon on the Public's Risk Agenda." *Science, Technology, & Human Values* 12(3&4):86–93 (Summer/Fall 1987).

NATIONAL RESEARCH COUNCIL. *Health Risks of Radon and Other Internally Deposited Alpha-Emitters (BEIR IV)*. Washington, D.C.: National Academy Press, 1988.

NERO, ANTHONY. "The Indoor Radon Story." *Technology Review* 89(1):28–41 (January 1986).

"RADON: TRACKING THE INVISIBLE THREAT." *The Morning Call*. Special Reprint, Allentown, Penn., October 28, 1986.

SMITH, V. KERRY, WILLIAM H. DESVOUSGES, ANN FISHER, AND F. REED JOHNSON. *Communicating Risk Effectively: A Mid-Course Evaluation*, EPA-230-07-87-029. Washington, D.C.: Environmental Protection Agency, July 1987.

U.S. ENVIRONMENTAL PROTECTION AGENCY. *A Citizen's Guide to Radon: What It Is and What to Do About It*, OPA-86-004. Washington, D.C.: Government Printing Office, 1986.

Appendix A: Major Issues and Alternative Goals in Assessing and Mitigating Radon in Structures, EPA, July 1985

	(1) Exposure and Risk Assessment	(2) Mitigation of Exposure in Existing Structures	(3) Prevention of Exposure in Future Construction	(4) Program Direction and Leadership
Goal A	Determine national distribution and identify high-risk lands.	Establish agency-led program to develop and implement mitigation techniques.	Establish agency-led program to designate high-risk lands and incorporate preventive techniques in new construction.	Issue federal guidance and recommendations on reducing exposure to radon in structures.
Goal B	Identify high-risk regions only.	Focus on developing state and private-sector capability to mitigate radon.	Focus on providing technical and policy guidance to states and private sector on radon prevention.	Develop agency recommendations for use by states.
Goal C	Respond to states' needs to assess known problem areas.	Promote private-sector development of mitigation techniques.	Promote development and implementation of preventive techniques in private sector.	Assist state-directed programs with technical information.
"Strawman" Strategy	Goal A	Goal B	Goal B	Goal A

APPENDIX B

Lung Cancer Deaths
Associated With Exposure
To Various Radon Levels
Over 70 Years

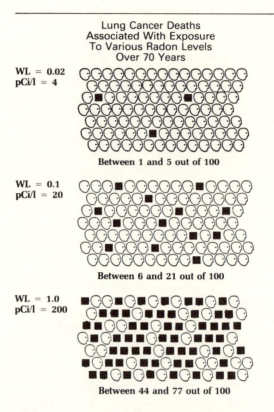

WL = 0.02
pCi/l = 4

Between 1 and 5 out of 100

WL = 0.1
pCi/l = 20

Between 6 and 21 out of 100

WL = 1.0
pCi/l = 200

Between 44 and 77 out of 100

If these same 100 individuals had lived only 10 years
(instead of 70) in houses with radon levels of about 1.0
WL, the number of lung cancer deaths expected would be:

WL = 1.0
pCi/l = 200

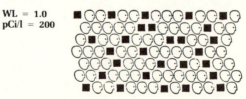

Between 14 and 42 out of 100

Source: United States Environmental Protection Agency, *A Citizen's Guide to Radon: What It Is and What to Do about It* (Washington, D.C.: U.S. Government Printing Office, 1986).

176

Dear Radon Test Customer:

Your radon sampler has been analyzed, and we've found the following results:

CODE #	107513	pCi/L	8.30
CODE #		pCi/L	
CODE #		pCi/L	

The abbreviation pCi/L means picoCurie per liter of air, the most common method of expressing radon/air concentrations. A picoCurie is a trillionth of a Curie, a unit of radioactivity. The following paragraphs explain what your result means, how you should interpret it, and what you might do to follow up on the test. If you have a problem interpreting your results, you may contact our technical service department at 704/891-5892 between 8:00 a.m. and 6:00 p.m. EST.

Variations in Radon Levels

During normal weather, indoor radon levels may rise and fall by a factor of two on a daily cycle—e.g., from 5 pCi/L to 10 pCi/L in 24 hours. During stormy weather, the levels will change even more drastically. The test results we've sent you are as they should be, an average of the radon concentrations in the area tested during the period the sampler was exposed. Bear in mind, if you are doing comparison testing or are averaging a series of tests, that any radon test is ONLY the average of the levels present during a specific period of time. Conditions during subsequent tests may not be the same.

(The following is condensed from the latest radon testing protocols being published by the EPA)

Conducting National Screening Measurements for Radon

Screening measurements should identify houses which contain high concentrations of radon that have the potential for causing increased risk of lung cancer. From a public health viewpoint, it is important to quickly identify such houses so that exposures can be reduced.

Screening measurements should be made in the lowest (potentially) livable area in the house. The highest concentrations of radon will usually be found in a room closest to the underlying soil. Therefore, a screening measurement should be made during closed-house conditions to ensure that the information obtained reflects the *potential maximum concentrations* to which the house occupants *may be exposed*. A measurement made under these conditions is less likely to miss a house where high concentrations may reach a living area. Conversely, if the results of a worst case screening measurement are very low, there is a high probability that the long-term average concentrations in the general living area of that house are also low. This is because average concentrations in living areas are about one-third of the level found in a basement (or other worst case measurement) during closed-house conditions.

If a house has, as a result of a screening measurement, been identified as having the potential for high exposures, *a follow-up measurement should be made before the homeowner makes any decisions* on the need for long-term remedial action.

Follow-up measurements can be used to estimate the long-term averages. Then these averages can be compared with the guidelines given in terms of safe annual exposure concentrations and used to determine the maximum health risk incurred by the occupants.

Using Your Screening Measurement Results

If the measurement results (from a test made under worst case conditions) are less than 4 pCi/L, there is a high probability that the annual average concentrations in the general living areas on non-basement floors of the house are less than 4 pCi/L. Exposures in this range (1 to 4 pCi/L) do present some risk, to the more sensitive occupants, of contracting lung cancer. However, reductions of concentrations this low may be difficult, and sometimes impossible, to achieve. If the screening measurement result is less than 4 pCi/L, a follow-up measurement is not necessary unless the room in which the test is made is used as a living area. This may be the case if there is a bedroom in the basement, or if the house has no basement and the test was made on the ground floor. Another test, conducted during cold weather, when continuous closed-house conditions are easily maintained, could be used to verify and average the current test results.

If the results of the test are between 4 and 20 pCi/L, the occupants should be aware that there is concern about the long-term exposure to these radon concentrations, but that there will probably be no large increase in risk with an additional 12 months of exposure. The occupants should consider making four measurements at three month intervals in order to determine the annual average and, therefore, their long-term exposure levels. Guidance for action to reduce radon levels as well as estimations of health risk are usually expressed in terms of annual average concentrations. (Measurements made to estimate annual averages should be made under normal living conditions in the major living areas.)

If the test results (of a worst case) are between 20 pCi/L and 200 pCi/L, a follow-up measurement, taken in the main living area, should be made within the next few weeks because the occupants may be exposed to dangerously elevated levels of radon. An annual measurement alone is usually not recommended because an additional 12 months of exposure could cause a significant increase in health risk.

If the results are greater than 200 pCi/L, the state health department or regional EPA office should be contacted for advice on immediate reductions in radon levels.

(Although the U.S. EPA is recommending that immediate action be taken at 200 pCi/L, some other agencies are considering 100 pCi/L as the level requiring immediate response. Call your state's health department or radiation control board for their recommendations.)

NOTICE: If your Radon test was conducted in a "potentially unlivable" crawl space or basement, the EPA says this test result SHOULD NOT be used to judge your likely exposure to RADON.

How To Select The Rooms To Be Used For Follow-Up Measurements

The test should be made in at least two main living areas (at two levels if more than one floor). Choose a frequently used room, such as a family room, den, playroom, or bedroom. A bedroom is a good choice because at least one occupant spends one-third of the year there.

If there are children less than 12 years old in the house, it is especially important to measure the concentrations in their bedrooms or in another area where they spend a lot of time. Evidence suggests that children may be more sensitive than adults to the effects of inhaled radon decay products.

The measurements should not be made in a kitchen, because of the likelihood that the range-exhaust fan system and the usually high traffic will cause abnormal changes in the radon levels. In addition, measurements should not be made in a bathroom, because relatively little time is spent in a bathroom. If radon in the water is thought to be one of a home's major radon contributors, then additional tests made in the bathroom could be used to determine a possible contribution from the water supply. Call 704/891-5892 for suggestions on a testing procedure that may locate water-bound radon. (NOTE: Radon is NOT usually found in community water systems where the water has been through a water treatment process. However, water from a private deep-well system may be suspect.)

If you need additional information on a radon problem, you may call your state health/radiation control department or your regional U.S. EPA office listed below. The EPA has published some comprehensive reports on radon and the various mitigation (radon reduction) techniques. You may also call Air Chek's technical service department at 704/891-5892; we maintain a full-time staff ready to assist you with general questions regarding radon. (NOTICE: AIR CHEK'S technical service number has been changed to: (704) 684-0893)

EPA REGIONAL OFFICES:

EPA Region 1
Room 2203
JFK Federal Building
Boston, MA 02203
(617) 223-4845

EPA Region 2
26 Federal Plaza
New York, NY 10278
(212) 264-2515

EPA Region 3
841 Chestnut Street
Philadelphia, PA 19107
(215) 597-8320

EPA Region 4
345 Courtland St., NE
Atlanta, GA 30365
(404) 881-3776

EPA Region 5
230 So. Dearborn St.
Chicago, IL 60604
(312) 353-2205

EPA Region 6
1201 Elm Street
Dallas, TX 75270
(214) 767-2630

EPA Region 7
726 Minnesota Avenue
Kansas City, KS 66101
(913) 236-2803

EPA Region 8
Suite 1300
999 18th Street
Denver, CO 80202
(303) 283-1710

EPA Region 9
215 Fremont Street
San Francisco, CA 94105
(415) 974-8076

EPA Region 10
1200 Sixth Avenue
Seattle, WA 98101
(206) 442-7660

FOR RADON INFORMA-
TION IN N.Y. or N.J.
IN N.Y. CALL 800-458-1158
IN N.J. CALL 800-648-0394
DO NOT CALL YOUR
"REGION 2" OFFICE
LISTED ABOVE!

For additional test kits or information on our open land radon flux test, call: 800-AIR-CHEK or 800-25-RADON or write For More Information Write P.O. Box 100—Penrose, NC 28766

Chapter 5

PLANT CLOSURE: THE ASARCO/
TACOMA COPPER SMELTER

Introduction

Community participation in regulatory decisions concerning the environment is a basic public policy dilemma. Defining criteria for community involvement has been an operational problem for environmental agencies since the 1970s. There has been considerable disagreement as to what community participation means. The following questions lie at the core of the participation issue: How should the public be involved? How much participation is desirable from the agency and community perspective? Is it possible to have technical and scientific discussions of risks and benefits in the context of long-standing emotional debates on these issues?

During the summer of 1983, the national headquarters of the U.S. Environmental Protection Agency, in conjunction with its Northwest regional office, carried out a highly publicized process of communicating the risks of industrial emissions to a neighboring community. The process grew out of EPA's role in setting a national standard for arsenic emissions. The standard under consideration was widely perceived to have a significant impact on the operation of the American Smelting and Refining Company (ASARCO) plant in the Tacoma, Washington, area—a copper smelter that emitted arsenic as an industrial by-product. The EPA had a dual role. First, it was involved in setting a national arsenic standard. Second, it played a major role in managing the risks of arsenic emissions from the ASARCO plant—a facility with a long history of environmental problems. This single plant was responsible for nearly 25 percent of the annual inorganic arsenic emissions in the United States and became a central focus of EPA's standard-setting process.

On one side of the regulatory equation regarding the ASARCO plant were the risks associated with the human exposure to arsenic from two major sources of emissions—fugitive emissions (high concentrations released at ground level during the manufacture of commercial arsenic) and stack emissions (lower concentrations released with other copper-processing by-products from the stacks). On the other side of the regulatory equation was the fact that strict control to a near zero level of both arsenic and sulfur dioxide (SO_2) emissions threatened the economic viability of the ASARCO plant—an important source of local employment and a major contributor to the tax base of Rustin, Washington.

The recently appointed EPA administrator, William D. Ruckelshaus, took a special interest in the ASARCO problem. The involvement in ASARCO/Tacoma gave Ruckelshaus an opportunity to capture national confidence in an agency that was rife with scandal early in 1983. It also afforded him an opportunity to put into practice some emerging ideas he had about the importance of risk communication in the scope of risk assessment and risk management function of federal agencies.

As part of its mandate to establish an arsenic standard, EPA initiated a new type of agency-citizen communication, subsequently referred to as the "Tacoma process." There were several unique aspects of this process. First, it captured the attention of the national press because of the special interest shown by William Ruckelshaus. Second, the intensity of EPA's direct participation in communicating the process of setting an emissions standard with a local community is uncharacteristic of the agency. Third, the Tacoma process case has been praised for having established an innovative and imaginative approach to agency risk communication.

The Tacoma process looms large in the EPA lore and among many who seek new ways of resolving intense controversies between polluting industries and local communities. By court order EPA was required to set an arsenic emission standard. Because of the age of the ASARCO plant and the protracted period of regulatory delays in controlling the emissions, the new standard was widely perceived as a threat to the preservation of the plant. The economic consequences of closing the plant were a major concern to certain segments of the Tacoma area. Many workers whose livelihood depended on the plant's operation believed the controversy was a fictional construct of environmentalists and misguided regulators. They were concerned that plant closure would cause emotional stress as a consequence of unemployment or a reduced standard of living for both plant employees and others whose livelihood depended on the smelter; more than one thousand people would be directly affected. Other residents in the vicinity of the plant decried the impact that arsenic emissions had on their community and feared for the health of their children and the quality of the environment.

In essence, EPA entered a long-standing local controversy and decided to open up its standard setting to the community. This was accomplished through a diverse public information strategy leading to a series of structured workshops held for members of the Tacoma community. In the workshops, EPA explained the dilemmas of regulation directly to the people most affected by the standards it was mandated to promulgate. It solicited inputs from residents, workers, and management, hoping to reverse a stereotype of "the faceless bureaucrat" in environmental risk management. A broad range of experts and community residents were involved, including regulators, administrators, scientists, the press, environmentalists, and workers. Although it retained its final decision-making authority, nevertheless, EPA tried something new.

A complex regulatory framework set the context for the Tacoma process risk communication. Five major state agencies had developed policies on SO_2 arsenic emissions from the ASARCO plant. The lead regional agency was the Puget Sound Air Pollution Control Agency (PSAPCA). Also involved were two locally based environmental groups solely devoted to solving the problems associated with the copper smelter, and six national public interest groups.

While a simplistic rendering of the debate in the national media characterized it as a "jobs versus health" issue, there were many complicating factors such as ASARCO's declining competitive position in the world copper market, obsolescence of the Tacoma plant, the costs of environmental cleanup from decades of pollution, and the slow transformation of Tacoma from a blue collar into a high-tech community.

Tacoma has achieved legendary status in the folklore of risk communication. Since the field of risk communication, as an object of study, is so new, the Tacoma experiment has set an important precedent for future risk management activities. This case study explores the myth and reality of the Tacoma process and examines whether the direct and high profile involvement of EPA headquarters in establishing a systematic link to the community at risk provides a generic model for risk communication of comparable events. The study also examines how different participants in the ASARCO/Tacoma controversy viewed the Tacoma process, what different stakeholders achieved by it, and how its outcome compares with the general goals envisioned by EPA. We begin the case with a rendering of the long and complex social and regulatory history of the ASARCO copper smelter and its relationship to the Tacoma community.

Historical Context

The American Smelting and Refining Company was formed in 1899 by the consolidation of several Guggenheim metallurgical firms. In 1912 the Tacoma facility, originally built as a lead smelter, was converted to a copper smelter.[1] Prior to World War II, Tacoma was the only smelter in the world capable of processing high-arsenic, high-sulfur ores from the Lepanto, Philippines, copper mine; consequently, Tacoma enjoyed a specialized operational niche. These ores were also a less expensive source of feedstock than those used by other smelters.

Similarly contaminated ores from northern Peru were added in the 1940s to comprise one-third to one-half of Tacoma's raw materials. Because there were no competing smelters for these high-arsenic, high-sulfur ores, ASARCO was able to pass through to the mines the full cost of processing, irrespective of world copper feedstock markets.

Following World War II, this type of custom smelting operation became increasingly marginal, the availability of feedstocks dwindled, and price controls were curtailed. To reposition itself in the market, ASARCO reduced custom smelting operations, consolidated most of its national operations in a new Amarillo, Texas, refinery, and upgraded its Hayden, Arizona, smelter. While these actions made ASARCO better integrated and less dependent on foreign markets, the added expenses of using comparatively low-arsenic copper feedstocks forced the company to reduce copper production to about 50 percent of capacity by 1974. This further eroded the earning power of the Tacoma plant and placed the smelter economically at risk for closing.[2]

Influence of International Copper Markets in the 1970s

The Tacoma plant became a more economically marginal operation to ASARCO. From the mid-1970s, investments in the plant failed to cover cumulative depreciation. This policy of corporate disinvestment in Tacoma was consistent with ASARCO's long-term management goals forced by changes in the availability of inexpensive feedstock and increased foreign competition. In 1975, for example, Philippine and Japanese interests announced construction of a plant to use the Lepanto copper ores. Loss of this captive base could only be replaced with more expensive, higher-quality ores with higher transportation costs. The Tacoma smelter would have closed at the time, except that continued operation was required to maintain production while ASARCO was completing its long-range corporate plan of capital redeployment and plant modernization elsewhere.

In 1980 ASARCO management estimated that $165 million was required for SO_2 emissions compliance with the Clean Air Act (CAA) in the Tacoma facility. The EPA estimated that only $65 million was required. Both parties could have been correct, as ASARCO's estimates included re-engineering the total Tacoma plant to make the capital investment required for the emission controls profitable in the long run; the EPA's estimate applied only to the costs of adopting the specific engineering improvements necessary for compliance. After more than a decade of resistance, ASARCO finally complied with the environmental laws, but it did so in its newly constructed, state-of-the-art plant located in Hayden, Arizona—not in the older Tacoma plant where such an investment seemed economically irresponsible to the company.

In 1981, EPA Northwest regional economist Robert Coughlin prepared a report which concluded that ASARCO/Tacoma had limited economic life, and that it probably would not continue to operate for more than five years.[3] The report utilized ASARCO's financial statements and standard secondary sources of information about refineries. The conclusions were that overcapacity in the copper industry in general and overcapacity within ASARCO in particular, and loss of high arsenic ore from the Philippines were the major factors that would cause plant closure by 1986 or 1987.

Early Regulatory Problems

When arsenic was listed as a hazardous air pollutant under Section 112 of the Clean Air Act (CAA) in 1980, no national standard regulating the atmospheric discharge of arsenic was proposed at that time. There was universal agreement in the scientific and regulatory communities that there were large gaps in the knowledge about arsenic particulates—such as its carcinogenic potential at relatively low levels of communitywide exposure.[4] On the national level, there were even more pressing air emission problems, namely sulfur dioxide, which commanded the attention of regulators and captured most of the funding. However, EPA did provide regulatory oversight for arsenic emissions, and it was involved in the design and approval of engineering control devices.

By 1983 a workplace standard was established by the Occupational Safety and Health Administration (OSHA). Several epidemiological studies strongly associated exposure to organic arsenic with lung cancer among workers in smelters and chemical plants. Follow-up studies confirmed a dose-response relationship (higher levels of arsenic exposure resulted in higher levels of risk); nevertheless, studies based on the Tacoma population found no excess morbidity and mortality related to arsenic exposure. Important scientific investigations and regulatory actions of arsenic were also conducted at the state level, particularly by the Puget Sound Air Pollution Control Agency (PSAPCA) and the Seattle-King County Health Department (S-KCHD). PSAPCA was established in 1968, and its initial efforts were directed to control SO_2 emissions from the ASARCO plant. It issued orders for compliance and was involved in protracted legal battles with ASARCO during the next fifteen years.

In the early 1970s field studies conducted by the state health department confirmed high arsenic levels in the urine and hair of children who attended grammar school near the plant. Contaminated soils, identified near the smelter, had arsenic levels of 1–30 ppm above background levels. The level of soil contamination declined with distance from the ASARCO facility, providing evidence of causality.[5]

In an 1980 agreement with PSAPCA, ASARCO agreed to install secondary hoods (an emission capture device) over the converter operations. One hood was in operation by 1983, and two more were scheduled for installation pending their recognition by EPA as the best available technology (BAT) to reduce arsenic emissions.

On June 22, 1983, EPA Administrator William Ruckelshaus delivered a major policy statement, entitled "Science, Risk and Public Policy," to the National Academy of Sciences. This speech, which stressed the importance of public involvement in regulatory decision making and carefully differentiated risk assessment from risk management, set the stage for the creation of the Tacoma process.[6] Less than one month later, on July 12, 1983, EPA issued its proposed arsenic emission standard. The proposed standard was established to regulate arsenic emissions specifically from copper smelters and glass plants. Other sources of arsenic emissions, such as arsenic chemical manufacturing plants, cotton gins, lead smelters, and zinc smelters, were not regulated by this standard.

Before adoption of the final arsenic standard, however, ASARCO announced that it planned to close the plant. All smelting operations ceased March 25, 1985, and arsenic production was terminated January 14, 1986. The risk of plant closure had been discussed for many years by workers in the plant and residents in the community. Since the mid-1970s ASARCO annual reports had made closure seem inevitable. Local newspapers had reported the probability of plant closure for nearly a decade.[7] The Chamber of Commerce and allied factions within the Tacoma community actually supported the notion of plant closure. These proponents of what was called the "New Beginning" program wanted to transform the smokestack image of the city into a high-technology center. Workers at the plant, however, because they faced economic dislocation, protested closure; the Steelworkers Union was particularly concerned with potential loss of jobs. Environmentalists, on the other hand, were pleased that the sources of contamination would be curtailed, but their concerns were immediately transferred to Superfund cleanup activities and follow-up health studies.

Legal and Regulatory Structure

The regulatory structure involved in determining the health risks of the ASARCO/Tacoma smelter was extremely complex.[8] There were five major state and federal agencies that had developed policies on arsenic and other emissions (primarily SO_2) dating back to the late 1960s. A key local agency was the Puget Sound Air Pollution Control Agency, which had reviewed numerous epidemiological studies and set regulations that required ASARCO to expend more than $40 million in air emission controls. From 1971 on, EPA played an increasingly important regulatory role, first in determining the strategy to control fugitive arsenic emissions in the ASARCO plant and later in developing the high-intensity risk communication activities around the proposed arsenic emission standards in 1983 that became known as the Tacoma process.

The ASARCO/Tacoma copper smelter introduced more arsenic in its feed material than the combined total from the fourteen other smelters operating in the United States. It was the only smelter in the nation that had this capacity, which defined its market niche. Because of this unique situation, it was estimated that the Tacoma facility generated almost 23 percent of the nation's total arsenic emissions. These special circumstances ultimately required EPA to propose a separate emissions standard for the ASARCO/Tacoma plant. A brief history of the regulatory risk assessments and communications leading up to the workshops in 1983 follows.

In 1968 PSAPCA was established and in the following year

adopted an enforceable SO_2 ambient standard and an SO_2 concentration standard. These were developed as part of a federally mandated state implementation under the Clean Air Act. In the summer and fall of 1972, three agencies conducted lead and arsenic exposure studies near the smelter: PSAPCA studied soil emissions: the Seattle-King County Health Department (S-KCHD) studied the soil and the hair, blood, and urine of children on Vashon Island; and the Washington State Department of Social and Health Services (DSHS) studied the hair, blood, and urine of children living near the smelter and analyzed vacuum cleaner samples taken from those homes. By year end, DSHS requested PSAPCA to adopt arsenic standards.

In February 1973, PSAPCA conducted a public hearing and adopted an arsenic standard. The next month EPA began a total environment study at the request of DSHS. By early April, ASARCO received the first compliance schedule for the control of arsenic trioxide emissions originating from handling and processing in the arsenic plant. Testing by PSAPCA and EPA disclosed the presence of arsenic in fugitive emissions from several sources.

In 1974 the EPA evaluated the risk of eating vegetables grown downwind from the smelter, and Washington State University test results confirmed the presence of heavy metal concentrations in the soil and in the vicinity of the smelter. To monitor fugitive emission control efforts from the furnace building, PSAPCA began collecting dustfall near the smelter. DSHS requested assistance from the Centers for Disease Control to draw blood leads.

Efforts to control arsenic emissions were intensified when an interagency task force was formed in July 1974. The task force, which included PSAPCA, ASARCO, EPA, DOE, DSHS, the United States Public Health Service (USPHS), and the Air Quality Coalition (a citizen group), decided to review all applicable studies and determine if additional studies were required. The task force later concluded that twelve additional research studies were needed. This activity led EPA to designate the study of ASARCO/Tacoma as its highest priority. To execute this task, EPA created its own in-house task force by year end 1974.

By April of 1975, EPA's Office of Toxic Substances decided to conduct an epidemiological study of airborne arsenic to determine if an ambient air quality standard should be set. On April 16, in an early public risk communication, Dr. John Beare of DSHS held a press conference to explain measures that should be taken to reduce exposure to cadmium and arsenic.

In May 1975, PSAPCA denied ASARCO's request for a variance from its arsenic emission standards. The agency then issued 349 civil penalties at $250 each for the emissions of arsenic-containing particulates. This action prompted EPA to conduct a public meeting in Tacoma entitled "Environmental Concerns and the Tacoma Smelter." During the remainder of this year, ASARCO became involved in multiple legal appeals to gain a variance.

In 1976 EPA became even more active when it requested information from ASARCO to determine whether air quality standards (for SO_2 and particulates) were being violated. The PSAPCA recommended that EPA require an Environmental Impact Statement (EIS). In early 1977, EPA requested that ASARCO submit economic data to ascertain if the company was financially able to comply with the State Implementation Plan (SIP). ASARCO declined to furnish these data. By April of that year EPA issued a Notice of Violation (NOV) of DOE and PSAPCA emission standards for SO_2 and particulates.

In August 1978, ASARCO filed a new variance application with PSAPCA over SO_2 emission standards, but PSAPCA refused to evaluate it until the EIS was completed. The company responded with more legal actions. Citizen groups reinforced PSAPCA's position by filing a supporting appeal with the Pierce County Superior Court. Also, the OSHA arsenic workplace exposure standard of 10 micrograms per cubic meter became effective at this time.

In an important regulatory-based risk communication, in 1980 EPA added inorganic arsenic to the list of hazardous air pollutants under section 112 of the Clean Air Act. One of the major sources of potential public exposure was primary copper smelters, and the ASARCO smelter in Tacoma became the focus of further studies. This listing increased the concern of local regulators and environmental groups about the health hazards of the smelter.

In 1980 ASARCO again applied for and received a one-year variance from SO_2 and particulate emission standards that were still the primary regulatory focus of PSAPCA. By year's end EPA had issued a notice of mainstack opacity violations to ASARCO and designated the smelter site as part of a Superfund (National Priority List) area known as Commencement Bay Near-Shore Tidelands.

In April of 1981, PSAPCA granted another one-year interim variance from SO_2 and particulate emission controls. After two public hearings were held on the Draft Environmental Impact Study (DEIS), ASARCO requested permission to install secondary hooding on the smelter converters to control fugitive SO_2 and particulate emissions. These hoods would also control arsenic emissions, but they were not specifically intended for this purpose. The PSAPCA requested EPA to promulgate national arsenic emission standards and regulations on dispersion techniques. EPA began negotiations with ASARCO to have the company voluntarily install and operate appropriate arsenic controls. By the end of 1981 the Final Environmental Impact Study, with an appendix about secondary hooding, was published. The PSAPCA then ordered installation of secondary hooding, instituted interim control requirements, and began a compliance program to target SO_2 emissions.

From 1982 onward, regulatory concern began to focus on the arsenic emission problem at ASARCO/Tacoma. The EPA followed a risk management approach consistent with section 112 of the CAA. According to EPA's interpretation of section 112, the law was not intended to eliminate *all* risks from airborne carcinogens but to provide an ample margin of safety to protect public health. Some small residual cancer risk could still be allowed. The agency's first step would be to evaluate how the current controls could be upgraded to the best available technology. If BAT was not in place on important emission sources, EPA would select an alternative based on economics, energy, and environmental impacts. The final step was to estimate residual risks beyond BAT control. In developing a proposed arsenic standard, EPA's Office of Air Quality and Standards had extensive communications with ASARCO, PSAPCA, and other interested parties.

In 1983 ASARCO and state agencies began to respond to higher arsenic levels as determined by a number of community studies. The development of pollution controls (BAT) at the ASARCO/Tacoma smelter involved extensive documentation and public meetings at a time when EPA was developing an approach to a national arsenic standard.

On July 11, 1983, the period of high-intensity risk communication began when EPA proposed national arsenic emissions standards. A series of three highly publicized workshops for Tacoma area residents were held in August. In November, public hearings were conducted in Tacoma and in Washington, D.C. These were the legal forums for presentation of testimony to establish the arsenic standard.

While EPA deliberated on setting a national standard, PSAPCA set a firm deadline of October 1, 1984, for installation of the secondary hoods. The Washington Department of Environment (WDOE) circulated a "community exposure standard" for arsenic in December, incorporating trigger levels for regulatory action.

In January of 1984, PSAPCA directed ASARCO to monitor arsenic emissions on a continuous basis rather than every sixth day. ASARCO announced its intention to evaluate a new smelter process, and the company requested a waiver from installing the secondary hoods. The state regulations for arsenic emissions were adopted by DOE on April 10, 1984, and by PSAPCA on May 5, 1984. In May, PSAPCA levied forty-nine civil penalties of $250 each on ASARCO for failure to monitor arsenic emissions during February and March. A year later in March 1985 ASARCO discontinued its smelting operation. As a result, the federal arsenic standard was never promulgated.

In summary, the regulatory context for the risk assessment and communication activities surrounding the ASARCO/Tacoma copper smelter occurred over a twenty-year period and involved national, state, and local agencies. The regulatory issues focused first on SO_2 and particulate pollution, and later on the arsenic exposure problem. Much of the risk communication activities prior to the proposed national arsenic standard were highly technical assessments of engineering approaches to emission control (the secondary hoods for fugitive emissions). The Clean Air Act provided the primary structure for EPA risk management activities, but the state-level environmental agencies and local activist groups all developed approaches to the issue of best available technology and the problem of providing an ample margin of safety to protect the health of Tacoma area residents. Differences between these groups would become the crux of risk assessment debates that will be described in the next section.

Risk Assessment

Risk assessment regarding the ASARCO/Tacoma smelter focused on estimating arsenic emissions into the air, evaluating the control technologies necessary to reduce emissions to a reduced level through BAT, and estimating the residual health risks to the exposed community given the implementation of BAT. The economic consequences to the smelter were important additional variables, as some control scenarios would clearly force plant closure. There was significant uncertainty in the risk assessment process for determining the health risks. This involved estimates of emissions, dispersion models, and expected cases of cancer in the exposed population. This section will focus on risk assessment activities leading up to the proposed national arsenic standard. Citizen involvement in the Tacoma process will be discussed in the following section.

The Problem in Context

On July 11, 1983, EPA proposed standards intended to reduce industrial emissions of inorganic arsenic.[9] These proposed standards were promulgated to comply with a New York District Court order issued in January 1983, resulting from New Jersey's objection to arsenic emissions from a New York glass plant. Of the estimated 1,200 million grams of inorganic arsenic emitted nationally in 1983, nearly 25 percent were identified as originating from the ASARCO/ Tacoma smelter, which used copper ores with an estimated average arsenic content of 4 percent (compared with the average of 0.6 percent in other U.S. smelters). Thus, the Tacoma smelter became a principal focus of the EPA's process for developing an arsenic standard.

The proposed standards were based on risk assessment studies conducted by EPA since arsenic was designated a hazardous air pollutant. The following were generally accepted standards concerning arsenic risk:[10]

- Inorganic arsenic was identified as a hazardous air pollutant listed under section 112 of the Clean Air Act. Arsenic caused or could contribute to "an increase in mortality or an increase in serious irreversible or incapacitating reversible illness."
- There is a high probability that inorganic arsenic is carcinogenic (cancer causing) to humans and that there was significant public exposure to the pollutant.

Inorganic arsenic, like most carcinogens, posed a risk at any level of exposure. Therefore, any emission was assumed to present some human health risk. The EPA's policy toward carcinogens that had no absolute safe level of exposure (no-threshold pollutants) was to provide a standard "at least to a level that reflects the best control technology available that is economically attainable."[11] This is the BAT standard described previously.

The EPA draft of the arsenic emission standard was widely circulated for comment about its technical accuracy and policy implications. This was, however, less than a month prior to the announcement of the Tacoma workshops.[12] A health document, prepared by EPA's Office of Health and Environmental Assessment, was developed in response to a request by the Office of Air Quality Planning and Standards. This comprehensive draft document contained a summary of scientific literature, evaluation of key studies, and evaluations of the toxicity of arsenic documented by the studies. Observed effect levels and dose-response relationships were placed in perspective with observed environmental levels. These were the two major risk assessment reports provided by EPA; they became important risk communication documents, even though they were intended to address technical and rule-making considerations.

Table 5–1 **Availability and Summary Assessment of Potential Control Options**

Potential Control Option	Category of Emissions	Availability	Assessment
Converter fugitive emissions controls	Fugitive	1984	Affordable
Equipment and work practice controls	Fugitive	1984	Affordable
Curtailment during malfunctions	Fugitive and process	1984	Affordable
More efficient stack controls	Process	Unknown	Ability of control devices to further reduce stack emission rates has not been demonstrated.
New smelting technology	Fugitive and process	Unknown	Technical feasibility of the use of flash smelting (or other technology) with high-arsenic concentrates has not been demonstrated. ASARCO will investigate the feasibility of flash-smelting Tacoma concentrates as required by Tripartite Agreement.[a] PSAPCA 1981 Board Order requires either FGD[b] system or new smelting technology by July 1987.
New arsenic trioxide production	Fugitive	Unknown	Bench-scale tests have been conducted. Pilot plant testing of several processes to begin July 1984.
Limits on arsenic content of concentrates	Fugitive and process	1984	Smelter likely to close if forced to use only low-arsenic concentrates.

Uncertainties in Health Effects

The relationship between arsenic emissions at copper smelters and excess lung cancers was well documented at occupational exposure levels, and it served as the basis for the need to set a communitywide arsenic emission standard. There was, however, significant uncertainty in estimating community risks from occupational exposure data. Early workplace studies showed that risks increased as cumulative exposures to arsenic increased. Lee and Fraumeni's study (1969) showed that workers had an overall significant increase in lung cancer mortality, and a dose response was evident.[13] Lubin et al. (1981), Lee-Feldstein (1982), and Higgins et al. (1982) studied the same smelter workers using different and combined periods of observation, and they verified the earlier findings.[14] Selected case reports and epidemiological studies of cancer or precancerous lesions associated with the Tacoma smelter are summarized in Table 5–1.

Enterline and Marsh (1982) completed a cohort study of 2,802 workers at ASARCO/Tacoma in 1982. This study of workers between 1940 and 1964 suggested that carcinogenic response to arsenic exposure can occur in about ten years. A Standard Mortality Ratio (SMR) of 168 was seen among employees who had worked between ten and nineteen years.[15] When all retired workers, including current workers and those retiring before age 65, were examined, a clear dose response was observed.

Near the ASARCO/Tacoma plant high levels of arsenic were discovered in soil samples, in plants, and in local children. Local epidemiological studies were not conclusive and added to the prevailing uncertainty. Alarmed by findings of high arsenic levels in the urine, hair, and fingernails of schoolchildren, Samuel Milham, a chronic disease epidemiologist, conducted a study near the Tacoma smelter in 1975.[16] His investigations suggested that, although high levels of arsenic were found in the schoolchildren, there was no evidence of any of the classical adverse health effects. The causal relationship between the level of arsenic emissions from the plant and adverse health outcomes was further clouded when two mortality studies of the Tacoma area population, conducted in 1978 and 1982, failed to document excess lung cancer.[17]

---

[a]Engineering and Compliance Plan for ASARCO/Tacoma smelter authored by ASARCO, Inc., the United Steelworkers of America, and Occupational Safety and Health Administration.

[b]Flue gas desulfurization.

SOURCE: Robert Ajax and Janet Meyer, "Policy Consideration in the Selection of National Emission Standards for Hazardous Air Pollutants for the Tacoma Smelter" (Paper presented at the annual meeting of the Society for Risk Assessment, Alexandria, Va., October 8, 1984).

A problem clearly suggested by both the local epidemiological studies and EPA's assessment of the data was that the estimated health risks were conservatively high and might not reflect actual risks in the exposed population. The risk assessment for residents living near the smelter was extrapolated from the cancer risks at a high occupational exposure using a linear model. Estimates of exposure in the population around the smelter were developed from population data and the projected ambient air concentrations calculated from dispersion modeling. It was estimated that in a population of 368,000 living near the smelter the excess lung cancer expected from exposure to ASARCO emissions ranged from 1.1 to 17.4 per year, before installing the hoods, and to .21 to 3.4 after this technology was installed. The assumptions made in the linear extrapolation model are considered conservative by most risk assessment experts; they viewed the mortality and morbidity estimates as the upper bound of cancer risk. Variation in the source emission data for the smelter compounded the inherent uncertainty of estimating cancer risks in the local population. Estimated risks are useful for comparing one substance with another and for setting regulation. However, to a lay audience such estimates are considered evidence of a clear risk to health. A "rough but plausible estimate of upper-level risk" to concerned lay audiences invariably became personalized to their own lives and those of friends and neighbors. Wicklund and Frost concluded: "The concern and fear is real even if the risk estimates are not. These concerns will not go away even with the resolution of the standard."[18]

ASARCO had conducted its own risk assessment. The company drew upon expert witnesses who disputed EPA's emission estimates, modeling assumptions, and no-threshold concept for arsenic. (See Figure 5–1.) ASARCO's major argument was that under prevailing economic conditions it could not afford additional control devices beyond the BAT recommended by EPA.

Different interpretations of risk were based on the risk calculations used as the basis for specifying the emission controls necessary to protect the public health. These arose from modeling assumptions, exposure assumptions, and unit risk numbers. These limitations can be described in the following manner:[19]

> Modeling Assumptions—*Measurement of air concentration of arsenic around the ASARCO plant have not been done thoroughly; however, the measurements that have been obtained indicate lower concentrations than those predicted by the dispersion model. Arsenic emissions data from the smelter used in the dispersion model are not precise. In many cases these emission rates were based on assumptions rather than actual emissions tests. This is especially true*

Source: Kelliher, *The West Seattle Herald* (Seattle, Wash.), August 3, 1983.

Figure 5–1

*for fugitive emissions, which are very important in calculating con-
centration yet are very difficult to measure. Also, estimates of how
these arsenic emissions mix with the ambient air are hard to determine because of the complex geography and lack of specific weather
data for the area around the smelter. These problems may explain
why the ambient monitoring around the smelter shows lower concentrations of arsenic than EPA's dispersion model predicts.*

*Exposure Assumptions—A principal assumption is that all per-
sons living within the 12-mile radius of the smelter will remain in the
same location for a 70-year lifetime and are exposed to a constant,
average concentration of airborne arsenic. This assumption could
result in large overestimates of arsenic exposure for those who spend
a lot of time away from their residences and in underestimates for
workers employed at the smelter. Additionally, exposure to arsenic
from resuspension of arsenic-bearing dust from city streets, empty
lots, and playgrounds has not been taken into consideration.*

*Unit Risk Number—Because arsenic is a carcinogen, it was as-
sumed that a linear relationship exists between exposure and risk.
Simply stated, this means that a person who inhales one microgram
of arsenic per cubic meter of air is one-tenth as likely to get cancer as
a person who inhales ten micrograms per cubic meter. If the relationship between exposure and risk is not linear, a different unit risk
number could result, which would in turn change the lung cancer*

Table 5–2 Summary of Case Reports and Epidemiologic Studies of Cancer or Precancerous Lesions in Persons Exposed to Arsenic

Media	Study Population	Author(s)	Type of Study	Results	Highlights and Deficiencies
Air	Smelter Workers—Tacoma, Washington (Analysis of deaths for 1946–60)	Pinto and Bennett (1963)	Proportionate mortality	No difference in lung cancer proportionate mortality between exposed and unexposed workers	Workers leaving plant before retirement were not included. In the classification of workers by exposure, the "non-exposed" group apparently were exposed since they also had high levels of arsenic in their urine.
	Smelter Workers—Tacoma, Washington (follow-up from 1950–71)	Milham and Strong (1976)	Cohort	40 observed lung cancer deaths versus 18 expected (p < 0.001)	Urinary arsenic levels of persons living around the smelter decreased with distance from the smelter.
	Smelter Workers—Tacoma, Washington (follow-up from 1949–73)	Pinto et al. (1977)	Cohort	32 observed respiratory cancer deaths, versus 10.5 expected (p < 0.05); Dose response seen by urinary arsenic levels and by duration and intensity of exposure.	Study consisted of only pensioners.
	Smelter Workers—Tacoma, Washington (follow-up from 1941–76)	Enterline and Marsh (1982)	Cohort	104 respiratory cancer deaths observed versus 52.5 expected (p < 0.01). Dose response found by intensity and duration of exposure.	Short-term high intensity arsenic exposures appeared to have a greater effect than did long-term low intensity exposures; sulfur dioxide exposure was found to have little or no effect.

SOURCE: U.S. EPA, Office of Health and Environmental Assessment, Draft Review. June 1983. EPA 600/8-83-021A. Table 501.

196

risk estimates made for the population around the smelter. It is
unlikely that the actual cancer risks would be higher than those
predicted by EPA, but they could be substantially lower.

EPA Alternatives and Their Evaluation

Some alternative methods to reduce arsenic pollution were discussed and included running the plant less often, using higher
grade feedstock, installing a different smelting technology, and adding supplementary control devices to the proposed hoods. EPA
officials concluded that these alternatives would be so costly to
ASARCO that the plant would definitely close. The agency did not
seriously investigate these factors as options or enhancements.

The EPA initially estimated that with its proposed arsenic standard in place, there would be four excess cancers among the 1,300
highest risk exposed populations, and 1.5 excess cancers among the
remaining 370,000 individuals residing within 18.8 km. (about 13
miles) of the plant. The EPA calculated the lifetime risk before and
after the proposed controls.

Later, when actual measured emission data were introduced by
both PSAPCA and ASARCO, EPA revised its risk estimates (see
Table 5-2). With this revised estimate, the proposed arsenic standard would have reduced the number of excess cancers from four to
two. The arsenic emission standards were delineated in the April
23, 1983, draft EIS, entitled "Inorganic Arsenic Emissions from
High-Arsenic Primary Copper Smelters—Background Information
for Proposed Standards."

The EPA's approach to risk assessment and management in the
ASARCO case is based on the requirements of the Clean Air Act.
This law is primarily technology-driven, in that it directs EPA to
impose specific levels of control, with a particular focus on technical and economic feasibility. Regulations from this law attempt to
control exposures by the installation of specific control devices. The
agency reviewed control technology that was in place at the
ASARCO plant and determined that only the addition of secondary
hoods over the converter process was required to have the best
available technology in place. These controls would only be 95
percent effective in capturing arsenic emissions, but the estimated
capital costs ($3.5 million to install and $1.5 million annual operating costs) would not be economically catastrophic to the smelter.

The residual health risks described earlier were high compared
to other hazardous air pollution solutions where BAT was implemented. The EPA assessment did not identify any technologies
that would reduce emissions to a near-zero level. Therefore, the

agency perceived that there were only two remaining options: to decrease ASARCO's production or require the plant to use more expensive, low-arsenic ore. The EPA considered that these alternatives would close the plant. As a result, the proposed standard did not go beyond BAT.

Economic factors played a large role in EPA's risk assessment. The EPA estimated that closure would eliminate an estimated 600 jobs at the plant and 500 jobs indirectly related to plant operations, a reduction in work force that represented about 1 percent of the civilian employment in Pierce County. The loss of $20 million in goods and services and $2 million in taxes was also anticipated. Since the Tacoma smelter was the only facility in the United States producing arsenic trioxide commercially, the plant's closure would also make the U.S. chemical industry totally dependent on imported arsenic.

Local Risk Assessment and Alternative Strategies

The EPA options seemed unpalatable to PSAPCA and the Vashon Island Community Council; they had prepared their own risk assessment and maintained that arsenic emission standards should be based on *health* risks rather than on specific equipment installed at the smelter. By establishing a different basis for setting standards, PSAPCA and the council shifted the burden of proof to the EPA to show that either the most exposed individuals, as measured by urinary arsenic levels, have an "ample margin of safety" with regard to adverse health consequences, or that the children as a group do not show excess arsenic in their urine. PSAPCA and the council wanted ASARCO/Tacoma to cease smelting high-arsenic ore, but from the perspective of EPA this approach would lead to closure of the plant and was rejected as a regulatory alternative.

The local risk assessment position was supported by nearly all environmental groups, local government officials, and citizen groups. They did not believe that such an approach to control would force plant closure. Only Friends of the Earth advocated total shutdown of the plant. Most Vashon Islanders favored strict controls regardless of economic consequences, as they lived closest to the plant.

The EPA Standard That Never Was

On June 27, 1984, ASARCO announced its plans to close down the Tacoma smelter within a year. This caused EPA to call off its scheduled promulgation of a standard specifically tailored for the Tacoma smelter.[20] According to Northwest regional policy analyst Randall Smith, the standard would have contained these requirements:[21]

- Use of hoods on converters.
- Use of leak-tight covers on certain equipment for handling and transporting arsenic-containing material.
- Cleanup of spilled arsenic-containing material.
- The regular maintenance and inspection of the equipment used to process, transport, and control arsenic.
- Steps to reduce arsenic emission during startup and shutdowns, and during malfunctions.
- Ambient monitoring of arsenic on and around the ASARCO complex.

Although the risk assessment activities did not lead to a final national arsenic standard, the issues raised in the Tacoma community and in EPA by the proposed standard had an important effect. The situation provided a setting for William Ruckelshaus to suggest application of a new approach to risk management where risk communication to citizens played a key role. The Tacoma process was the most public risk communication activity and will be discussed in the next section.

Risk Communication

Overview

The characteristics of the Tacoma risk communication process were determined by the statutory requirements facing EPA in the development of a national arsenic standard under section 112 of the CAA. The regulations were proposed just prior to the time of the workshops, and a period of commentary and public hearings was required. The workshops and other extensive risk communication activities organized by EPA were established in addition to the mandated hearing process. Through these community workshops, EPA attempted to educate the public about the decision process of setting a national standard prior to the formal hearings.

There were three major phases of risk communication associated with the health hazards of the Tacoma smelter. The first period began July 1, 1967, with the establishment of PSAPCA, and it lasted until July 11, 1983, when EPA announced that it would conduct a series of workshops to obtain input from the public most directly affected by the proposed arsenic emissions standard. During this period communications were generally limited to technical written legal directions from the regulators to ASARCO officials, informal and often acrimonious verbal exchanges between the parties, and low-key press coverage, usually within the context of economic news. Occasionally, news items appeared about environmental groups that sought to educate the public about their perspective. Editorial coverage was sparse and limited to the local area, and comments were mild in tone.

The second period, referred to as the Tacoma process, took place between the EPA announcement of the proposed arsenic standard and the conclusion of the public hearing. At this time national attention was focused on ASARCO/Tacoma. The EPA administrator played a visible role in this phase, although he was not a hands-on participant in the workshops themselves. Communications were numerous and intense, and they originated from many sources— EPA headquarters, EPA Northwest Regional Office, ASARCO, and a wide variety of local health and environmental groups. National print and electronic media covered the events. Many of the most influential newspapers throughout the country reacted to the process in editorials. Coverage was expanded to include network TV news and feature presentations. National interest was concerned with how the EPA operated and the symbolic jobs versus health issue. Media discussion of the technical aspects of the proposed arsenic emission standard itself was minimal. The formal hearings for setting the standard following the public workshops did not generate the extensive media interest that the workshops had.

The third period began when the hearings were concluded on November 2, 1983, and continued through the closure of the ASARCO/Tacoma smelter. The smelter closure and the cessation of arsenic production received only mild reactions from the media. Coverage once more became a local issue. Tacoma began to attract new high-tech industry, and public attention and loyalty gradually transferred to new possibilities of economic well-being. Environmentalists switched their activities to solve the problems associated with the Commencement Bay Superfund site. Follow-up health studies continued, but they were viewed more as routine research.

This section will focus on risk communication events in the second and third periods, which cover the rise and fall of ASARCO as a national issue. The EPA placed great weight on encouraging the citizens of Tacoma to participate in setting their arsenic emission standard because the potential risk would be high, even after the best available technology was applied. EPA Administrator Ruckelshaus was quoted as saying:

*In essence, the citizens will have an opportunity to share with EPA
their reactions to managing the risks involved. We must ask them if
they are willing to accept certain risks associated with exposures to
low levels of arsenic. . . . We must also educate them as to the health
risks involved and the options available to EPA.*[22]

Section 112 of the Clean Air Act frames the issue of residual risks
to human health after the installation of best available technology as
one of determining *acceptable* risk given the economic value of
keeping a plant open. In the ASARCO case, the level of residual
risk (.2 to 3.4 cases per year) was larger than in previous cases. The
risk management strategy of Mr. Ruckelshaus was to open the
process of risk assessment to the affected population. A series of
technical questions would be posed to the public: Will the secondary hoods reduce arsenic emission to an acceptable level? Are
current levels of arsenic emission a problem? What are the long-
term health consequences of arsenic exposure at the levels predicted to exist after the installation of the hoods?

The Tacoma process, as described by Mr. Ruckelshaus's an-
nouncements in the local and national press, would bring in direct
contact the separate worlds of the risk assessment experts and the
affected public. In theory, better risk management decisions and a
better informed public would result from this new approach. In
practice, the process proved to be more complicated.

Mr. Ruckelshaus spoke to the editorial boards of some of the
major area newspapers in Washington State prior to the workshops, but he did not participate directly in any of the Tacoma area
public meetings. Our analysis of the television reporting on Tacoma revealed no interviews or statements from Mr. Ruckelshaus.
In newspaper interviews he stressed that he was "eager to hear
other suggested approaches to arsenic emissions, including additional technical efforts industry can make."[23]

Susan Hall, a public participation consultant hired by Northwest
regional EPA to assist in planning the workshops, reported that
press coverage in the beginning fostered more confusion than understanding. Administrator Ruckelshaus's desire to consult with
the public was interpreted by some elements of the press to be an
abrogation of his decision-making role. The public, in turn, was
uncertain of its role; one frequent response to the invitation to
participate in the workshops consisted of calling up EPA and asking
where to go to vote on the standard. Both TV and print media
emphasized the citizen participation as the "newsworthy" part of
the story, and relegated the important statutory requirement that

EPA would make the final decision to a brief comment at the end of a report or story.

The EPA headquarters flew in an entourage of its experts for informal meetings with interested parties prior to the workshops. Included among EPA experts were Elizabeth Anderson, director of the Office of Health and Environmental Assessment, Robert Ajax, chief of the Standards Development Branch of Research Triangle Park, and members of their staffs.

The EPA was not ready for the press onslaught; when the press started asking hard questions, some members of the EPA staff acknowledged that they did not agree on the value of the process. Unit risk numbers and modeling assumptions, they knew, would be difficult to explain: "They felt that risk assessment was a function of experts, and they were afraid that soliciting wider input might make the public afraid and angry."[24]

Northwest Regional Administrator Ernesta Barnes recalled:

The national interest was a surprise. It only became an issue because of Ruckelshaus' interest in risk communication. It was a misinterpreted issue. I don't recall any involvement of the Region with the issue prior to the arrival of Mr. Ruckelshaus at headquarters.[25]

Both national and regional EPA staff were required to develop an approach in a hurry, as there was only one month between the July 11, 1983, announcement of the proposed standard and the first workshop on August 10, 1983. The staff had to design a program to communicate the complexities of the risk assessment process in a manner that would establish a relationship of trust between the agency and the local community and that could survive the emotionally charged comment period for the proposed arsenic standard. While the standard-setting process requires citizen input at hearings and a public docket of all testimony and background materials related to the proposed standard, the regional office developed a more extensive and comprehensive procedure.

Beginning with a press release on July 13, 1983, on the proposed standard, the public involvement activities included distributing informational materials to all local libraries, compiling a mailing list of 1,000 names, developing fact sheets on the technical issues, providing numerous interview opportunities for the press with key EPA staff, conducting small meetings with interested groups such as the Steel Workers Union, the Lung Association, the ASARCO manager, and others, and setting up an information hot line to answer telephone requests. The activity that generated the most attention from citizens and the media was the public workshops.

The Goals of the Public Workshops

The Tacoma process was accomplished by a series of workshops at which scientific and technical experts and members of the general public were invited to give input about how to determine an appropriate arsenic standard. From EPA's perspective, comments were received with the understanding that suggestions were advisory but not binding on EPA. The EPA revealed its plans for risk management of the copper smelter and shared state-of-the-art scientific and technical data on the health effects and smelter operations. The public was also educated about how EPA assesses trade-offs involving reduction of risk and what investment decisions are required to achieve this goal. Thus, the workshops were designed to inform the public about the regulatory process, the proposed arsenic standards, risk assessment, and risk management. Questions and comments from the public would be used to prepare testimony for the public hearing.

The EPA did not seek to develop a consensus or to change public opinion (except to correct misinformation and to provide new information). However, the public did not appear to fully understand or appreciate the rationale behind EPA's objectives. Some community participants wanted debates, and some expected on-the-spot problem resolution. Conflict and obstructionism were minimized by the EPA's carefully orchestrating the agenda while providing an opportunity for all participants to speak. The EPA also had security personnel present on the scene in one meeting, but found that there was no need for crowd control.[26]

Most of the official communicators desired to correct misperceptions and provide new information. However, there is a very fine line between attempting to fix public misperceptions and trying to alter the beliefs and values of the public. It is understandable that during and after the workshops participants would differ in their interpretation of EPA's basic goals.

The workshops were held on August 10, 16, and 19, 1983, in the evenings in public schools in the community. The forums consisted of introductory remarks by Regional Administrator Ernesta Barnes, who chaired each of the meetings. Then Elizabeth Anderson and Robert Ajax from the national EPA presented background information on the health effects and the proposed standards. Fact sheets which attempted to describe in clear terms the range of technical issues bearing on the standard-setting process were distributed.

After the formal presentations, the audience was divided into three smaller groups of about 75 persons, where discussion was facilitated by both EPA and outside consultants. Barnes, Anderson, and Ajax rotated among the groups, spending about one-half hour in each group. There were also members of various regional and state agencies in the audience who were able to respond to questions. A summary session followed the question-and-answer period. At the first two workshops, two professors from the University of Washington (Kai Lee and Gilbert Omenn) presented closing remarks on more general issues of risk assessment and risk management.

The planning effort for the workshops was divided between regional and national EPA staff. The public relations firm handled logistical details, advised the EPA staff on how to improve and clarify their prepared statements for the workshops, and conducted a dry run of the workshop presentations. This collaboration was generally viewed as successful, although there was some tension between regional and national EPA staff concerning who was leading this effort. Ernesta Barnes recalled,

> *Regions normally don't get involved in the standard-setting process.*
> *While this was a joint effort in communication and education, head-*
> *quarters was too involved. The short sheets [fact sheets] were pre-*
> *pared by headquarters, and our material was reviewed, at least in*
> *part, by headquarters.*[27]

Social Context of the Public Workshops

Risk communication strategies are part of a larger social context. While science and technology might have formed the basis of the risk assessment information presented in the workshops, the public had been exposed in addition to a lengthy and often emotional risk communication process from other sources. Much of the risk communication in the ASARCO/Tacoma case was of an informal nature and occurred before the workshops. Parties with opposing views sent a range of messages that reflected the cultural and value issues that were prevalent in the community.

Local news media played an important role as a communicator of risks concerning the smelter. In 1981 papers ran stories on the risk of plant closure ("Tacoma Smelter Has Dim Future, Says Economist") and reported on citizen anger at public hearings conducted by PSAPCA ("Angry Citizens Sound Off At Smelter Hearing"). Reporting at this time indicated that a high level of community concern and conflict was already in place:

> *How many deaths will occur while we wait for ASARCO to clean up*
> *its emissions, asked Marjorie Williams, Secretary of the Seattle chap-*
> *ter of the International Association of Cancer Victims and Friends.*[28]

ASARCO officials denied that emissions were a threat to the public's health, while at the same time Greenpeace representatives accused ASARCO of "reaping profits, poisoning the land, and causing lung cancer." The most dramatic early message was given by a representative of Greenpeace, who in October 1981 scaled the tall stack and unfurled an 80-foot banner which read:

After the last tree is cut, the last river poisoned, the last fish dead, you will realize you can not eat money.[29]

Many local citizen groups formed between 1981 and the time of the EPA-sponsored Tacoma process. The Clean Air Quality Coalition, Clean Air for Washington, Americans Protecting the Environment (APE), and Tahomans for a Safe and Healthy Environment were just some of the groups that evolved in response to a public perception of a health threat from the smelter. They joined more established organizations like the Lung Association and the League of Women Voters in a broadly based citizen coalition. On the other hand, ASARCO, the Steelworkers Union, and some local elected officials generated community support for the smelter.

The infamous "green undershirt" is the best-remembered visual image associated with Tacoma's arsenic emissions. It was recalled by nearly all of the interviewees contacted in the preparation of this case. The green undershirts belonged to John C. Larsen, who worked at the smelter for thirty years and lived within the shadow of its tall stack. Larsen believed that he became immune to arsenic. He spoke at one meeting of the Steelworkers Union prior to the workshops and wrote this description to a local newspaper:

The first twenty years I worked there [ASARCO], I ate so much arsenic that when I perspired my white undershirt would turn green from the arsenic that came out of my body. No, the arsenic did not kill me, and I don't believe I will die from cancer. The human body builds up a resistance to arsenic so that after a while you become immune to it. The last 10 years that I worked, the smelter cleaned up the air so that I no longer had green streaks in my underwear. The main reason I retired early was because of a regulation saying we had to wear a respirator the whole eight hour shift.[30]

National news began to focus on the ASARCO/Tacoma smelter issues after January 1983, when the court order to develop a national arsenic standard was issued. Most stories clustered around the issuance of the proposed arsenic standard and Mr. Ruckelshaus's comments on risk management and community input discussed previously. National television focused on the workshops and did not play a large role in influencing the local context of risk communication prior to the workshops.

The Dynamics of Workshop Participation

The public workshops generated the greatest interest. Three sessions attracted a total of more than 750 people. The first sessions, held on Vashon Island on August 10, had more than 350 attendees, including members of the press. The two other workshops, held in Tacoma, attracted about 550 people each, but neither of them had large attendance from the media. The questions recorded at the workshops documented that the public was not monolithic. Susan Hall identified three distinct segments—the concerned general public, the informed public, and the technical public.[31] Each group had its distinct perspective and language of communication.

The Concerned General Public.   The concerned general public was pragmatic and anecdotal. They wanted to know the specific adverse health consequences that arsenic emissions would have on them personally, the severity of the problem, and actions that could be taken to eliminate all dangers. The general public was frustrated and angered by complicated technical answers. They believed that technology can provide the answers, and they wanted immediate action. Some members of this group were open to arguments regarding the limits of present technology and cost-benefit trade-offs. Others were suspicious of scientists, the bureaucracy, or ASARCO.

One example of the general public were the union leaders who advised their rank and file to "keep it cool" but to have a visible presence at the workshops. The members were advised not to cause disruptions because that might alienate neutral citizens and weaken their cause. A representative of Tahomans for a Safe and Healthy Environment sported a sandwich board claiming, "Give a hoot, don't pollute." Given this mix of participants, the potential for significant confrontation was clearly present. Some of the questions posed by this group were:

- "Why isn't anyone from ASARCO or the city in front [with EPA] answering questions?"
- "The only risks discussed tonight are long-term risks, such as lung cancer. What about the effects on our health right now? How does arsenic affect common ailments such as asthma?"
- " 'Ample margin of safety' seems like a balance between health and loss of jobs. Safety and health is far more important [to me] than jobs. How can I relate this for formal public testimony?"
- "How much money is available for Vashon under Superfund? What is included in this amount (it seems awfully low)? Will EPA Superfund test fish and vegetables?"

These comments were more emotional and dramatic, and became part of television coverage of the workshops.

The Informed Public. The informed public were familiar with the issue. Some were members of the local community environmental groups and shared a perspective. Others represented only themselves but had followed the issues closely for many years. Most of this group had opinions about how all the responsible parties should respond. They were concerned not only with health and economics but also with the long-range implications of EPA's response.

Employees of ASARCO considered the smelter and its payroll as a blessing, not a curse. The value of cleaner air was a remote and abstract "extra." Their comments in the meeting expressed this position. Workers presented data from epidemiological studies of the effect of job loss on health status and calculated that more lives would be lost from the stress of plant closure than would be lost from community exposure to arsenic at the levels estimated in the EPA models. Some questions from the informed public perspective were:

- "Why isn't arsenic in soil included in the risk assessment? What will EPA do about arsenic in soil and water?"
- "Is it safe to eat produce from Vashon gardens? Where can citizens take produce to be tested for contamination?"
- "What amount of arsenic was previously emitted by the smelter (10–40 years ago) and what effects did this have on the population?"
- "Does EPA consider economics in setting standards? What economic considerations are taken into account? It seems like EPA is only concerned about plant closures and loss of jobs. What about other economic factors such as health costs to the affected public, insurance rates, property values, etc.?"

This group challenged EPA to discuss the implications of technical data in lay terms. They wanted answers to practical questions.

The Technical Public. The technical public attended the workshops to debate the technical nuances of the problems. Their questions focused on interpretations of research data, state-of-the-art knowledge, the validity and protocols of research, and other technological points. Members of the public sought legitimacy by speaking in scientific terms; many of these participants had scientific backgrounds themselves. Brian Baird, president of Tahomans for a Cleaner Environment, was perhaps the best example of this group. He received permission to distribute a questionnaire to workshop participants that would be used as the foundation for his doctoral dissertation. He was highly visible and very articulate. Media representatives often selected him to present the environmentalist point of view.

Here are some examples of technical questions in the workshops:

- "A major assumption being made by EPA is that airborne arsenic is the greatest risk from arsenic emitted by the ASARCO smelter. What is the basis for eliminating *ingestion* from the health risk assessment, especially with prevailing winds that will cause arsenic deposition on soil and possibly contaminate vegetables and drinking water in the area?"
- "Ajax said tall stack emissions do not have effects at ground level. EPA wants to control converter emissions which EPA said has no ground level effects and that 'other' ground level sources cause most problems. Is this true? What is EPA doing about other (fugitive) emissions?"
- "We have two aquifers on Vashon. I have seen no information on levels of arsenic in water. Is Superfund looking at this? Are we addressing only arsenic, or cadmium, too? Cadmium came from the smelter years ago."

There was a range of responses to the carefully prepared EPA presentation. The examples presented suggest that there was a difference in the technical logic of the EPA presentation and the cultural logic of many of the participants' questions. The EPA had come to discuss the assumptions of the health risk assessment studies, the dispersion models, the details of a best available technology standard under the Clean Air Act, and other technical questions. Many of the participants' questions focused on personal issues and equity considerations. An important challenge from the audience was its questioning of the cost-benefit framework as an acceptable regulatory strategy.

Gilbert Omenn, one of the two academic commentators at the workshops, observed:

> *The personal nature of the questions they [EPA] encountered made a striking counterpoint to the presentation of meteorological models and health effect extrapolations the agency presented.*[32]

EPA officials had to contend with hostile responses from participants for providing too little and too much protection, depending on the perspective of the questioner. Workshop participants communicated their messages to EPA through posters and lapel buttons, placards, and other informal risk communications. Brief slogans can often convey a strong message. Some of the placards displayed at the workshops, the hearings, and various faction meetings included the following symbols of the protest:

Heavy metals are forever.
Make love, not arsenic.
Pass the regulatory buck.

Some workers appeared at the workshops wearing a button that said "Jobs." Downwind residents sported buttons that said, "Health." At the next meeting many people formerly representing both widely divergent positions appeared with a new button that read "Both." The new button expressed both a position on the issue and a change in perspective, indicating that the process had served a useful purpose.

Risk communications are sometimes used to sway the uninvolved and uncommitted public in the hope of obtaining increased support on an issue. To a neutral audience, the line between hero and villain was often ill-defined. The conflict of the jobs versus health issue was a clearly defined message that played well on network TV news. A supporter or detractor of ASARCO who had a visual "hook" or a pithy one-liner could become the focus of media attention.

Network news coverage of the workshops stressed the jobs versus health trade-off and the different risk perceptions of smelter workers and fearful residents. These media images presented a community anguished and divided over a tragic and unfair choice. (See Figure 5–2 for the results of one community poll.) Interviews with citizens, the local mayor, the plant manager, and Ernesta Barnes were used to represent the diversity of opinions on the risks of the smelter. Video footage from the workshops focused on questions that challenged EPA or presented emotional and personal stories of fear.

The announcement of the workshops and the public's input in the final EPA decision on the ASARCO smelter had raised expectations for the Tacoma process. Local papers suggested that the citizens had "the onus of deciding 'acceptable' risk." The Steelworkers Union sponsored a door-to-door survey to poll residents' opinions on closing the smelter. Some residents expected the Tacoma process to operate like a town meeting where the public's will would determine a policy decision. EPA officials intended the workshops to be an educational effort directed toward increasing the public's understanding of arsenic-related health risks, the control options that were available, and the cost consequences of alternative decisions. With such a difference in the interpretation of the basic goals of the risk communication process, it is not surprising that there were different evaluations of the success of the intervention.

Source: *The Morning News Tribune* (Tacoma, Wash.), July 24, 1983.

Figure 5–2

EPA Evaluations

From the perspective of the EPA staff who conducted the workshops, the Tacoma process was a qualified success. Most believed that the goals of the workshops had been clearly stated and that the agency had carefully explained that while the process accepted input, EPA did not relinquish decision-making authority. As Ernesta Barnes explained at the time of the workshops:

> *This is not a vote, it is not like the Ted Mack Amateur hour. . . . We are not taking a poll because this is not a polling issue.*[33]

The regional office staff was energized by the efforts required to prepare for the workshops. They were prepared to play a significant role in this approach-to-risk-management event, although the timing was stressful. Ernesta Barnes recalled:

> *Region 10 was way out in front on their true commitment to the public before this thing ever happened. If this had happened in some other region, I'm not so sure that it would have been so successful. The Pacific northwest was way out ahead in real participatory government and public involvement. People really expected to be consulted.*[34]

Throughout the process the regional staff thought the agency gained credibility with the press, environmentalists, and local industry.

An evaluation of the workshops was conducted by Susan Hall, the public relations consultant hired to assist in the development of the workshops. This largely descriptive assessment was quite positive regarding EPA's role in the process:

> *The workshops were in themselves a successful public involvement activity. All of EPA's goals and expectations for the workshop were met. The workshops were well planned, well organized, ran smoothly and on schedule. Conflict and obstructionism were kept to a minimum, while respect was shown for all viewpoints, concerns, and need for information.*[35]

The evaluation pointed to several specific success factors; the open and honest personal styles of Ruckelshaus and Barnes, the EPA's willingness to involve local health experts and environmentalists to help design the workshops, the careful attention to logistics by the staff, and the presence of a knowledgeable "technical public" whose familiarity with the issues helped to inspire confidence in other citizens. According to EPA policy analyst Randall Smith, the formula for success is to be "as open as possible, as soon as possible, and to be as clear as possible." The use of scientific and technical jargon makes "people often think you are trying to trick them. . . . The process proved terrifically costly and time consuming. We clearly don't have the funds or the staff to tackle such a project every time out of the blocks," he concluded.[36]

The EPA staff also recognized that there were some problems with the process:

> *The most misunderstood part was the Carcinogen Assessment Group. They are not normally interactive. The people wanted to hear about all of the pollutants not just lung cancer from arsenic. The emission controls discussed at the workshops were too narrow. . . . We probably didn't do too good of a job of educating. The people didn't know about the risk. They wanted to tell us about their fears. The health groups didn't have time to review the documents and PSAPCA didn't agree with the EPA data.*[37]

In general, the EPA participants in the workshops considered that they had done well considering the situation. The agency was in an early stage of understanding risk communication. The key belief shared by EPA risk communicators was that the purpose of their risk management process was not to eliminate *all* risks but to set a standard that allowed some acceptable residual risk to health. There was some disagreement with this basic premise among other groups who participated in the process.

Public Evaluations

Presenting detailed technical information in the workshops had a mixed effect on the lay audience. The process simultaneously inflamed emotional debate, as it offered a more science-based framework for discussion of the issues. Kai Lee, one of the academic commentators at two of the workshops, stated:

> The workshop experiment was a suboptimized success—it raises questions about how this kind of effort can work. EPA was excellent in describing a clear decision process for an unclear decision. The audience understood the message, but given their divided interest in the outcome, however, understanding what EPA is up to and up against does not make them feel more confident in the outcome. That is the suboptimal part.[38]

The interests of Tacoma residents were very specific and simply stated. Would they get cancer? Would there be any jobs? Owen Gallagher, plant worker for forty-three years and mayor of Ruston for twenty years, spoke for supporters of ASARCO. Dismissing the threat of arsenic emissions, Gallagher concluded, "When you're in a bakery, you expect to get some flour on you."[39]

Those residents who desired stricter controls and were less concerned with the potential loss of jobs became quite alarmed about arsenic and cadmium in their home gardens. Residents of Vashon Island, which was downwind of the smelter, expressed a different perception of the risks from the smelter workers and were outraged that there was communitywide exposure from plant emissions. (See Figure 5–3.) Their position, as expresssed in the workshops, was that ASARCO should absorb the costs of zero-level emissions.

Some community members were quite critical of the workshop process. Janet Chalupnik of the Lung Association had advocated for stricter controls on ASARCO emissions for more than ten years. In her view the workshops

> tended to polarize people. Things got crazy for a while. I felt that more background preparation was needed for the community. The whole thing just seemed to kind of drop on the community. The people perceived that EPA is ready to let some people die. EPA is going to say, "Here is an acceptable risk." Not much education was going on; there were too many people for discussion.[40]

A stronger statement was expressed by Alfred M. Allen of the Seattle-King County Health Department:

> How did EPA perform? I'll be cavalier. We knew ahead of time "that's not the way you do it.". . . They gave a vast amount of scientific information to a lay audience. . . . It was an absolute error

Source: Horsey, *The Post Intelligencer* (Seattle, Wash.), July 14, 1983.

Figure 5–3

in communication. Scientifically and politically it was a failure. The presentors didn't understand the community—this is a blue collar mill town.[41]

There were many opinions in the community regarding the success of the workshops. One environmentalist tried to measure if the workshops had changed any underlying beliefs about the risks of the smelter for the participants. Brian Baird, president of Tahomans for a Cleaner Environment, compiled data from 347 questionnaires (representing approximately 80 percent of the citizens who attended the workshops). Results indicated that citizens' informal risk estimates and risk tolerance were associated with these factors:

- judged environmental attitudes
- perceived benefits of the hazard source
- evaluations of exposure
- voluntariness
- acceptance or denial of vulnerability

He summarized his findings:

Neither factual knowledge of formal risk estimates and proposed standards or residential distance from the smelter was found to be closely related to risk tolerance or informal risk estimates.[42]

After the presentations at the workshops, many of the attendees were still unfamiliar with the basic information; factual knowledge was not a very useful predictor of risk tolerance. Even when information is made available to the public, few individuals will study it carefully. Of those who study the information, many will fail to retain it. Of those who retain it, many will disagree about its validity. Tolerance for risk, based on the Tacoma workshop questionnaires, appears not to be derived either from formal risk estimates or from specific details of risk management proposals.

The Baird study suggests that the Tacoma process of informing the public about risk may not change underlying perceptions of risk tolerance. It documented that respondents on both sides of the issue continued to believe that the general population was in agreement with their respective positions. Consequently, whichever decisions the EPA made, one side will believe that the regulatory actions are an inappropriate response to their assessment of the risk and that of the majority of the public.

The mayor of Tacoma, Douglas Sutherland, played a visible role in both local and national risk communications about the smelter. He characterized ASARCO as a "good neighbor" on ABC's "Good Morning America." He evaluated the EPA workshop efforts this way:

> *If I were to give them a grade, I'd imagine I'd give them about a "B." They found themselves on the short end of a very difficult stick. They had to get together a great number of people to begin a public process that involves a great deal of emotion and a great amount of uncertain information.*[43]

After the EPA workshops, the mayor moderated a municipally televised workshop at the Tacoma Dome. An invited panel of local and national experts, including several from EPA, participated in a day-long roundtable. The audience did not participate directly, but they could submit questions in written form. The audience was small and orderly. There was some local media coverage of this event, but no national coverage. The presentations were highly technical and focused on the emissions model and the estimates of cancer deaths.

Press Evaluations

Both national and local media attention focused on the process and outcome of the EPA workshops. The local press gave detailed coverage to the substantive issues, addressing both the economic and health effects problems. Community residents were quoted often, and the area papers provide a rich chronicle of how the EPA message was embedded in the local history and culture. Some of the local reporting translated EPA's expressions of scientific uncertainty in the workshops as an error in the agency's analytic procedures. For the media, such uncertainty was translated as an "error" in the scheme.

The inaccuracy of the EPA measurement of arsenic output from ASARCO/Tacoma became a public controversy, and it was reported in a newspaper article, "Weighing the Risks: The Smelter Debate." Two undocumented comments in the local media allegedly summarized the attitude of the EPA and its administrator:

Agency insiders, however, admitted the agency itself was embarrassed by its makeshift numbers and welcomed the chance to take more time to take another look at the arsenic emissions at the plant. . . . Ruckelshaus himself reportedly was appalled at problems with the risk assessment model and ordered a reworked model himself. The EPA administrator reportedly told EPA staffers that he was faced with making a decision with "25 percent of the facts."[44]

While such quotes are not official statements, they suggest how scientific uncertainty is perceived by the media and public.

Local and national press focused on evaluating the fairness of the media-characterized jobs versus health decision facing the community. The *Journal-American* of Bellevue, Washington, took the editorial position before the workshops that the ASARCO/Tacoma problem does not lend itself to "resolution by plebescite":

Although the 650 individuals whose jobs are on the line are known, no one knows the additional person who will contract lung cancer because of the smelter. That person is a statistical addition to the projected cases of lung cancer caused by other factors. The Journal-American *asks, "Who will buy funeral bouquets in gratitude for his sacrifice on behalf of all of us here at the smelter?"[45]*

Letters to the editor, expressing a range of viewpoints, appeared in all the Tacoma area newspapers. One reader, Sharon Sauve, a Vashon Islander, wrote her local newspaper before the workshops, exhorting the press to play a more active role:

Because I've worked as a newspaper reporter, I know that the press has power—and certain doors are open to a reporter that would not open to the average person on the street. I'll get polite evasive answers in return, when what I want is hard answers to hard questions. Come on, Editor—you've got clout, if you just use it. Get your staff out into the field(s).[46]

The *Washington Chronicle* defended the process and the administration. Its editorial, entitled "Tacoman's Involved in a Risky Decision," concluded before the workshops that there is no easy answer for Tacoma:

In asking the residents of Tacoma how safe they want to be, Ruckelshaus is only following the democratic tradition of popular participation in government. The benefits of this tradition are many, but they have never come without risk.[47]

The *Boston Herald* concurred. In an editorial entitled "A Risky Choice," they concluded:

In sum, he [Ruckelshaus] has proposed that they weigh benefits against risks and choose how safe they [Tacomans] want to be, in order to help EPA decide what it should do to resolve the issue. That's not a case of Ruckelshaus passing the buck; it's participatory government in action—and it's good.[48]

"Mr. Ruckelshaus as Caesar," a highly critical editorial in the *New York Times*, criticized the administrator for imposing an impossible task on the citizens of Tacoma. It quoted Ruckelshaus as saying, "I don't know what we will do if there is a 50-50 split," while contending that he was not taking a referendum, just testing for consensus. The *New York Times* stated that he "has no business doing either"—his job is to protect the public from hazardous air pollutants "with an ample margin of safety." The editorial concluded with this statement:

The Roman Caesars would ask the amphitheater crowds to signal with thumbs up or down whether a defeated gladiator should live or die. Mr. Ruckelshaus, sitting in Washington, D.C. or Washington state, shouldn't need to ask the public whether a 5 percent extra risk of cancer is acceptable.[49]

The Public Hearings

The public hearings mandated in the CAA were held in November of 1983. Meetings were held in Tacoma and in Washington, D.C. At the Tacoma meetings more than 143 persons presented testimony, which EPA officials considered a very good turnout. One effect of the workshops was evident at the public hearings: the different interest groups had developed more precise statements of their respective positions. The testimony presented was quite focused and centered on specific options for setting the arsenic standard.

There was opposition to the EPA's proposed standard. Harvey Poll, representing PSAPCA, testified that requiring hoods to control fugitive emissions was a good approach but insufficient to protect the health of Tacoma residents. He proposed a monitoring system that would define an operational standard for emissions from the plant. When monitors determined that arsenic levels were too high, the plant would be required to cut back on operations. PSAPCA also wanted to link the problem of SO_2 controls to the technologies that the smelter would be required to use.

There was a mixture of opinions expressed at the hearings. Most of the environmental groups supported the action-level approach and also criticized EPA's use of ASARCO's data in determining the economic impact of alternative control technologies. "This places the regulation of a carcinogen in the hands of corporate accountants," one group said. The ASARCO representatives, on the other hand, refuted EPA's modeling assumptions and challenged the no-threshold notion for arsenic. The union supported the EPA proposal and suggested that further research should be conducted on future controls. Residents who lived near the smelter presented emotional testimony on the social costs of arsenic pollution as they had at the workshops. They were not concerned about the smelter closing; they simply wanted to stop being exposed to arsenic. There was some local media coverage of the hearings, but there was no national coverage of this part of the process.

Robert Ajax, of the Air Quality Standards Division of EPA, read more than 650 comment letters on the proposed standard during the public comment period after the hearings. He found that a "slight majority" of all commenters recommended that EPA adopt only hoods. This suggests that nearly half of the letters supported other positions ranging from "no regulation is needed" to "the smelter should be closed." These comments, along with revised computer models of exposure and risk estimates, were factored into developing a final regulation on arsenic emissions by the summer of 1984.

The regulations finally proposed by EPA headquarters were "significantly different" from the recommendations prepared by Ernesta Barnes and her staff.[50] The Northwest regional office had recommended that a boundary limit for arsenic emission levels be established around the smelter. If arsenic controls inside the smelter were adequate, the excessive emissions would be indicated by monitoring beyond the smelter site boundary. The boundary limit concept was brought to the EPA's attention during the public comment-public hearing process, but it was not incorporated in the promulgation EPA was to have made in August 1984.[51] Robert Ajax recalled that such an approach was not accepted because the effects of the regional proposal were not quantifiable and would have likely caused plant closure.[52]

Reactions to the Smelter Closing

Reactions from Public Officials. When the smelter closing was announced, emotions ran from anger, shock, and dismay to quiet resignation and even optimism. County Labor Chief Clyde Hupp explained the closure this way: "Asarco's troubles have been known for many years. . . . The community has been in an anticipatory mood, but it's still bad news. It's a shock."[53]

The mayor, who had characterized ASARCO as "a good neighbor" on the television program "Good Morning America," was characteristically upbeat:

> *It's another challenge for the city. . . . It's a part of the transition from what Tacoma was to what Tacoma will be. . . . The timing is bad in the short term. . . . It's going to be tough, but it's something gutsy cities like Tacoma are able to handle.*[54]

The destruction of the smokestack image might have been a positive accomplishment—except that it was achieved by the elimination of 550 well-paying jobs. This loss and its attendant ripple effect were compounded by Pierce County's preexisting 11 percent unemployment rate. The economic jolt was somewhat softened by the formation of the Tacoma-Pierce County Development Board and the city's "New Beginnings" revitalization program. But this was of no immediate solace to the unemployed smelter workers.

The media interest reawakened for the plant closing story. There was some national television reporting and extensive local news coverage. The Tacoma *News Tribune* described the plant closing this way:

> *A new skyline is taking shape, one that need not depend on the giant smokestack for nurturing. New jobs are being created, as are new opportunities for business. The mourning for the smelter's closing may be filled with sadness for many, but there is no need for dirges of despair.*[55]

"Its time has more than come," concluded the *Seattle Times* in its June 28, 1984, editorial post-mortem, "Symbol of Pollution":

> *No matter how contradictory through the years the pronouncements of industrial chemists on the one hand and environmentalists on the other, we could never buy the notion that pumping hundreds of tons of arsenic from the smelter's soaring stack every year did not somehow pose health hazards. Meanwhile, the plant has continued to hinder efforts to upgrade Tacoma's civic image.*

The final decision to close the copper smelter was made by ASARCO, not by EPA or the public. Contrary to popular belief, the decision was not really influenced by the Tacoma process. While the jobs versus health slogan made catchy headlines for editorials, it failed to reflect both the statutory obligation to reduce arsenic emissions and the economic realities of the international copper market. Internal management considerations, including the shift of ASARCO production and the prohibitive cost of installing additional sulfur dioxide controls in an obsolete plant, reinforced the impact of market changes. The general public and the national press linked the plant closing with the high cost of arsenic control devices, but the controls required for sulfur dioxide were estimated to be more than ten times more costly than the EPA's proposed arsenic controls, and the combined effect of external and internal forces would probably have forced plant closure, irrespective of the consequence of arsenic emissions control requirements. The regulatory environment and the public controversy were less important variables in causing the smelter to shut down.

After the copper refinery operation ceased, the ASARCO/Tacoma plant continued to process commercial arsenic for nearly a year. During this time, the smokestack emissions, the visible sign of the health risk, were no longer present. As the arsenic processing operations continued, the fugitive emissions problem was not abated, but active protest and press coverage regarding this risk subsided. The national press considered the case closed, and the local press focused on the economic impact of the plant closure and the possible public health implications of high arsenic concentrations in the soil and in the urine, hair, and fingernails of exposed individuals, particularly children.

In 1988 there continue to be significant problems related to ASARCO for the Tacoma community. The ASARCO plant grounds have been incorporated into the Commencement Bay Superfund site, one of the nation's top-ten toxic waste sites. When portions of the old plant were dismantled in early 1978, asbestos removal was added to the list of ASARCO's environmental and health concerns. The entire site, including the access road, was cordoned off to protect the public from possible danger during removal operations.

Important epidemiological studies of arsenic pathways are still being conducted to determine the health consequences of arsenic exposure. The plant is no longer operating, but the effects on the environment and public health will continue to be studied.

Conclusion

The ASARCO/Tacoma case is both simple and complex. At one level it represented an exercise in a more extensive community involvement in standard setting than that usually required under the CAA. While this was a resource-intensive activity and required the efforts of both regional and national EPA officials, it was an extension of an established process. The complexities arise when we consider the symbolic and cultural aspects of the Tacoma process, which assumed large, if fleeting, prominence in the national media and continues to play a prominent role in EPA lore.

The proposed arsenic standard set in motion a regulatory process that the agency had been through many times before in other section 112 regulatory initiatives. The necessity to develop a standard in response to a lawsuit, however, placed EPA in a reactionary position from the start. Add to this a lengthy local and regional regulatory history regarding the Tacoma smelter and a more complex risk communication problem emerges. The EPA-initiated citizen participation process was a late entry in an ongoing local problem.

Should EPA have developed some communication process with the residents of Tacoma before the summer of 1983? From the agency perspective, the lawsuit set the timing for discussions with the community about the risks of the smelter. From the community perspective, there had been concern about arsenic exposure for years accelerated by the placing of arsenic on the list of hazardous air pollutants in 1981. Timing is a key variable in risk communication, and communities and agencies sometimes differ in their determination of when risk communication activities should begin.

The involvement of William Ruckelshaus in defining the goals for the Tacoma process in a highly visible manner created the symbolic complexities in this case. National media characterized the Tacoma process as a bold new initiative and implied that this particular case was representative of a new and more open EPA approach to regulatory decision making. The drama of the jobs versus health dichotomy carried the media story to prominence on evening news broadcasts, but placed quite a burden on the EPA staff who had front-line responsibilities for implementing the Jeffersonian ideals underlying Mr. Ruckelshaus's approach to risk management.

It is important to distinguish between factors that led to the smelter closing and the factors that led to the proposed arsenic standard and the Tacoma process. The EPA was required to develop a standard, but the controls proposed as best available technology were carefully selected to avoid increasing the risk of plant closure. The ASARCO smelter had experienced severe pressures in the international copper market, and the economic future of the plant was uncertain from these factors alone. In the national and local media, however, EPA was portrayed as controlling the destiny of the plant, and this image placed additional burdens on the local risk communication process.

Considering the controversies and complexities surrounding the proposed arsenic standard, we must ask, Was the Tacoma process a good approach to risk communication? Should it be replicated in similar situations in the future? The answer depends on one's goals for risk communication. If the objective is to have community participants accept EPA definitions of a risk problem and respond to environmental hazards from a technical perspective, the Tacoma process did not reach this goal. Risk controversies are only partially addressed by presenting technical information in a clear and accessible manner as the workshops attempted. Fundamental value differences between participants must also be considered. Risk assessment involves a high percentage of scientific judgment and a relatively small percentage of "hard" science, and the public focuses on the areas of uncertainty. When the uncertainties influence community health and local economic growth, the public will treat risk management as a political process.

If the goal is to have an inexpensive and low-profile risk communication, the Tacoma process will not meet this objective. The process intensified the frequency and level of interaction between EPA headquarters and the regional office. Public participation was costly in terms of the number of high-level administrators and scientists involved and the duration of their involvement. Setting standards in public can raise expectations, attract great media attention, and be misunderstood.

On the other hand, if a risk communication goal is to increase the visibility of the agency and to communicate a message that risk managers respect community input, the Tacoma process was fairly successful. A Roper report survey documented that 58 percent of local residents agreed that EPA should have sought public input.[56] Interviews for this study also suggest that many Tacoma area citizens appreciated EPA's effort to involve the public. However, much of this goodwill was directed toward the regional EPA. Because the Washington office actually developed the proposed standard, it became the lightning rod for a broad range of citizen frustration. Also, appreciation of EPA's efforts did not mean that the community accepted the basic assumptions of the risk management decision. Many disagreed with the proposed standards and the fundamental notion of an acceptable cancer risk at the high levels predicted in the arsenic exposure models.

Former EPA Administrator Ruckelshaus termed the process a "qualified success." There are other situations like that faced in Tacoma, and he suggests that environmental regulators have a special responsibility toward those communities at risk:

Tacoma shows that we have to prepare ourselves for other Tacomas. Environmental stress falls unevenly across the land and we

*have a special responsibility to people in communities that suffer
more than their share. We are prepared to make an extra effort in
those communities.*[57]

Developing approaches to risk communication that incorporate the
goals of citizen participation present a challenge to all parties involved in these controversies. To the environmental regulators,
citizen participation means input into a process defined by officials
in technical terms. For the public, participation often means gaining greater control of both the process for making decisions and the
language of risk. The ASARCO/Tacoma case represents a first chapter in an ongoing public policy dilemma.

Endnotes

1. The ASARCO/Tacoma smelter processed copper ore concentrates, precipi-
   tates, and smelter by-products from multiple foreign and domestic sources. At
   full smelt, 320 Mg. (353 tons) of anode copper were produced each day. The
   smelter site included 10 roadsters, 2 reverbatory furnaces, 3 converters, 2
   anode furnaces, and the only arsenic production facility in the United States.
   The arsenic processing operation consisted of 3 roasters, 3 settling kitchens,
   storage facilities, and an arsenic plant.
2. Robert Coughlin, *Factors Leading to Closure of the Tacoma Smelter* (Seattle:
   U.S. EPA, Region X, 1985).
3. Ibid., p. 21.
4. U.S. Environmental Protection Agency, Office of Air Quality Planning and
   Standards, *Inorganic Arsenic Emissions for High-Arsenic Primary Copper
   Smelters: Background Information for Proposed Standards*, EPA-450/3-83-
   009a (Research Triangle Park, N.C.: U.S. EPA, April 1983).
5. D. Davoli and M. Johnson, "Arsenic Emissions from the ASARCO Smelter"
   (Paper presented at U.S. EPA Region IX Air Toxics Conference, September
   13, 1983).
6. William Ruckelshaus, "Science Risk and Public Policy" (Speech presented at
   the National Academy of Sciences, June 22, 1983).
7. "Tacoma Smelter Has a Dim Future, says Economist," *Seattle Times*, October
   6, 1981, p. E7.
8. Discussion and documentation of legal and regulatory events is highly reliant
   upon records and sources provided by Janet Chalupnik of the American Lung
   Association of Seattle.
9. EPA Washington, *News Release*, "EPA Proposes Standard for Inorganic Arsenic Emissions," EPA Press Release, Contact: Robert L. Ajax, Standards Development Branch, U.S. EPA, Research Triangle Park, N.C., July 12, 1983.
10. Davoli and Johnson, "Arsenic Emissions from the ASARCO Smelter."
11. EPA Northwest Region, *News Release*, Contact: Bob Jacobson, July 12, 1983.
12. U.S. EPA, Office of Air Quality Planning and Standards, *Inorganic Arsenic
    Emissions for High-Arsenic Primary Copper Smelters*.
13. A. M. Lee and J. F. Fraumeni, Jr., "Arsenic and Respiratory Cancer in Man:

An Occupational Study," *Journal of the National Cancer Institute* 42:1045–52 (1969).

14. J. H. Lubin, L. M. Pottern, W. J. Blot, S. Tokudome, B. J. Stone, and J. F. Fraumeni, Jr., "Respiratory Cancer Among Copper Smelter Workers: Recent Mortality Statistics," *Journal of Medicine* 23:779–84 (1981); A. Lee-Feldstein, "Arsenic and Respiratory Cancer in Man: Follow-up of an Occupational Study," in W. Ledeor and R. Fensterheim (eds.), *Arsenic: Industrial, Biomedical and Environmental Perspectives* (New York: Van Nos Reinhart, 1982); I. Higgins, *Arsenic and Respiratory Cancer Among a Sample of Anaconda Smelter Workers*, Report submitted to OSHA in the comments of the Kennecott Minerals Company on the Inorganic Arsenic Rule-Making (Exhibit 203-5), 1982.

15. The Standard Mortality Ratio (SMR) indicated was 329. The SMR is the observed number of deaths divided by the expected number (expected based upon the age and sex of the exposed population). The normal death rate for a group from a specific cause is 100. Any number above that indicates an increase above the normal rate. For example, for workers with exposures of 580 ug/cubic meter, an excess risk of 445–567 percent was observed. By contrast, workers exposed to 290 ug/cubic meter experienced an excess risk of 150–210 percent.

16. Dr. Milham had also conducted other smelter-related studies. See: S. Milham and T. Strong, "Human Arsenic Exposure in Relation to a Copper Smelter," *Environmental Resources* 7:176–82 (1974), and "Studies of Morbidity Near a Copper Smelter," *Environmental Health Perspectives* 19:131–2 (1977).

17. Fred Hutchinson Cancer Research Center study of Pierce County and King County cancer cases diagnosed between 1974 and 1976. Two surrogates used for exposure were the distance from the smelter and the concentration of sulfur dioxide (which serves as a tracer) over background level. In 1982 Hartley et al. conducted another mortality study which yielded the same results.

18. Christine Wicklund and Floyd Frost, "Asking the Public to Decide Upon an Acceptable Risk: The ASARCO Copper Smelter Near Tacoma, Washington" (Unpublished paper, 1983).

19. EPA *Fact Sheet*, "Uncertainties and Risk Calculations," Distributed in the public workshops held in the Tacoma area, August 1983.

20. R. Coughlin, *Factors Leading to Closure of the Tacoma Smelter* (Seattle: U.S. EPA, Region X, 1985).

21. U.S. EPA, Northwest Region, *News Release*, Contact: Randy Smith, November 2, 1984.

22. U.S. EPA, Washington, *News Release*, Contact: Robert L. Ajax, July 12, 1983.

23. Ibid.

24. Susan Hall, "Arsenic and Old Smelters: Public Involvement in Risk Assessment" (Unpublished paper, 1983).

25. Ernesta Barnes, interview, March 19, 1987.

26. Susan Hall, "Public Involvement in Standard Setting for Arsenic Emissions in Tacoma, Washington," September 1983.

27. Ernesta Barnes, interview, March 19, 1987.

28. Blaine Schultz, "Angry Citizens Sound Off at Smelter Hearings," *Seattle Times*, October 23, 1981, p. B4.

29. David Hooper, "City at Bay: The Poisoning of Tacoma," *Pacific Northwest*, December 1982, p. 37.
30. John C. Larson, "You Become Immune to Arsenic," *Tacoma News Tribune*, June 28, 1984.
31. Hall, "Public Involvement in Standard Setting for Arsenic Emissions in Tacoma, Washington," p. 8.
32. Henry Lee, "Risk Management and Public Perception: The Curse of the ASARCO Smelter" (Unpublished paper, January 1984).
33. Ernesta Barnes, interview, March 19, 1987.
34. Ibid.
35. Hall, "Public Involvement in Standard Setting for Arsenic Emissions in Tacoma, Washington," p. 20.
36. Barnett N. Kalikow, "Environmental Risk: Power to the People," *Technology Review*, October 1984, p. 61.
37. Ernesta Barnes, interview, March 19, 1987.
38. Lee, "Risk Management and Public Perception," p. 7.
39. John Gillie, "Pro: Many Rustonites Cite Smelter Benefits," and "Con: Vashon Residents: Fallout Won't Let Us Enjoy Our Land," *Tacoma News Tribune*, August 21, 1983.
40. Janet Chalupnik, interview, March 18, 1987.
41. Alfred Allen, interview, March 18, 1987.
42. Brian N. R. Baird, "Tolerance for Environmental Health Risks: The Influence of Knowledge, Benefits, Voluntariness, and Environmental Attitudes," *Risk Analysis* 6:425–35 (November 4, 1980). This research was conducted as a doctoral dissertation in clinical psychology from the University of Wyoming. Partial financial assistance was provided by the Tacoma-Pierce County Health Department.
43. Douglas Sutherland, interview, March 17, 1987.
44. Gillie, "Pro: Many Rustonites . . ." and "Con: Vashon Residents."
45. David Hooper, "Airborne Arsenic," (Bellevue) *Journal American*, July 13, 1983.
46. Sharon Sauve, "Letter to the Editor," (Bellevue) *Journal American*, July 13, 1983.
47. "Tacomans Involved in a Risky Decision," (Centralia) *Washington Chronicle*, August 19, 1983.
48. Editorial, "A Risky Choice," *Boston Herald*, August 26, 1983.
49. Editorial, "Mr. Ruckelshaus as Caesar," *New York Times*, July 16, 1983.
50. Hall, "Public Involvement in Standard Setting for Arsenic Emissions in Tacoma, Washington."
51. U.S. EPA, Northwest Region, *News Advisory*, Contact: Randy Smith, November 2, 1984.
52. Much of the discussion of setting the arsenic standard draws upon Robert Ajax and Janet Meyer, "Policy Consideration in the Selection of National Emission Standards for Hazardous Air Pollutants for the Tacoma Smelter" (Unpublished paper presented at the Society for Risk Analysis, annual meeting, Alexandria, Va., October 8, 1985).
53. Richard Syphen, "So Long Smelter," *Tacoma News Tribune*, June 28, 1984.
54. Ibid.

55. David Hooper, "Asarco Decision Closes an Era," *Tacoma News Tribune*, June 28, 1984.
56. Lee, "Risk Management and Public Perception," p. 14.
57. Charles Robers, "The EPA Now Trades Lives for Jobs," *High Country News*, December 26, 1983.

Sources

Interviews

ALFRED M. ALLEN, M.D., M.P.H., Tacoma Pierce County Health Department, March 18, 1987.

SUSAN H. HALL, Hall & Associates, 2033 Sixth Avenue, Seattle, March 16, 1987.

SUSAN MCNALLY, Production Coordinator, City of Tacoma Municipal Television, March 17, 1987.

ERNESTA BARNES, Northwest EPA (Region X), Administrator, March 19, 1987.

GIL OMENN, Dean of Public Health, University of Washington (Telephone Interview), March 15, 1987.

DOUGLAS (DOUG) SUTHERLAND, Mayor of Tacoma, March 17, 1987.

DANA DEVOLY, Northwest EPA (Region X), Technical Specialist, March 17, 1987.

CAROL JOHNSON, Northwest EPA (Region X), March 16, 1987.

BOB JACOBSON, Northwest EPA (Region X), March 16, 1987.

JANET CHALUPNIK, Washington Lung Association, Seattle, March 17, 1987.

DOUG PIERCE, Tacoma Pierce County Health Department, Solid Waste Manager, March 18, 1987.

GREG GLASS, EIS Author for Battelle, March 17, 1987.

JEFF WETHERSBY, Tacoma News Tribune, Reporter, March 17, 1987.

DOUGLAS PIERCE, Tacoma Pierce County Health Department, March 18, 1987.

Television Coverage

"RUSTON, WASHINGTON," Off-Air News Excerpts, ABC, NBC, CBS, August 10–12, 1983.

"JOBS OR CLEAN AIR," *Good Morning America*, Bill Ross, KOMO, August 10, 1983.

"CENTRALIA, PENNSYLVANIA, AND RUSTON, WASHINGTON, RESIDENTS MUST MAKE CRUCIAL DECISIONS," *Nightline*, ABC News, August 11, 1983.

"RISK ASSESSMENT WITH DR. GOLDSTEIN," U.S. EPA, September 7, 1984.

"Jobs and Health," *Good Morning America*, Bill Ross, KOMO, August 10, 1983.

"ARSENIC AND TACOMA," *Nightline*, ABC News, Lynn Sherr, August 11, 1983.

"TOUGH FOR ALL CONCERNED," *MacNeil/Lehrer Report*, August 11, 1983.

ARSENIC AND HEALTH EFFECTS:

"ASARCO Smelter Hearings," Tacoma Municipal Television, Channel 12, Tacoma Dome, September 1984.

EPA: David Patrick, Pollutant Assessment Chief
Ted Torslund, Carcinogen Assessment Group
Clark Gaulding

Dana Devoly
Chuck Kleburn
Washington State, Department of Ecology:
John Spencer
Hank Droghy
Jim Crowell
University of Washington
Gil Omenn
Lincoln Pollitzar (Fred Hutchenson Cancer Institute)
Karle Motte, Environmental Pathology
Washington State, Department of Social and Health Services
John Beare
Sam Milham
PSAPCA: Art Damkoheler
James Knowland
ASARCO: Larry Lindquist
John Newlands
Mike Varner
Tahomans for a Healthy Environment
Brian Baird
Tacoma General Hospital
Rick Branchflower
Centers for Disease Control
David Sokol
Georgetown University: Health Policy Analysis
Edward Berger
Tacoma Pierce County Health Department:
Kim Lowry
Bud Nicola
DEIS (Battelle Research Institute):
Greg Glass, Consultant to Tacoma Health Department
"CHOICES IN RUSTON, WASHINGTON," *MacNeil/Lehrer Report,* PBS News, October 3, 1983.
"ARSENIC WORRIES," NBC News, May 21 and 22, 1984.
"WAS RUCKELSHAUS THE ENVIRONMENTAL KNIGHT?" *Nightline,* ABC News, December 30, 1984.
CITY OF TACOMA PROMOTIONAL VIDEO, Tacoma Municipal Television (no date given).
"ASARCO PLANT CLOSING," ABC News, March 24, 1985.
"ASARCO CLOSES," Sam Donaldson, March 24, 1985.

Chronology*

1890

The Tacoma smelter begins operation.

----

*This chronology is adapted from a document prepared by Janet Chalupnik of the American Lung Association of Seattle.

1905

ASARCO purchases the Tacoma smelter.

1950

The smelter installs a sulfuric plant to recover up to 18 percent of the SO_2.

1967

July 1: The Puget Sound Air Pollution Control Agency (PSAPCA) is activated by state law.

1968

March 13: PSAPCA adopts Regulation I, an enforceable SO_2 ambient standard and an SO_2 stack concentration standard.

July 11: After numerous violations are observed, PSAPCA orders compliance with Regulation I by July 31, 1968. ASARCO files suit declaring Regulation I unconstitutional. PSAPCA countersues for a restraining order.

Fall–Winter: Legal suits are held in abeyance while ASARCO and PSAPCA attempt to develop a method of compliance.

1969

July: PSAPCA assesses civil penalties for violations of Regulation I.

December 31: ASARCO applies for a SO_2 ground-level standard variance to be in effect until May 1, 1972.

1970

February 2: PSAPCA denies the ASARCO variance, which includes a plan to construct a 1,100 foot stack.

March 23: ASARCO petitions the Thurston County Superior Court for judicial review of the PSAPCA decision.

June 10: As a result of PSAPCA's public hearings on amendments to Regulation I, ground-level SO_2 emission standards are tightened, and the stack concentration monitoring exemption is deleted.

August 12: PSAPCA establishes a 90 percent SO_2 emission control standard and rejects ASARCO's five alternatives to the immediate adoption of the amendments to Regulation I.

December 9: PSAPCA conducts a hearing for ASARCO's request to obtain a five-year variance to construct a local SO_2 plant designed to control 51 percent of the converter's emissions.

1971

January 13: PSAPCA grants a three-year variance with annual renewal options to 1975 for the installation of the liquid SO_2 plant.

November 23: Washington State Pollution Control Hearings Board (PCHB) hears the ASARCO appeal. An estimated $20 to $35.9 million is required to meet a 90 percent standard. PCHB remands the variance to PSAPCA.

1972

January 12: PSAPCA adopts an amended variance which allows ASARCO to amortize its investment in the liquid SO_2 plant, and it extends the time for ASARCO to decide if it will meet the 90 percent standard.

May: Professor William Rodgers requests PSAPCA to study arsenic and lead emissions near the Tacoma smelter.

Summer–Fall: Three agencies conduct lead and arsenic studies near the smelter. PSAPCA studies soil emissions. The Seattle-King County Health Department (S-KCHD) studies soil and the hair, blood, and urine of children on Vashon Island. The Washington State Department of Social and Health Services (DSHS) studies the hair, blood, and urine of children living near the smelter and vacuum cleaner samples.

November 13: DSHS requests PSAPCA to adopt arsenic standards.

1973

January: PSAPCA completes a Hi-Vol Air quality study. The heavy metals are analyzed by the Washington State Department of Environment (WDOE).

January 10: DSHS requests PSAPCA to conduct a total environmental study. PSAPCA forwards their request to EPA.

February 14: PSAPCA conducts a public hearing and adopts an arsenic standard.

March: EPA assigns Robert Carson to coordinate the ASARCO study.

April 10: ASARCO receives the first compliance schedule for control of arsenic trioxide emissions from handling and processing in the arsenic plant.

May 9: PSAPCA collects pasture grass samples, and EPA-NERC in Corvallis analyzes them. Significant concentrations of arsenic, lead, cadmium, and mercury are found.

June 28: ASARCO appeals PSAPCA's civil penalties before PCHB.

July 26: PSAPCA source test shows the presence of arsenic in fugitive emissions from larry cars, converters, and reverberatory furnaces.

September 24: Robert Statnick (EPA) conducts a source test of ASARCO emissions. PCHB determines the civil penalties for the ambient SO_2 violations.

Fall: PSAPCA collects house dust, vegetable, and soil samples. DOE and EPA analyze them.

December 12: PSAPCA grants ASARCO a variance extension to April 30, 1974.

1974

January 3: GCA, a private firm under contract to EPA, completes the analysis of Hi-Vol samples of sulfates and heavy metals.

February 14: Construction is begun on the pilot baghouse to determine the best available technology (BAT) for main stack particulate emissions.

Fall–Winter: DSHS continues to take samples of children's urine.

March 22: PSAPCA asks EPA for technical assistance to evaluate the risk of eating vegetables grown downwind of the smelter. Washington State University (WSU) Research Center presents results of heavy metal concentrations in the soil and vegetation near the smelter.

April 15: EPA gives PSAPCA the evaluation of risks associated with the consumption of vegetables grown downwind from the smelter.

May 1: PSAPCA starts collecting dustfall near the smelter to monitor fugitive emission control efforts from the furnace building. DSHS requests assistance from Dr. Phliip Landrigan to draw blood leads.

May 6: PSAPCA grants ASARCO an extension on its variance until June 30, 1974. On June 20 a further extension is granted until July 31, 1974.

July 31: PSAPCA, ASARCO, EPA, DOE, DSHS, United States Public Health Service (USPHS), and the Air Quality Coalition form a task force on smelter emissions.

September 18: The task force decides to collect all applicable studies, and to determine if additional studies are needed.

November 20: The task force determines that twelve additional research studies are needed.

December 9: ASARCO announces that it will not comply with the 90 percent standard by December 31, 1976; consequently, the variance terminates on that date.

December 18: EPA assigns a study of the Tacoma smelter as its highest priority, and it creates its own in-house task force.

1975

January 22: EPA proposes four alternative actions to the task force.

April 2: Jim Everts of EPA reports to the task force that the Office of Toxic Substances will conduct an epidemiological study of airborne arsenic to determine if an ambient air quality standard should be set. EPA and HEW will study heavy metals in the hair, blood, and urine around 28 nonferrous smelters. The three smelters with the highest risk will be studied more intensively.

April 16: At a press conference Dr. John Beare of DSHS explains measures which should be taken to reduce exposure to cadmium and arsenic.

May 17: PSAPCA denies ASARCO's request for a variance on visible emissions (sec. 9.03) and visible emissions of arsenic-containing particulate (sec. 9.19). PSAPCA issues 349 civil penalties at $250 each for emissions of arsenic-containing particulate.

May 21: EPA conducts a public meeting, "Environmental Concerns and the Tacoma Smelter," in Tacoma.

June 25: ASARCO appeals to PCHB for relief from enforcement of the visual standard and the arsenic emission standard. ASARCO also seeks reversal of the denial of a variance from the arsenic emission order.

July 17: PSAPCA petitions Pierce County Superior Court (PCSC) to restrain ASARCO from violating arsenic emission standards.

August 14: ASARCO petitions PCSC to prevent PSAPCA from issuing civil penalties for arsenic emissions while the PCHB hearing and decision is pending.

September 3: PCSC issues a verbal order restraining PSAPCA from issuing further civil penalties and notices of violations pending the PCHB decision on ASARCO appeal.

October 31: PCHB upholds the denial of ASARCO's variance application.

November 1: The National Institute of Occupational Safety and Health (NIOSH) contracts with PSAPCA for testing of an improved method to measure total arsenic in the ambient air near the smelter by January 31, 1976.

November 19: A PCSC motion to vacate the preliminary injunction granted for the pendency of ASARCO's appeal is denied.

December 5: ASARCO files for variances from the SO_2 emission standards (9.07[b][c]), visual standard (9.03[b]), and requirement to use BAT to limit arsenic emissions (9.19[c]).

December 8: ASARCO's petition to the PCHB for clarification of its October 31 decision is denied.

1976

February 19: PSAPCA grants a variance to ASARCO for periods ranging from two to five years, depending upon the control program and the pollutant. ASARCO is to provide SO_2 and arsenic trioxide emissions monitoring, support a urine sampling program, and cooperate with the state environmental assessment requirements.

March 18: PCHB issues a proposed order denying PSAPCA's variance application heard in May 1975. ASARCO settles PSAPCA's civil penalties.

March 19: Thirteen individuals, citizen groups, and PCHB appeal the ASARCO variance.

March 29: ASARCO files a petition in PCSC to compel adoption of a variance resolution in the form they drafted rather than the one signed by PSAPCA.

May 5: ASARCO's petition is granted by PCSC.

June 28: Four citizens petition King County Superior Court (KCSC) to vacate ASARCO's variance and to grant other requests for relief. The order for dismissal becomes final.

July 13: DSHS reports no abnormal urinary levels of arsenic in samples taken from Vashon Island residents.

July 28: PSAPCA determines that an Environmental Impact Statement (EIS) is required by the EPA. ASARCO appeals this action to the PCHB.

October 1: DSHS reports on urinary levels of arsenic in 133 samples taken in 1976. Arsenic levels are about the same for comparable groups tested four years previously.

October 8: DSHS requests worker mortality and arsenic exposure data from ASARCO.

October: ASARCO applies for tax credits on the purchase of a variety of air pollution controls and air monitoring equipment.

November 18: EPA requests information from ASARCO to determine whether air quality standards are being violated.

December 1: DSHS urine sampling of eight children records slightly lower levels than in the previous sampling.

1977

January 1: ASARCO pays penalties for its violations of SO_2 standards and then drops its appeal.

February 2: EPA requests ASARCO to submit economic information to determine if it is financially able to comply with the State Implementation Plan (SIP), and to evaluate the variance granted in February 1976. ASARCO declines to furnish these data.

March 30: Based upon the finding that an EIS is necessary, PCHB vacates the variance and remands it to PSAPCA.

April 5: EPA issues a Notice of Violation (NOV) of DOE and PSAPCA emission standard for SO_2 and particulates to ASARCO.

April 18: ASARCO petitions PCSC for a judicial review of the PCHB order, and petitions the PCHB for a stay until a judicial review can be held. ASARCO obtains a 90-day stay.

June 16: PCHB orders that ASARCO's appeal not to prepare an EIS should be held in abeyance until the judicial review is complete.

July 25: PCSC grants an extension to the stay of the PCHB order to October 25, 1977.

August 12: ASARCO files a new variance application with PSAPCA.

August 18: PSAPCA orders ASARCO to prepare an EIS to complete the variance application.

October 11: PCSC conducts hearings on ASARCO's petition for judicial review of the PCHB order, and it requests briefs from the parties.

1978

January 1: PCSC judge vacates and sets aside the PCHB order. This decision supports the variance granted to ASARCO.

February 3: The Citizens Group appellants challenging PSAPCA's variance to ASARCO file a "Motion for Reconsideration" of the PCSC decision to vacate the PCHB order of March 30, 1977.

March 24: Judge denies the "Motion for Reconsideration."

April 27: ASARCO files an appeal on the SO_2 nonattainment designation (No. 78-1929, 9th Circuit Court).

May 1: ASARCO petitions for reconsideration and a stay of designation of nonattainment for SO_2.

May 5: PSAPCA submits comments on the Draft EIS to Dames & Moore, the ASARCO consultants.

May 11: ASARCO advises postponement of the Draft EIS until the A. D. Little report from the EPA on the Tacoma plant economics and the Nonferrous Smelter Final Guideline is available.

June 12: A $5 million roaster baghouse begins its operation at ASARCO.

June 14: Supporting affidavits and a brief are submitted by the Citizen's Group appellants who oppose granting a variance to ASARCO to modify the ruling of the PCSC (Cause No. 254512).

June 26: The supplementary ASARCO/PSAPCA contro system (SCS) data link begins operation.

July 11: ASARCO submits its annual report on Variance No. 157 (Resolution No. 359).

August 1: The OSHA arsenic standard of 10 micrograms/cubic meter becomes effective.

September: The Citizen's Groups file an appeal in PCSC which supports the PSAPCA action in granting the variance (Cause No. 254512).

October 12: In an emergency temporary regulation the Washington Industrial Safety and Health Association (WISHA) adopts the OSHA arsenic standards and compliance dates.

October 13: PSAPCA receives the EPA's SO_2 attainment status records at ASARCO through August 31, 1978.

November 14: A public hearing is conducted on a permanent WISHA standard for arsenic.

November 27: ASARCO files a brief on the variance and provides comments on the State Implementation Plan (SIP) SO_2 strategy (PCSC Case No. 45508).

November 29: ASARCO withdraws its request for a Nonferrous Smelter Order (NSO) and its application for a variance.

December 8: WISHA reapplies the emergency arsenic standard regulation.

December 14: PSAPCA approves the SIP, including the SO_2 strategy.

1979

January 3: ASARCO requests an extension for completion and shakedown of the anode furnace emission control project until June 30, 1979.

January 9: ASARCO applies for a one-year variance to conduct a tall stack test program to resolve the causes of the opacity violations. ASARCO requests DOE to redesignate the vicinity of the smelter as an attainment area.

January 12: The proposed tall stack regulation is published in the *Federal Register* (FR 44, No. 9).

January 15: DOE requests EPA to redesignate the vicinity of the smelter as an attainment area. ASARCO ceases operation of its electrolytic refinery.

January 23: WISHA files a permanent arsenic regulation.

January 31: Proposed regulations for NSO's are published in the *Federal Register*.

February 6: Citizens file rebuttals to variance case (Supreme Court case No. 45508).

Feb. 14–15: ASARCO roaster baghouse source is tested.

February 22: WISHA arsenic regulation becomes effective. DOE provides PSAPCA with a draft of the SO_2 control strategy for Tacoma.

March 8: PSAPCA approves extending the anode furnace control program to June 30, 1979.

March 8–9: EPA conducts an informal public hearing about the NSO regulation.

March 21: EPA proposes regulations implementing section 120 of the Clean Air Act (CAA) for the "assessment and collection of noncompliance penalties." ASARCO files a brief with OSHA for an administrative stay of enforcement pending the outcome of litigation in the Ninth Circuit Court of Appeals.

March 27: PSAPCA receives the ASARCO research proposal for the tall stack opacity.

April 27: DOE releases SIP, which includes the SO_2 maintenance plan.

June 7: PSAPCA grants ASARCO a one-year extension of the variance for the tall stack (Resolution No. 446).

June 8: EPA advises DOE that the SO_2 area near the stack should be designated as unclassifiable.

June 19: The Ninth Circuit Court stays positions of the arsenic standards for ASARCO, nationally.

June 30: The anode furnace control is completed, and its operation begins undergoing a shakedown.

July 23: Expansion of the Supplementary Control System (SCS) for SO_2 nonattainment designation is completed.

August 30: WISHA makes minor revisions to its arsenic regulation (FR 44, No. 152).

September 18: ASARCO No. 2 Reverbatory Furnace burns out.

October 11: The State Supreme Court rules that the ASARCO variance should have been preceded by an EIS. The ruling remands the variance to PSAPCA until an EIS is prepared.

November 30: EPA redesignated ASARCO SO_2 area as "unclassifiable" (FR44, No. 232, sec. 107[d][1][D] of the CAA).

1980

February 25: ASARCO applies for a one-year variance on the 1976 variance application or for an extension until the State Environmental Policy Act is completed.

April 10: PSAPCA grants ASARCO an interim variance ending on or before April 10, 1981, to provide time for completion of the EIS required by the Supreme Court in Washington.

June 24: EPA issues rules establishing the content of the initial NOSs and the criteria and procedures to be used to evaluate NSOs issued by the state.

July 1: A labor strike at ASARCO stops the smelter operation. SO_2 in the ambient air is greatly reduced.

August 11: ASARCO requests permission from PSAPCA to increase the size of one converter to reduce the SO_2 emissions. Permission is granted.

August 12: ASARCO applies to PSAPCA to extend the 1976 variance, which had been invalidated by the State Supreme Court, until December 31, 1982.

September 11: ASARCO is notified that its NSO should be submitted to PSAPCA rather than to a federal agency.

September 30: ASARCO receives approval for its Notice of Construction (NOC) to replace the open conveyor system with a closed system.

November 5: EPA issues a NOV to ASARCO for mainstack opacity violations (sec. 113 of the CAA).

December: ASARCO/Tacoma site is designated as part of a Superfund (National Priority List) site known as Commencement Bay Near-Shore Tidelands.

1981

February 4: ASARCO requests an interim variance from Regulation I SO_2 control (Sec. 9.07[b]) for one year or until the final decision is made on the variance application of December 5, 1975.

March 12: PSAPCA grants an interim variance to end no later than April 9, 1982.

March 23: Draft Environment Impact Statement (DEIS) is published on ASARCO variance application of December 5, 1975 and subsequent updates.

April 21: Two DEIS public hearings are held in Tacoma.

April 29: ASARCO submits a NOC to install secondary hooding on smelter converters to control fugitive SO_2 and particulate emissions.

May 8: PSAPCA requests EPA to promulgate arsenic emission standards and to promulgate regulations on dispersion techniques.

June 10: EPA replies to PSAPCA that regulations will be proposed by the end of the year and final rules will be promulgated before December 1982.

June 19: EPA informs PSAPCA that it plans to negotiate an agreement with ASARCO to install and operate appropriate arsenic controls.

July 10: PSAPCA approves ASARCO's NOC for secondary hooding on smelter converters.

September 30: Final Environmental Impact Study (FEIS) is published. An appendix of correspondence with EPA about secondary hooding is attached.

October 22: PSAPCA holds a public hearing on the ASARCO variance request.

November 12: PSAPCA denies ASARCO's variance, orders installation of secondary hooding, and grants variances from the SO_2 standard. ASARCO is ordered to follow interim control requirements and begin a compliance program.

December 7: Civil penalties are issued for arsenic-laden dust deposited on cars driving near the plant.

1982

January 26: ASARCO agrees to participate in a study to determine the arsenic emission rate from slag dumping.

February 8: ASARCO hires a contractor to make repairs and to increase the power on the electrostatic precipitators (ESPs).

February 12: EPA approves federal funding of slag emissions study.

March 19: ASARCO explains that discrepancies in the monthly SO_2 emissions reports to PSAPCA are due to the use of two different methods of reporting. ASARCO concludes that the sulfur balance report is a better method than the Hastings Flow Meter method.

March 26: ASARCO completes the hood enclosure on Converter #4.

May 24: PSAPCA submits samples to Battelle Pacific Northwest labs for arsenic specification.

May 28: ASARCO files a NOC to replace the baghouse on the Godfrey roasters and metallic arsenic furnaces.

June 8: DSHS collects urine samples from Ruston Elementary School children.

June 25: ASARCO submits report on the effectiveness of meteorological curtailment.

July 14: ASARCO reports ambient arsenic average over 1.0 $\mu g/m^3$. One station near the main stack measured over 10 ug/m3 on three consecutive days.

August 5: ASARCO adjusts the stack control system and high emissions are eliminated.

October 29: To improve performance ASARCO repairs the ESPs and increases the power to them.

November 17: DSHS reports that arsenic levels in the urine samples of children living near the smelter have not fallen significantly since the early 1970s, even though the arsenic emissions have declined during the same period.

December 6: PCHB upholds ASARCO's appeal of civil penalties for emissions of dust containing arsenic.

December 9: PSAPCA calculates that 1.3 tons of arsenic trioxide per year is emitted from dry slag dumping, but no arsenic is emitted from wet dumping.

December 30: EPA issues a Notice of Violation to ASARCO for the violation of opacity standards during slag handling and dumping.

December 31: To meet variance requirements, ASARCO reports the results of its testing and data collection.

1983

January 13: A secondary converter hood-capture efficiency test is conducted by an EPA consultant.

February 7: The new replacement baghouse becomes operational.

March 3: ASARCO explains that high arsenic measurement anomalies were due to the demolition of the old baghouse in August, fugitive dust from dust pulling in October, and meteorological conditions in November.

May 2: Washington Department of Ecology (WDOE) is designated the lead agency to investigate the ASARCO Superfund site.

May 24: ASARCO concludes that moderate to strong surface winds carry reentrained dust from buildings and other surfaces to the sampling stations, thereby producing high ambient arsenic.

May: ASARCO provides details of downtime for the liquid SO_2 and acid plants including meteorological curtailment, malfunction, upset and breakdown, and scheduled breakdown.

June 30: DSHS collects children's urine samples from North Tacoma, Vashon Island, and Olympia.

July 11: EPA proposes national arsenic emission standards.

August: EPA holds workshops about the proposed national emission standards on Vashon Island on August 10, and in Tacoma on August 16 and 18. The public hearing is postponed from August 30 to November 2.

September 1: DSHS reports the results of urine sampling taken in June 1983.

September 19: The Vashon Island Community Council objects to the BAT standard, and it proposes that a new standard protecting public health is more appropriate.

October 10: The mayor of Tacoma holds a conference on the health effects of arsenic. Participants include regulators, other officials, academics, and citizens.

October 19: PSAPCA informs ASARCO that the October 1, 1984, deadline to install and operate secondary hoods is firm.

October 20: EPA issues revised arsenic emissions. The new estimates conform closely to those made by PSAPCA.

November 15: ASARCO informs PSAPCA that it could meet the October 1, 1984 construction deadline, but that it wishes to have the EPA regulations finalized before it proceeds.

December 3: In order to meet slag-handling standards for arsenic emissions, PSAPCA recommends replacement of "pancaking" slag with the use of water granulation.

December 14: The Washington Department of Environment (WDOE) circulates an interim "community exposure standard" for arsenic. The proposal requires evaluation of causes of arsenic levels of 2 µg/m³ of arsenic in a 24-hour period and levels of an annual average of 0.3 micrograms.

December 30: ASARCO seeks permission to establish a new smelter process. ASARCO announces that it does not wish to install FGD systems, and that it wants PSAPCA's requirement waived.

1984

January: EPA agrees to fund a consultant to study the feasibility of developing a chemical mass balance model for arsenic emissions from ASARCO.

January 9: EPA advises ASARCO to proceed with installation of the hoods at the earliest possible time. PSAPCA defers ASARCO's request for waiver for an FGD system, and requests assurance that the new smelting technique (INCO flash smelting) is technically feasible.

February 29: ASARCO is notified to monitor for arsenic on a continuous basis, as required by PSAPCA Order No. 53, rather than every sixth day as they had been doing.

March 3: ASARCO responds to PSAPCA stating that they will resume continuous sampling.

April 10: DOE adopts the proposed state regulation relating to arsenic standards. ASARCO submits a NOC to PSAPCA for the water granulation system and asks that they be relieved from penalties during the construction period.

April 20: PSAPCA approves the compliance schedule for construction, but it is silent on penalty relief.

April 30: PSAPCA sends ASARCO a NOV for the 49 days in February and March which were not monitored.

May 5: PSAPCA adopts state regulations for arsenic emissions.

May 31: PSAPCA levies 49 civil penalties of $250 each for new reporting violations.

1985

January: The Ruston-Vashon Superfund investigation begins. It will determine the routes of exposure responsible for elevated urinary arsenic levels in children.

March 24: The ASARCO smelting operation stops production.

1986

January 17: Arsenic production at ASARCO is stopped.

Acronyms

APE	Americans Protecting the Environment
ASARCO	American Smelting and Refining Company
BAT	Best Available Technology
CAA	Clean Air Act
CWA	Clean Water Act
CERCLA	Comprehensive Environmental Response, Compensation Liability Act
DEIS	Draft Environmental Impact Statement
DOE	Department of Energy
DSHS	Department of Social and Health Services (Washington State)
EIS	Environmental Impact Statement
EPA	Environmental Protection Agency
ESP	Electrostatic Precipitators
FEIS	Final Environmental Impact Statement
FR	*Federal Register*
GCA	GCA Corporation (A private consultant reporting to EPA)
KCSC	King County Superior Court
NAAQS	National Ambient Air Quality Standards
NERC	National Environmental Research Center
NESHAP	National Emissions Standards for Hazardous Air Pollutants
NIOSH	National Institute of Occupational Safety and Health
NOAA	National Oceanographic and Atmospheric Administration
NOC	Notice of Construction
NOV	Notice of Violation
NPDES	National Pollutant Discharge Elimination System
NSO	Nonferrous Smelter Order

OSHA   Occupational Safety and Health Administration
PCHB   Pollution Control Hearings Board (Washington State)
PCSC   Pierce County Superior Court
PSAPCA   Puget Sound Air Pollution Control Agency
RCRA   Resource Conservation and Recovery Act
S-KCHD   Seattle-King County Health Department
SCS   Supplementary Control System
SEPA   Washington State Environmental Policy Act
SIP   State Implementation Plan
SMR   Standard Mortality Ratio
TPCHD   Tacoma Pierce County Health Department
TSP   Total Suspended Particulate
USA   United Steelworkers of America
USPHS   United States Public Health Service
WDOE   Washington (State) Department of Environment
WISHA   Washington (State) Industrial Safety and Health Association

Selected Bibliography

ARTHUR D. LITTLE, INC. *Economic Impact of Environmental Regulators on the U.S. Copper Industry*. January 1978.

BAIRD, N. R. "Tolerance for Environmental Health Risks: The Influence of Knowledge, Benefits, Voluntariness, and Environmental Attitudes." *Risk Analysis* 6(4):425–35 (1986).

CHASON, JACK. "Sunset for the Smelter." *The Weekly*, August 24–30, 1983, pp. 29–40.

COUGHLIN, R. *Factors Leading to the Closure of the Tacoma Smelter*. (Washington: U.S. EPA, March 1985).

DAMES & MOORE. *Environmental Impact Statement*. September 1981.

FROST, FLOYD, SAM MILHAM, LUCY HARTER, AND CHERYL BAYLE. *Tacoma-Area Lung Cancer Study*, January 30, 1984, pp. 1–20 unnumbered.

GISKE, MARK. "Proposed Arsenic Emission Standard: Was It Worth the Wait?" *Environmental Law Quarterly—Puget Sound School of Law* 5(2):1–5 (1985).

HALL, SUSAN. "Arsenic and Old Smelters: Public Involvement in Risk Assessment." Paper presented at Hazardous Wastes and Environmental Emergencies Conference, March 1984, pp. 1–7.

———. "Public Involvement in Standard Setting for Arsenic Emissions in Tacoma, Washington." September 1983, pp. 1–10, Appendix with summary of Public Comment.

HOOPER, DAVID. "City at Bay: The Poisoning of Tacoma." *Pacific Northwest*, December 1982, pp. 26–59.

JOHNSTON, MICHAEL, AND DANA DAVOLI. "Arsenic Emissions from the ASARCO Smelter: Tacoma, Washington." Paper presented at U.S. EPA Region IX Air Toxics Conference, San Francisco, California, September 13–14, 1983.

KALIKOW, BARNETT N. "Environmental Risk: Power to the People." *Technology Review*, October 1984, pp. 55–63.

KALIKOW, BARNETT N. "Shadow of a Smelter." *View*, January 1984, p. 12.

LEE, HENRY. *Risk Management and Public Participation: A Case Study of the ASARCO Smelter*. Cambridge: Harvard University, 1984.

MILHAM, S., JR., "Studies of Morbidity Near a Copper Smelter." *Environmental Health Perspectives* 19:131–32 (1977).

MILHAM, S., JR., AND T. STRONG. "Human Arsenic Exposure in Relation to a Copper Smelter." *Environmental Resources* 7:176–82 (1974).

NATIONAL ACADEMY OF SCIENCES, COMMITTEE ON MEDICAL AND BIOLOGICAL EFFECTS OF ENVIRONMENTAL POLLUTANTS, Washington D.C., 1977. Docket Number (OAQPS 79-8) II-A-3.

RUCKELSHAUS, WILLIAM D. *Science, Risk and Public Policy*. National Academy of Sciences, June 22, 1983, pp. 1–20.

U.S. ENVIRONMENTAL PROTECTION AGENCY, OFFICE OF HEALTH AND ENVIRONMENTAL ASSESSMENT. "Health Assessment Document for Inorganic Arsenic," EPA-600/8-83-021A. Research Triangle Park, N.C.: U.S. Environmental Protection Agency, June 1983.

U.S. ENVIRONMENTAL PROTECTION AGENCY, OFFICE OF AIR QUALITY PLANNING AND STANDARDS. "Inorganic Arsenic Emissions from High-Arsenic Primary Copper Smelters—Background Information for Proposed Standards." EPA-450/3-83-009a. Research Triangle Park, N.C.: U.S. Environmental Protection Agency, April 1983.

U.S. ENVIRONMENTAL PROTECTION AGENCY. "Health Assessment Document for Inorganic Arsenic," p. 2-1.

WICKLUND, CHRISTINE, AND FLOYD FROST. *Asking the Public to Decide Upon the ASARCO Copper Smelter Near Tacoma, Washington*. Unpublished background paper, 1984.

YUHNKE, BOB. "EPA's Risk Assessment Process . . . A Critique." *The Environmental Forum*, July 1985, pp. 19–24.

Chapter 6

A HAZARDOUS WASTE SITE: THE CASE OF NYANZA

by John Duncan Powell

Introduction

The Nyanza hazardous waste site in Ashland, Massachusetts, became known in 1981 as one of the top ten sites on the Environmental Protection Agency (EPA) Superfund priority list. As a risk event to be studied for problems of risk communication, it is representative of a class of industrial sites found on the outskirts of metropolitan areas such as Boston, which emerged in the 1970s and 1980s as a focus of concern about toxic wastes and public health. A number of aspects of the Nyanza case make it particularly promising as a source for lessons to be learned in improving risk communication.

First, as an industrial site in a small town, it lies in close proximity to an exposed residential population, making the assessment and communication of health risks a prominent (and problematic) factor. But its historic presence also means that, in terms of employment, taxes, and the cultural identity of the town, the transformation of the residents' image of the dye plant from a place to work (even if a source of nuisance odors) to a potentially life-threatening menace was a major reversal in perceptions and definitions of reality. For neighbors of the plant and for other townspeople and some local officials, Nyanza has been a rude awakening to the risks of modern American life. The local media have played a prominent

Acknowledgments are given to Professors S. Krimsky and A. Plough for their important contributions to this chapter and to Sharon Becker for her research assistance.

role in the situational redefinition and the related assessment of health risks, a role that has called forth both praise and censure.

The second aspect of the Nyanza situation that makes it a rich case for study and analysis involves the complexities of the historic waste disposal practices on the site and the wide variety of environmental and potential health impacts. Liquid wastes—raw, partially and fully treated—and dye waste sludges were for sixty years routinely spilled, accidentally or deliberately, and discharged into wetlands and brooks adjacent to the various buildings and sheds in which the dyes and intermediaries were manufactured. In addition, organic liquid and inorganic wastes and mixed sludges were either pumped or barreled and hauled up Megunco Hill above the plant and then dumped on the ground, into wetlands, or into waste pits, a standard disposal practice at the time. Thus the (still active) light industrial site itself is contaminated in a variety of specific locations by a wide range of toxic materials, and the waste disposal hill area suffers from surface, subsurface, and wetlands contamination. Both surface and groundwater contamination have moved offsite, polluting downgradient water courses, wetlands, and neighborhood basements during seasonal high water table periods. The physical complexities of the site and its contamination impacts have provided ample material for risk communication conflicts during the process of site assessment, the selection of remedial alternatives, and the assessment of health risks.

Third, the Nyanza case involves an intricate pattern of intergovernmental relations concerning a single site. Historically, the local Board of Health responded to neighborhood complaints of nuisance odors and waste discharge impacts on Chemical Brook. By the early 1960s, the state Division of Water Pollution Control (DWPC) took a series of steps to force the dye company to upgrade the partial waste treatment system that had been operating since 1919. The division and later the state Department of Public Health (DPH) were gradually drawn into dealing with the disposal practices on Megunco Hill. When the JBF Scientific Research Corporation, under contract to EPA, traced mercury contamination of the Sudbury River to its source at the Nyanza site in 1970, the Metropolitan District Commission (MDC) became involved because Reservoirs No. 1 and No. 2 were standby water supplies for the MDC distribution system (Boston and some forty surrounding communities). A short time thereafter the Division of Fish and Wildlife (DFW) became involved, since mercury contamination of river and reservoir fish was encountered. In 1975 a number of state environmental agencies were combined into the Massachusetts Department of Environmental Quality Engineering (DEQE), which under the 1979

Hazardous Waste Management Act (Massachusetts General Laws [MGL], Chapter 21C) took an aggressive role in dealing with the Nyanza site. The Environmental Protection Agency, in setting up its Superfund Program as mandated in 1980, placed the Nyanza site on its National Priority List in October 1981, and after an abortive attempt to deal with the responsible parties (begun by DEQE), the agency took over the cleanup of the site in August 1982. Such a pattern of complex intergovernmental involvement affects risk communication significantly. While it brings a wide range of expertise and experience to a common problem, it complicates and slows down the process of managing the site and remediating its impacts. In turn, delay erodes local confidence in the process and eventually may produce a skeptical or even adversarial posture on the part of concerned local citizens, officials, and the media. This erosion of confidence and trust amplifies the already existing uncertainties and technical complexities of the site itself. In turn, this intensifies disagreement over the assessment and remediation of risks.

Fourth, under such circumstances dealing with the local public can be a communications minefield. The initial media reporting of the Nyanza site as among the top ten on the National Priority List (NPL) established a perception of the site as a dreaded menace. This perception, reinforced by a talented editorial cartoonist, became—and remains—embedded in the public mind.

When the Nyanza site gained notoriety, a citizen group, concerned about health issues, conducted a door-to-door sampling of health problems from among the immediate neighbors to the site. It found eight cases of cancer and assorted ills among the 47 households it covered. This fact was communicated by several means: a public meeting called by the group, letters from responsive local legislators, and guest appearances on local radio shows.

Local reporters quickly learned of and reported an additional twelve cases of cancer in the town. State and local health authorities were skeptical, a technically reasonable reaction to the findings; from an epidemiological standpoint the data were inconclusive because of the small numbers, the lack of precise endpoints and the absence of direct exposure measurements. However, from the perspective of effective risk communication, their response was badly flawed. The cancer scare in the early summer of 1982 generated a legacy of distrust in governmental agencies and officials and an explicit dissatisfaction with a process that elevates the principles of a science above the concrete health concerns of citizens. This legacy remains perhaps the greatest obstacle to the successful communication of health risk information as the assessment and remediation process continues at Nyanza.

Finally, the Nyanza case speaks beyond itself by offering lessons to participants in similar situations. This is not to say that the Nyanza case can teach participants how to *avoid* difficulties in risk communication. At best, risk communication regarding hazardous waste sites is a difficult, delicate process. But understanding the dynamics of the Nyanza case can help risk managers of hazardous waste sites to anticipate, understand, and patiently deal with the most problematic aspects of the situation. This case study is designed to illuminate those dynamics by addressing a number of questions: For older hazardous waste sites such as Nyanza, how do different sectors of the impacted community (i.e., residents, business operators, and local officials) construct the risk problem in the Superfund process? How are locally derived problem-solving efforts usually viewed by Superfund officials? Once Superfund takes control of a local cleanup effort, why are its remedial plans usually considered inadequate by the host community? What role do the media play? What role do local citizen groups play in the assessment and communication of the risks posed by the site? What forms of informal and formal risk communications and public participation programs are implemented at Superfund sites? How well do they work? What can be done to improve them? These questions will be addressed later in the study, but first we examine the evolution of the Nyanza Superfund site.

Historical Context

In this section particular attention will be paid to the history and nature of the operations at the Nyanza plant and how they led to the contamination of the air, water, and soils on and around the site. Equally important to understanding the risk communication process under Superfund are the views of neighbors of the plant—those who worked there, whose children played on the site, and who complained of its odors—and the content of the Nyanza folklore. Tracing the historic process by which the plant's operations and its environmental impact were brought to the attention of and dealt with by local, state, and finally federal government officials over a twenty-year period is essential to an understanding of community response. The legislative-regulatory framework within which the local hazardous-materials business operated changed drastically over this twenty-year period, and a significant part of the Nyanza story is the local business community's struggle to cope with this changing framework. Finally, once Nyanza was designated a National Priority List Superfund site in 1981, a formal bureaucratic process began to unfold in terms of assessing, containing, and resolving Nyanza's problems. This process, which is still in progress, will only be described briefly in this section, since many details are addressed in subsequent sections on site assessment and risk communications.

The Dye Plant and Its Neighbors

Construction by U.S. Color and Chemical Company, a manufacturer of textile dyes and intermediaries, began on the site at the foot of Megunco Hill in 1917. Over the years since then the company facilities grew from one or two buildings to well over a dozen. As a result of changes in corporate ownership, the company name changed several times, becoming Nyanza Incorporated (hereafter Nyanza) in 1965 (see Figures 6–1 and 6–2). Dye manufacturing was a good business in the United States for much of this period, but by the end of the 1960s German and Japanese competition was taking a heavy toll on the U.S. industry (much of which, like Nyanza, has since disappeared from the marketplace).

The dye plant had installed a primitive waste treatment system for its operation by 1919, and without major changes, the same arrangement continued into the 1960s. Basically, a series of drainage pipes from a number of the buildings conveyed liquid wastes, most of them highly acidic and laced with organic and inorganic compounds, including arsenic, lead, mercury, and chromium, to a collecting sump at one end of the operation. From this pit the raw wastes were pumped to a treatment tank, neutralized, and then pumped into a series of settling basins/lagoons, whose outflow went directly into a wetland brook. This wastewater brook then joined Trolley Brook, which flowed along the plant boundary toward the center of town 500 feet away, where it was conveyed by a culvert and dumped into the Sudbury River.

Additional wastes were also produced: process solid wastes (dye sludge), sludges that accumulated in the waste treatment settling ponds, bad batches of products, recovered phenols, solvents, and mercury, and the distillation residues from solvent recovery processes. Solid wastes were either barreled and hauled or pumped up Megunco Hill above the site, where they were disposed of by dumping them on the ground, into pits, or on the hillside, where they migrated downward into a wetland pond bordering the railroad track on the northern boundary of the Nyanza site. This wetland in turn discharged into a ditch which ran along the track toward the center of town, where it eventually joined Trolley Brook before entering the culvert into the river. Very frequently

Figure 6–1   Location Map: Nyanza Chemical Site, Ashland, Mass.

Base map is a portion of the U.S.G.S. Framingham, Mass. Quadrangle (7.5 minute series, 1965 photo revised in 1979). Contour interval is 20′; scale is 1″ = 2000′.

Prepared by NUS Corporation, a Halliburton Company. Source: Massachusetts Department of Environmental Quality Engineering, Nyanza Preliminary Site Assessment Report, October 1980.

(nightly during its last years of operation), the raw waste collection sump of Nyanza overflowed into this ditch, as did floor drains conveying spills and cleanup water from the buildings on the site not serviced by the treatment system. These wastes were on various occasions blue, red, or purple. The colors and the odor of the raw waste stream that overflowed into the ditch gave rise to the name Chemical Brook.

Several years after Nyanza closed its operations in Ashland, when news of the site's listing on Superfund's priority list reached the local media, *Middlesex News* reporters located and interviewed some former plant workers. Dominic Riano, who had worked at the Nyanza plant for thirty-nine years, told of driving tractors up the hill to dump the solid wastes from the plant: "They buried everything up there. Paste. Trash. Everything," Riano commented. He told of the liquid wastes dumped into Chemical Brook: "It would just run down the railroad track and into a brook [Trolley Brook]. . . . In the winter they'd let it run all over the place." Another employee who worked at the site from 1936 to 1972, Kenneth Pentheny, remembered the drums of waste that were hauled up the hill for disposal: "They just put the chemicals in open drums and left them up there. . . . I always knew that someday, when those drums start leaking, there is going to be a problem. I know there is," said Pentheny.[1]

But if Nyanza in the last decades of its existence was a "dirty" operation, as one gathers from these eyewitness accounts, it was also a place of fascination for the children of the neighborhood across the railroad tracks, along Pleasant Street, according to the *Middlesex News* interviews. Some children used to practice target shooting at liquid-filled glass containers half buried on the site. One recalled years later that "we'd watch the squirrels drop out of the trees when the [red vapor] would hit them."[2] The neighborhood children would often return home with their shoes and clothes stained green, blue, or purple. Sometimes they would come into contact with sludges and blister their hands and arms. Others, and their pets, frequently played in, occasionally swam in, and may even have drunk from the wetlands ponds and brooks on and flowing off the site without suffering more observable damage than coming home green, blue, or foul-smelling. "It was the best spot. You could swim and we had forts up there," said a neighborhood resident who came forward in 1985 to pinpoint for investigators where a large number of barrels had been buried on the hill. "It was a place we roamed in our childhood," he said. According to these eyewitnesses, some thirty to forty of their former playmates regularly visited the site. Traditionally, the hill was also used as a

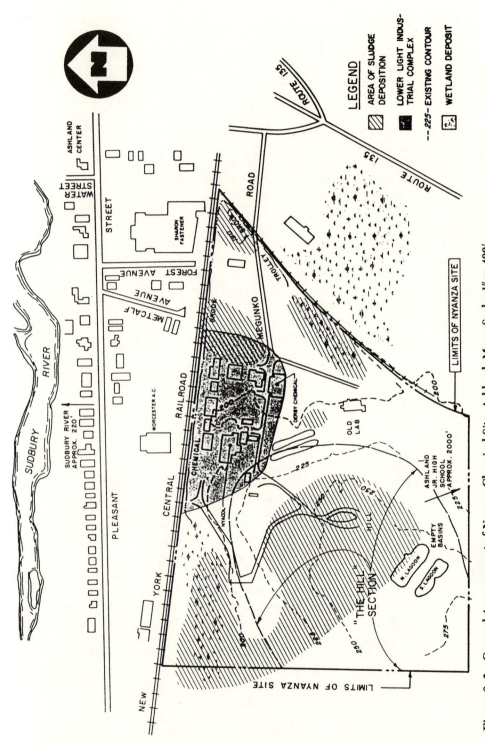

Figure 6–2 General Arrangement of Nyanza Chemical Site, Ashland, Mass. Scale: 1″ = 400′.
Prepared by NUS Corporation, a Halliburton Company. Source: Massachusetts Department of Environmental Quality Engineering,
Nyanza Preliminary Site Assessment Report, October 1980.

246

shortcut to the Ashland High School, which lies 2,000 feet to the south of the hill. Eyewitnesses also reported that the multicolored wetland ponds froze into colored ice in the winter. "I had a friend who was skating one day, and he fell through the ice. He was purple when we pulled him out," said one.[3] And many remember the day of the blue snow.

The parents of these neighborhood children often complained about the nuisances caused by the dye plant. In addition to the problems of dealing with the dyed clothing and skin of their children and pets who came home multicolored and dripping wet sludge, homeowners regularly complained to the Board of Health about the various odors coming from the group of buildings across the tracks. They even had to evacuate their homes for several hours one evening in the 1960s because an accident at the Nyanza plant released a fog of sulfuric acid which drifted over the neighborhood. Sometimes they complained about pitted or blistering paint on their houses or automobiles and about the colored water that entered many basements each spring. But mainly they complained about the odors. One elderly woman recalled years later that she had to give up her canary because of the fumes from the plant.[4] While some residents remember Nyanza as having been a nuisance in the 1960s and earlier, almost no one thought of the site as a health hazard.

Government Action

The first layer of government that addressed the Nyanza problem was the local Board of Health, under the authority of Massachusetts General Law, chapter 111. Among the sections of this statute are the traditional common law nuisance provisions. Odors are often caused by spills or accidents during routine manufacturing operations, which allow the fumes from noxious substances to escape. By the time a complaint is made and the Board of Health arrives at the plant, the episode is usually over. Nevertheless, episodes continue to happen; neighbors continue to complain; and elected Board of Health members continue to try to deal with the problem in a variety of ways.

The Ashland Board of Health requested assistance from the State Department of Public Health on a number of occasions. In 1963 they sent a letter to the DPH regarding Nyanza's discharge of waste into Trolley Brook. The DPH responded in its reply that "the Department is still of the opinion that the neutralized industrial waste may properly be discharged into the brook"[5] and that Ashland should seek to expedite the completion of its municipal sewer system; Nyanza's waste could then be discharged into its tie-in, thereby eliminating the nuisance! Until the closing of the Nyanza operation, and even afterward with other chemical manufacturing entities on the Megunco Road site, the Board of Health continued to deal with the frustrating reality of odors, spills, and noxious episodes.

During the mid-1960s the plant's discharge of wastewater into the waters of the Commonwealth had come to the attention of the Division of Water Pollution Control of the Water Resources Commission. In August 1967, the DWPC issued an order for Nyanza to comply with a specified Pollution Abatement Program, which required the rebuilding of the original 1919 waste treatment system.[6] A number of drains from additional buildings were tied into the waste stream. The old sump (which regularly overflowed into Chemical Brook) was bypassed and boarded over and a new one installed. Two new holding and settling lagoons took the waste from the neutralization tank and trapped sediments prior to discharge into the same wetland brook outlet as before.

However, Nyanza was ordered, under this program, to tie its entire waste stream into the municipal sewer system, which was moving toward completion in the Pleasant Street neighborhood across the railroad tracks from Nyanza. This hookup was supposed to be completed prior to the end of 1969. Although inspectors from the DWPC took note of the solid waste and sludge disposal practices on Megunco Hill, regulatory attention was primarily focused on the waste treatment system and the chemical nature of its final effluent.[7]

It was the discharge of Nyanza's partially treated waste stream into the sewer lines that brought the firm to the attention of the next pair of government regulators, the Ashland sewer authorities and the Metropolitan District Commission, the area-wide Boston sewer and water authority. Because of the DWPC's involvement in restructuring Nyanza's treatment system, it was drawn into a conflict about two issues: the pH level and the level of mercury in the waste stream. The Ashland Sewer Department was concerned about the effect the pH level would have on the condition of its new sewer system pipes; the MDC was concerned about the effect of the high pH levels on its treatment plants in Boston Harbor. The MDC regulations called for a maximum level of pH 9.0 for permitted discharge into its systm, and the Nyanza effluent regularly exceeded the level.

Local and state sewer authorities had stipulated that Nyanza's treatment and metering equipment must be automated and improved as a condition of discharge.[8] After hookup to the sewers in

July 1971, monitoring by these authorities found that limits were exceeded. When this happened, wastewater was temporarily diverted and discharged the historic way—by spilling it into the Trolley Brook wetland. But this violated the DWPC order to cease discharging into the waters of the Commonwealth. During this period, from 1967 through about 1973, Nyanza seemed to make regular efforts to comply with the various requirements of the regulatory community but also began to indicate that these efforts were becoming an increasing burden in a situation of declining demand for its products.[9]

In the summer of 1970, following up on an EPA contractor's study of mercury contamination of the Sudbury River, a federal water quality official visited the Nyanza corporate offices in Lawrence, Massachusetts, to discuss the mercury used in the Ashland plant's manufacturing processes. The official was told by the Nyanza executive vice-president that the Ashland plant used perhaps twenty pounds of mercury a day in certain of its dye operations, that none of it was recovered, and that he assumed it went out with the waste water discharge. Two months later, after being notified of this fact, the DWPC ordered the Ashland plant to reduce the mercury content of its waste to 0.08 pounds per day.[10] Subsequent actions followed the same pattern as happened with the pH levels: Nyanza attempted to comply, but monitoring showed that it failed; further production and waste treatment changes were ordered; the effluent would come into compliance for a period but then exceed the standards once more.

During periodic plant inspections, DWPC agents began to pay more attention to the dumping practices on Megunco Hill. In June 1971, a DWPC official visited the site, observed a 35-foot-deep sludge pit on the hill, and noted that its overflow near the spring was contaminating Trolley Brook at the foot of the hill. He also observed that it was contaminating the groundwater, since the sludge went below the water table.[11] The DWPC began testing surface water for mercury levels, and as a result in September 1971 sent a telegram to Nyanza ordering it to cease discharge into the waters of the Commonwealth from all sources on its site.[12] Decades of careless waste disposal practices had produced on the site more than 100,000 cubic yards of contaminated sludges, soils, and debris, most of them in either constant or periodic contact with ground and surface water. Although the precise amount was not known in 1971—in fact, it was not established until a decade later—every time testing was done subsequent to this order, more indications of contamination turned up.[13] In the spring of 1972 DWPC ordered Nyanza to meet a specific list of abatement conditions. Nyanza hired an engineering firm to develop a compliance proposal, but DWPC rejected the company's proposed solutions, noting that they failed to meet many of the conditions specified. By the fall of 1972 Nyanza's attorneys were corresponding with DWPC, seeking to delay the remedial actions required by the state.[14]

At the local level, the Ashland Board of Health, with the assistance of the town counsel, had taken the odor problem to the Metropolitan Boston Air Pollution Control Commission in November 1972. The commission ordered Nyanza to submit an air pollution abatement program within thirty days.[15] Utilizing a special consultant (a former plant manager), the Ashland Board of Public Health had by December 1972 reported Nyanza to the Department of Public Health for dumping barrels of expended phenols on Megunco Hill without a permit. The DPH cited the Nyanza plant for violations of the state's public health code (MGL, chapter 111, section 150A), and required a statement of compliance with landfill regulations.[16]

This affirmative action on the part of the Board of Health is particularly significant in light of an effort by the Ashland Industrial Commission the previous summer to slow down enforcement of landfill regulations. The new members of the commission had tried to persuade the town to back off from its pressures on Nyanza, which was one of the largest industrial facilities in a town that saw itself as being bypassed by economic progress. The manufacture of clocks by Telechron and other corporations, for which Ashland was historically noted, had already fallen on bad times (in fact, all these plants were destined to close). Members of the Industrial Commission attended the Board of Health meeting on June 3, 1971, and told the board that they had already met with the selectmen (the town executive board) and had spoken their piece. Now they wanted other town officials to know their view: "Ashland is dead. All due to bad publicity. The banks, the real estate people, and the lawyers all know it. The Selectmen need cooperation among the town Boards. The Town has to bend a little."[17] Notwithstanding opposition of the business community, the Board of Health continued to enforce the law.

In this same watershed year of 1973, the DWPC asked Nyanza for further engineering studies to address the problems of the site. The DWPC had contracted its own assessment of mercury contamination from the same EPA contractors who discovered the Sudbury River mercury contamination in 1970; it received the grim results in July 1973: from 1940 to 1970 nearly 5,000 pounds of mercury per year were discharged from the plant with no effort at recovery. Some 3,500 pounds annually went into the brooks, and hence the Sudbury River Reservoirs now contained 90,000 pounds of mercury. Megunco Hill held 40,000 pounds more, which continued to leach into ground and surface water migrating off the site.[18]

The reports by Camp, Dresser, & McKee (CDM), the Boston engineering firm hired by Nyanza to address the total plant problem, had also been received by the summer.[19] As a result of these two reports, the DWPC ordered Nyanza to remove all metals from its wastewater discharge, reduce sulfates to allowable standards, dispose of sludge according to the DWPC's new hazardous waste regulations, abate the infiltration of groundwater into the waste treatment system, and carry out all plant process improvements specified in the CDM engineering report. Subsequent to negotiations among the attorney general's office, DWPC, and Nyanza attorneys, a Suffolk Superior Court decree of September 27, 1973, recorded a consent agreement detailing the terms for Nyanza's compliance with the DWPC conditions.[20]

By 1975, when DWPC was absorbed into the state's new Department of Environmental Quality Engineering, it was clear that no compliance was taking place. In a series of maneuvers that seemed to be designed to stave off financial ruin, Nyanza attorneys in 1975 turned to several state and federal agencies seeking financial relief from the remedial requirements of the consent agreement. These requests were turned aside as inappropriate. For example, the Water Resources Commission was asked to purchase the contaminated Megunco Hill parcel under a state statute designed to protect water quality in the Commonwealth. In his letter responding to this request, the commission director concluded that "the property owned by Nyanza, Inc. in Ashland is not a feasible water impoundment site for future water resource needs of the Commonwealth."[21]

Similarly, the regional EPA office was requested to provide some financial assistance in meeting the remedial requirements imposed on the site. Nyanza argued that not only were local taxes and jobs at stake, but that if it were financially ruined, federal taxpayers, not private parties, would have to bear the greatest burden of protecting the environment. EPA Enforcement Director Michael Deland referred the Nyanza attorney to the Small Business Administration for possible financial assistance.[22] The period of stalling was terminated in the fall of 1976 when the DEQE and the state attorney general's office filed for contempt proceedings against Nyanza. A contempt citation was issued early in 1977.

Business Strategies

When their initial strategies of shifting the burden of costs and using legal measures failed, Nyanza's principals turned to other business means to deal with a bad situation. They sought ways to cut their losses. If dyes were not manufactured, then the expensive waste treatment conditions would not be needed. If Nyanza were to cease to exist in Ashland, then the corporation's noncompliance with the consent agreement would become moot.

Six months after receiving the contempt citation, Nyanza sold three of its industrial lots, buildings, equipment, and easements to Nyacol Products, Inc. (hereafter Nyacol), which continued to manufacture colloidal silica on the site. Nyacol's property did not include the hill dumpsite. According to Nyacol President Robert Lurie, $50,000 of the purchase price paid to Nyanza was placed in an attorney general's escrow account to be used for a cleanup of the sediments in Chemical Brook and the contaminated wetland patch along the railroad, as called for in the consent agreement.[23] According to the remedial plan, these cleanup wastes were to be taken to Megunco Hill, above the already contaminated areas, buried in two large pits dug for this purpose, and then covered over. Thus, the Nyacol principals believed that they had acquired an already sound manufacturing business from a company that was liquidating due to contamination problems, without inheriting any of those problems themselves.

Several months after Nyanza had disposed of parts of its plant and property to Nyacol, a local small-time industrial speculator named Edward Camille bought the remaining buildings and property, including the contaminated hill. A few years earlier Camille had bought two small parcels of the Nyanza property, dismantled the unused building, graded over the surface contamination (which was endemic to the entire site), and set up other light industrial uses. According to an EPA counsel's notes (gathered from the Cambridge Registry of Deeds, on December 19, 1978):

> Nyanza sells the property to Edward Camille (individually) for $170,000. To pay for it, a trust called Megunco Trust is created (President of Nyanza is the trustee) and the trust gives a mortgage to Camille. This mortgage was never discharged.[24]

Camille hired a local engineer to finalize the original cleanup recommendations. Impermeable material (Volclay) would be worked into the bottom of the receiving pits on the top of Megunco Hill (since they were permeable). Further, a diversion trench would be constructed upgradient of the pits to divert surface water runoff and lower the groundwater elevation in the vicinity of the pits.

Sludges on the site would be dug out and deposited in the two pits and covered with an impermeable cap to prevent infiltration from precipitation. Camille's engineers assured DEQE's water pollution experts that these measures would solve the problem on the site. And since the cleanup included Nyacol's stretch of Chemical Brook and the contaminated wetland at the foot of the hill (for which the $50,000 cleanup escrow account had been established), the cleanup would be done at no cost to Camille, according to his engineer's calculations.[25]

In the mind of a small businessman, this must have seemed a sensible, thrifty, and practical way to move forward and turn a problem into an investment opportunity. There was, however, one large problem, which Camille had not anticipated. The original 1975 cleanup scheme fell far short of the more rigorous environmental standards of the 1980s.[26] While local businessmen had been cutting deals and attempting to cope with the contamination problem as they saw it, a series of events radically altered the regulatory environment with regard to hazardous wastes. In 1975 DEQE was created and later absorbed DWPC from the Water Resources Commission. During this period DEQE came under strong environmental leadership. Within a few years the Love Canal story exploded (in 1977 and 1978), and in 1979 a Division of Hazardous Wastes was added to DEQE as part of a strong legislative mandate to attack the assessment and cleanup problem in the commonwealth (MGL, chapter 21C). In 1980 the Superfund program was established under EPA (40 CFR, part 300).

The interplay of these four events brought Nyanza under new scrutiny. As EPA took steps to implement the Superfund program, an early task was to formulate a National Priority List of major sites. The EPA turned to the states for information and suggestions on top priority problem sites. By virtue of the ranking methods used, the Nyanza site received an extremely high score (due to the weight given to the 3.1 million consumers of MDC water who were potentially exposed to the Framingham Reservoirs standby).[27] Just when Camille was attempting to carry out his bargain-basement solution to the Nyanza contamination problem, the local, state, and federal bureaucracies were redefining the Nyanza site. From their perspective, it was one of the worst hazardous waste sites in the country; it has been referred to as "among the top 10," "number 11," and "number 13," depending on which iteration of the Mitre Corporation's model was used. For each step Camille took to get the cleanup moving, state or federal environmental engineers inspected the site, reviewed the plans, and determined them to be inadequate.

In February 1981, a more powerful business interest formally entered the picture. The MCL Corporation, owned by a New Hampshire man, purchased Camille's Megunco Hill property. (MCL also bought Nyacol's property on the site, then leased back the buildings to Nyacol for their continued operation).[28] From indications given in a guarded appearance before the Ashland Planning Board, MCL's business strategy was to take over the floundering cleanup effort from Camille and incorporate the Megunco Hill property into its plans for building condominiums.[29]

Superfund Cleanup

The MCL's dreams of condominiums on Megunco Hill were not realized. By October 1981, the same year MCL purchased the property, the Nyanza site was listed on Superfund's first National Priority List of hazardous waste sites. The MCL and its local engineering consultants tried, but failed, to formulate assessment and remedial plans to the satisfaction of DEQE engineers. The EPA took responsibility for the site in August of 1982, and in 1983 a scope for a Remedial Investigation/Feasibility Study (RI/FS) was drawn up and a contractor selected (the NUS Corporation of Pittsburgh). This RI/FS, Phase I of the Superfund process, was aimed at source containment or the identification and disposal of all the sources of contamination on the site itself. Later, a Phase II study would determine off-site groundwater impacts and remediation, and still later, a Phase III study would deal with the Sudbury River water and sedimentation problem.

Ironically, this logical, step-by-step Superfund process—identifying the sources of contamination, containing them, and addressing the routes of off-site impacts—was the basis for some of the most negative local reaction to Superfund. From the beginning of the process, residents were more concerned about the off-site impacts on their health than about the details of site assessment required to contain the chemicals on the site. The townspeople assumed, incorrectly, that the contamination on the site would be cleaned up and carted off somewhere else for disposal. Local officials assumed, again incorrectly, that they would have a decisive role in the selection process of remedial action. In 1982 a local citizen's group played an active role in focusing attention on off-site health impacts; the issue received strong press coverage (discussed in detail later in the chapter). From 1983 through 1985, local officials and the citizens' group struggled to achieve their goals through the Superfund process, which gave them an opportunity to comment on EPA's proposals; all decision-making authority, however, rested with Superfund officials. (This process, which affected risk communication significantly, will be addressed more fully later in the chapter.)

Phase I field work was conducted by the NUS Corporation during the spring and summer of 1984. A draft RI/FS appeared that fall, and the final RI/FS and supplement were finished in the spring of 1985. The EPA made its choice of a cleanup option for Phase I in July 1985 amidst intense resistance from some segments of the community. The EPA published its Record of Decision in November 1985.

According to the Phase I plan, the on-site contaminants will be buried on the site in a secure landfill. When completed, the fenced-in site would encompass about 14 acres, with a 3- to 4-acre cap in the center. This remedial action is estimated to cost up to $9.8 million and, according to EPA, will prevent any additional health or environmental impacts from the site.

Design work for the construction of the Phase I remedial action was begun in 1986 and was nearing final completion in July 1987. Upon completion of the construction design plan, the EPA will execute an Inter-Agency Agreement with the Army Corps of Engineers. The Corps will then take responsibility for advertising for bids and selecting a contractor (anticipated in the fall of 1987) and then oversee the construction, expected to take one year. The Phase I bid opening is scheduled for February 1988. Phase II will begin in the spring of 1988, and the work plan for Phase III is currently being developed, with field work to be completed during the summer of 1988.

In addition to these actions, a partial cleanup of the original Nyanza waste sump took place in the fall of 1987 under the direct supervision of the Region I Emergency Response Team On-Site Coordinator. The old waste sump and contiguous contaminated soils, boarded and covered over in the 1960s' conversion of Nyanza's waste treatment system, were found to contain extremely high amounts of nitrobenzene, a semi-volatile toxic. The sump, known as "The Vault," measures 77 feet in length, 49 feet in width, and 8 feet deep. An inflatable portable structure was positioned over the vault. The hazardous materials that spilled from the vault were fed by conveyor belt into a portable incinerator. The materials in the vault itself were found to be too liquid to treat in this manner. They will be dealt with in a later stage.

In the five years since the Superfund process began at Nyanza, little concrete remedial action took place. The DEQE fenced off the perimeter of the hill site in November 1982, augmented the fencing in 1985, culverted Chemical Brook, and posted warning signs to fishermen in 1986, citing that the Sudbury River and the

Reservoirs are contaminated with mercury. The DEQE also covered a residential yard with one foot of clean material as an interim
remedial measure. But Superfund *site containment* actions (Phase
I) had been five years in the making. Why did the process take so
long? Why, especially when delays breed frustration among many
of the local participants in the process and thereby become part of
the risk communication problem for Superfund managers, could
events not move faster?

In the case of any Superfund site, the early stages of a remedial
investigation often uncover previously unknown or unsuspected
aspects of a site and its impacts, calling for an expansion of the
scope of the study. This happened at Nyanza in several instances,
including the "discovery" of the Vault and the contamination of a
wetland on the far side of Trolley Brook. Moreover, at the insistence of local people, the seasonal groundwater, which enters the
basements of the Pleasant Street neighborhood, was tested several
times for the presence of hazardous materials. But the site complexity generic explanation of why Superfund moves at a slow pace is
only a partial answer at Nyanza.

The political turmoil at EPA during the winter of 1982 and spring
of 1983, which saw Rita Lavelle fired, William Ruckelshaus returned to Washington as EPA administrator, and the Region I
administrator replaced, certainly slowed things down at the Nyanza
site (see Figure 6–3 for evidence of local interest in these events).
In addition, Superfund's expiration and the reauthorization im-
passe in Congress forced the administrator to suspend funds already earmarked for the Nyanza cleanup in August 1985. The Superfund Amendments and Reauthorization Act of 1986 (SARA) was
not passed until December, after which progress could once again
be made. Everything considered, five years may not represent
slow progress.

Risk Assessment and Remedial Actions

To set the stage for the analysis of formal risk communications,
three aspects of the Nyanza risk will be addressed: (1) the sources of
contamination and their routes of exposure to human populations;
(2) the struggle between the site owner's engineering consultants
and state and federal regulatory officials; and (3) the specific remedial measures selected by the EPA in its Record of Decisions that
grew out of the Remedial Investigation and Feasibility Study. At
various times it will be necessary to refer back to events described
in the previous discussion.

Granlund—

Source: Dave Granlund, *Middlesex News* (Framingham, Mass.), August 13, 1982.

Figure 6–3

Sources of Toxic Waste Exposure

The risks posed by the Nyanza operation may have originated within the old Nyanza buildings, but as we have seen, waste disposal practices over time, in conjunction with the contingencies of nature, resulted in a large volume (estimated 100,000 cubic yards) of hazardous materials in and around the industrial site. They exist on and adjacent to the hill above Megunco Road, in wetlands and streambeds in and downgradient of the site, in surface waters and sediments in the Sudbury River and its reservoirs, and in a groundwater plume that moved under the site toward the river. Human exposure to these toxic materials can occur through direct contact, through the air, or through surface or groundwater. Thus "risk" can be defined as the likelihood of injury, disease, or death from coming into contact with one of these sources on-site or with toxics carried by environmental transport media off-site. To assess this likelihood, one must identify the toxic substance, measure the exposure level of a human to the substance, and then apply toxicological principles and data on the chemical to arrive at an estimate of "likelihood of injury, disease, or death." This kind of assessment normally takes one of two forms, qualitative and quantitative. In the case of Nyanza, risk assessments have been almost exclusively qualitative. Moreover, they have occurred in piecemeal fashion and have been communicated in an episodic and informal manner.

The first and most obvious risk posed by such sites arises when one comes into direct contact with toxic materials. This was pointed out in DEQE's 1980 Preliminary Site Assessment, which cited hazards to workers in the industrial area and to neighborhood children who played on the site or used it as a shortcut to the high school.[30] There are temporary, direct ways to deal with this problem, such as covering the toxic deposits with gravel or topsoil. When Camille purchased his first industrial parcels from Nyanza in 1976, he employed such measures in and around the industrial buildings on Megunco Road. There was no permit and no plan; the logic was simply to cover up the hazardous material so that there would be no contact. When Camille purchased the hill property in 1978, he tried to do the same thing, to a certain extent. Thus, the possibility and degree of direct contact exposure had been, to a small degree, addressed temporarily prior to the state and federal risk management efforts of the 1980s.

But a large volume of exposed toxic materials remained, and neighbors provided, through the press and public meetings, much information concerning children playing on and cutting through the site. Since it was a simple and obvious thing to do, DEQE fenced off the hill portion of the site in November 1982 and posted the fence with warning signs; this direct action helped to minimize but did not eliminate the direct contact risk associated with the site. Later, as the Remedial Investigation data provided a more complete picture, DEQE, using state funds as part of its Superfund matching requirement, extended the fencing to close off the high school shortcut over the bed of the old trolley line and installed a culvert to carry Chemical Brook along the railroad tracks. This alleviation of the exposure opportunity for direct contact with toxic materials was completed in the fall of 1985.

Exposure to risks also occurs when airborne contaminants are carried off-site. This concern in the Nyanza case was raised in the 1980 DEQE Preliminary Site Assessment, at various meetings, in study documents, and in news articles. Private engineers, DEQE, and EPA field technicians have all measured air contaminants on, adjacent to, and downwind of the toxic materials on and about the site. Even in hot weather, when volatilization of organics and metal-laden dust particles might be expected, few toxic contaminants other than traces well below state, federal, and international (World Health Organization) standards have been detected.[31]

After direct contact, the second most serious route of toxic exposure from Nyanza was through consumption of fish contaminated with mercury. Once again, this fact had been known and communicated within governmental regulatory circles since the early 1970s, when studies were completed for EPA by the JBF Corporation. Over the years multiple tests of reservoir sediments and mercury-contaminated fish have been done by EPA, DEQE, MDC, and DFW. The test levels for the mercury-contamination varied from well below the federal Food and Drug Administration action level of 1.0 ppm to as high as 12.0 ppm.

Groundwater in leaking basements in the neighborhood is the third route of exposure from the Nyanza site. Many residents complained about the potential for risks to their health should the annual spring high water in their basements contain contaminants. Tests done in 1986 indeed revealed low levels of a number of organics and some inorganics and metals in groundwater.

Private versus Public Standards

With the sources and nature of the risks to human health posed by the Nyanza site as a backdrop, certain problems that arose during the site assessment process at Nyanza are more easily understood. The first of these concerned the private assessment and cleanup plans of the MCL Corporation.

As the initial premise, both the state and federal Superfund programs rely on responsible party cleanup of contaminated sites. That is, the property owner or the corporate entity operating on the site is expected to pay for the assessment and remediation of the problem. Private parties act out of a mixture of financial self-interest, responsibility, and legal liability. If they are unwilling or unable to perform to the satisfaction of the regulatory agency and Superfund monies are used for assessment and cleanup, the government can seek to recover up to three times its costs. This naturally puts pressure on the responsible parties to perform, as the legislation had intended. But private parties tend to be very cost-sensitive, while regulatory agencies, at least in the site assessment stage, tend to be relatively cost-blind. In the minds of the regulatory community, the priority concern is assessing the potential threat to the public health and the environmental resources impacted by a given site. Thus a built-in difference of perspective and concern exists between private and public interests, and it is frequently expressed in terms of standards.

At Nyanza the gap between private and public standards was most glaringly evident during the time when Camille was attempting to clean up the hill. His local engineer, Connorstone, Inc., estimated that it would cost $10,000 to prepare the disposal pits near the top of the hill, $20,000 to excavate and place the site contaminants in the pits, and $5,000 to clay-cap the pits; the groundwater diversion trenching could be done for the balance of the original $50,000 escrow account held by the attorney general for cleanup.[32] Eight years later, federal Superfund expenditures for site assessment and initial remedial measures have amounted to nearly $2 million. The first phase remedial measure, including excavating site contaminants, burying them on-site, capping the burial pits, and diverting groundwater around the pits with backup monitoring, are estimated to cost in the vicinity of $10 million. Thus, the same general concept of the on-site solution could be done to the satisfaction of private standards for $50,000 (1980 dollars), whereas meeting federal standards will cost $12 million (1985 dollars). Once Camille realized that state officials would not accept his standards, he sold the business to MCL.

Under MCL, remediation was still guided by the basic disposal concept first developed in 1974. A more adequate program of air, soil, and groundwater site assessment studies was developed by Jerome Carr Research Laboratories, Inc. As testing continued, DEQE raised issues concerning field and laboratory techniques. In August 1981, Carr submitted a draft copy of the Phase 1 on-site assessment results to DEQE.[33]

Additional phases were intended to address off-site water contamination impacts, but MCL's strategy was to complete Phase 1 assessment and cleanup, thereby opening the way for the development of the Megunco Hill property. However, by the time the draft Phase 1 assessment study was submitted, DEQE had begun to work with EPA in evaluating the site for possible inclusion in the National Priority List. With state and federal engineers characterizing Nyanza as one of the worst sites in the country, and MCL's Carr/Connorstone team thinking in terms of a quick, practical cleanup job, the stage was set for a clash of conflicting standards.

The Carr report of August 1981 revealed a very conservative, minimalist approach to the site assessment. Specific conductivity was the primary measure used to establish the groundwater contamination plume and assess the surface water off-site impacts. Testing had been conducted for heavy metals and inorganic and organic compounds, including mercury, chromium, iron, lead, zinc, cadmium, antimony, phenol, sodium, ammonia, copper, selenium, and silver. Carr, arguing partly on the basis of drinking-water standards (no standard had been established for most of these chemicals) and practicality, proposed dropping all but mercury, chromium, and selenium from future testing requirements. Carr followed the same line of reasoning when the organic contamination results were subsequently finalized. By September 18, 1981, Carr forwarded the final results to DEQE, indicating that "the new materials show that there is no hazard from dust or gases on the site and that the list of organic pollutants can be reduced to three chemicals."[34] The three chemicals included only trichloroethylene, trans 1,2-dichloroethylene, and chlorobenzene. Carr then requested a meeting within two weeks to review these findings and obtain approval to proceed with the Carr/Connorstone remedial plan; MCL, Carr, and Connorstone hoped to finish the assssessment and remedial work before the end of 1981. Was this approach elegantly simple and practical; or was it merely quick and dirty?

The DEQE and EPA engineers and their special consultants were not satisfied with using conductivity as a surrogate measure for the long list of inorganic contaminants and heavy metals on the site, nor were they satisfied that all subsequent testing should be narrowed to only three inorganics.[35] Thus, private interests were arguing for a narrow and inexpensive approach, whereas government engineers and decision makers believed that the situation required a more comprehensive analysis of threats to the public health and environmental resources. The disagreements about standards were at times narrowly technical (what kind of seal to put on water sample containers) and at times philosophical in nature (given limited private resources, the more spent on testing, the less left for cleanup; how can this be justified?).

Within EPA, legal counsel and engineers were trying to decide whether to let the private Nyanza cleanup efforts proceed or to take over the assessment and remediation process under Superfund auspices.[36] To this end, Camp, Dresser & McKee, which had performed the 1974 cleanup study, was hired to review all of the data and produce a proposed Remedial Action Master Plan (RAMP) for the Nyanza site. In the meantime, further testing by Carr/Connorstone was approved by the DEQE.[37]

These delays and disagreements apparently began to take their toll on Jerome Carr. After reviewing detailed responses to his August 1981 Phase 1 Report, Carr lectured DEQE's engineers on their errors in judgment and fact:

We feel the request for extensive pH monitoring will only increase the costs of field time with little or no benefit to the study. . . . We find the requirements contain unnecessary detail and are unnecessar-

ily restrictive. . . . We will resist the use of a cleaning agent that
could interfere in any way with our sampling program. . . . We feel
this is too much sampling and is too restrictive in terms of tech-
nique. . . . Thus the entire request for bedrock water should be
eliminated. . . . We insist on the right to use the best methods. . . .
The requirement for pumping the permeable strata is also extremely
questionable.[38]

This document, seven single-spaced pages in length, contains extensive criticisms of DEQE's response to Carr's report. It cites
obscure documents in footnotes, an apparent reflection of Carr's
doctoral training in geology. Failing to receive a satisfactory abatement of DEQE's requirements to upgrade his procedures, three
months later, in May 1982, Carr wrote once again to DEQE's site
engineer (under the heading *"The Bottom Line"*):

I hope this letter places these minor issues in perspective. If you still
have problems with the field techniques, then I must insist that the
problems be documented along with the rationale behind your con-
cerns. It is only after you have fully thought out and organized your
concerns that I can address them at the required level.[39]

Despite these efforts to tutor government engineers in proper standards, Carr faced another setback late in July when EPA released
the results of CDM's Remedial Action Master Plan. The report
concluded that the Carr/Connorstone plans for cleanup were "insufficient and incomplete," citing "several data gaps" and "deficiencies" in testing. CDM also criticized the lack of information on
potential health impacts on residents living near the site.[40]

In August, DEQE and EPA regulators met with MCL representatives and the Carr/Connorstone team in an attempt to negotiate a
binding consent agreement which would delineate a testing and
remediation program to the satisfaction of public officials. But
MCL had already expended $150,000 in consulting fees to assess
the site and plan its cleanup. Even so, the results were disputed by
other consultants, as well as state and government engineers. Furthermore, the RAMP recommended another $225,000 to $300,000
in studies in order to meet the standards of the regulatory agencies,
which was more than MCL would accept. The private effort to
assess and clean up the Nyanza site collapsed, and henceforth EPA
took the lead role. Jerome Carr responded to his antagonists in the
public sector at a Framingham meeting of the League of Women
Voters in September 1982:

The Ashland site could have been 60 to 80 percent cleaned up
before the start of the 1982–83 high water season if the EPA hadn't
interfered with the cleanup. . . . They came in February. I haven't

*been to the site since, and I'm the chief investigator. A lot could
have been done in that time. . . . DEQE will admit they are un-
derstaffed. Nyanza is number 10 in the country, number three in
the state, yet the engineer assigned to monitor the site has zero
months experience in hazardous waste. He's only been there 14 to
15 months, period. . . . People don't understand that there are
many criteria in deciding whether something is a hazardous waste.
They shut off the thinking mechanism. They lose their heads, and
the subject becomes emotional. . . . People see the word "benzene,"
and they stop thinking.*[41]

Superfund Risk Management

Whether or not Jerome Carr was correct in his characterization of
how people think about hazardous wastes, matters were taken out
of his hands in August 1982, when responsibility for site assessment
and remediation came under Superfund. As described earlier, a
Remedial Investigation/Feasibility Study was performed by the
NUS Corporation, and a remedial alternative was selected by EPA
in its November 1985 Record of Decision. The basic Phase I site
containment measures selected by EPA are designed to manage
the risks posed by the on-site contamination, providing reasonable
protection to the public. These steps include:

1. Excavation of all outlying sludge deposits and contaminated
    soils and sediments associated with these deposits.
2. Consolidation of this material with the hill sludge deposits.
3. Capping of the hill area in conformance with the technical
    requirements of the Resource Conservation and Recovery
    Act (RCRA).
4. Construction of a groundwater and surface water diversion
    system on the upgradient side of the hill.
5. Backfilling the excavated areas to original grade and revegetat-
    ing the wetland areas.
6. Construction of a more extensive groundwater monitoring
    network to enable future evaluation of the effectiveness of the
    cap.[42]

These measures are only the first in a series of risk management
steps to be taken by Superfund and are aimed solely at site containment. Additional risk management steps, as described above,
include assessment and remediation of off-site groundwater contamination and assessment and remediation of the off-site mercury
contamination in the sediments and fish found in the Sudbury
River in its downstream reservoirs.

Risk Communication

Risk communication is often thought of by experts as a one-way transmission of knowledge to lay people in a manner that will foster the acceptance of risk management decisions.[43] From a more comprehensive perspective, risk communication is a multi-party, multi-directional process that takes place over time. In this section the advantages of this comprehensive definition of risk communication will be illustrated by focusing on the activities of a local citizen's group and the media in developing and communicating information about the Nyanza site. Following the analysis of local, health-oriented risk communication, a description and analysis of the informal and formal risk communications that originated in the Superfund process will be presented.

Local News Media

Virtually all public officials interviewed for this study—local, state, and federal—were negative in their characterization of the press. The strongest terms used were inaccurate, misleading, and sensationalist. This attitude toward the media seems to be widespread among government scientists and managers dealing with hazardous waste issues and the public reaction to them. The press, in effect, is perceived as provoking in the public the anger and fear that scientists and managers so resent having to confront. Occasionally, individuals representing government regulators or industry or business interests are emotionally battered, especially at large-scale public hearings, but it is highly questionable whether the media can be said accurately to cause public anxiety about hazardous wastes and other environmental hazards such as radiation. Media coverage may precipitate or exacerbate public reaction, but it doesn't create it.

In the Nyanza case, furthermore, the nature of coverage by the local media, primarily the *Middlesex News*, cannot be considered unique or highly parochial. Indeed, the *News* coverage of Nyanza fits exactly into the national pattern of media coverage illuminated in the Media Institute's recent study, *Chemical Risks: Fears, Facts, and the Media*. The Media Institute found that journalists who report on chemical risks tend not to be scientifically trained. They rely heavily on government sources for information about a specific event, including its risks. Media reports emphasize information about high risk as opposed to low-risk or no-risk information. Moreover, the media do not generally solicit reaction from scientists in industry or government about risk events. Rather, the typical pattern of reporting is completed by seeking reaction from the "man in the street."[44]

While the Nyanza reporting fits this national pattern, several characteristics of the coverage account for some of the complaints by government officials. The *News*, which is heavily read in Ashland (of 5,300 households, 2,689 subscribe to the *News*, 788 to the *Boston Globe*, and 347 to the *Herald American*), has a medium-sized circulation (60,000 daily). The paper tends to hire young generalists as local reporters, who are assigned to the smaller towns in the circulation area where they can gain some writing experience and then move on or up. Thus, periodically, new reporters unfamiliar with the Nyanza situation become responsible for covering events with a complicated history, accounting for some of the complaints about "inaccurate and misleading" reporting. To the extent that the sensationalism charge can be sustained, close examination of the 165 articles on Nyanza published from October 1981 through July 1987 reveals that during the first two years of coverage the fault lay with the *Middlesex News* headline writers rather than the beat reporters. Another specific feature of the *News* that distinguishes it from the average pattern is the talent of its editorial cartoonist, Dave Granlund, who on several occasions so brilliantly captured and symbolized the hopes and fears of the people of Ashland with regard to the Nyanza story that his images remain indelible to this day. One dimension of this symbolism that is repeatedly articulated in the public and the media—and repeatedly qualified or put into a different perspective by government officials—is the status of Nyanza as one of the top ten worst sites in the nation.

The first *News* story on Nyanza carried the October 1981 announcement by EPA of its first National Priority List of hazardous waste sites. Of the 114 initial sites, Nyanza was, indeed, ranked in the top ten due to the weight that one model (developed by the Mitre Corporation) gave to the potential for exposure of the 3.1 million MDC water customers to mercury-contaminated water in Framingham Reservoirs Nos. 1 and 2. The reason for Nyanza's ranking was not carried in the initial release of information by the government. By November, DEQE and EPA spokesmen were taking great pains to inform *News* reporters that they should put the ranking in perspective, pointing out the *potential* risk was highly unlikely to occur, while the *actual* risks, on and near the site, were rather modest. Yet the top ten label remains lodged in the public mind, and officials dealing with the site still have to make repeated efforts to address the negative image suggested by the description.

Is the *Middlesex News* responsible for perpetuating an unrealistic level of fear about the Nyanza site? Since the initial story, a total of 165 stories on Nyanza (as of July 1987) were published in the *News*. In 1981 all stories mentioned the top ten status of the site (as determined by EPA); during 1982 and 1983, the years of heaviest coverage (57 and 30 articles), mention of the ranking held at 54 percent; and in the years since, the frequency dropped to and held steady at about 33 percent. Furthermore, since 1983 citations of the site's status have been reduced to a set formula, provided toward the end of the article, which incorporates the "perspective" element desired by government officials: "In 1981, the EPA declared Nyanza one of the 10 worst chemical dumps in the nation because it is 220 feet from the Sudbury River, eyed by Boston as a future water source."[45]

Yet, while its ranking as one of the ten worst sites is still mentioned frequently in public discourse about Nyanza, it seems clear that it is not the status *per se* that arouses anxiety in the mind of the public but the association of the site with risks to the health of its neighbors. And no risk, it seems, is more dreaded in the public mind than cancer. It was not the general reporting of the *Middlesex News* on the Nyanza site, including its NPL status, that precipitated a deep-seated fear of the Nyanza site but the spring 1982 cancer scare and its aftermath. Two unpublicized sets of events, one carried out in public, the other occurring within local and state government, intersected in May 1982. These events indicated that public concern about (if not fear of) Nyanza-related cancer predates any media coverage of the topic by several months.

Organized public interest in the Nyanza site centered in the Social Action Committee of the Ashland Federated Church. Leaders of this committee, after reading of the Nyanza site in the *News*, decided that this would be a good topic on which to focus.[46] Love Canal, the Superfund Act, and the general problem of hazardous wastes had all received extensive national media coverage prior to the publication of the NPL list. While it originated within the Social Action Committee, local organized interest in hazardous waste in general and in the Nyanza site in particular grew, by late summer 1982, into a larger citizen's group, Ashland Advocates for a Clean Environment (AACE). AACE is in most respects typical of local citizen groups that have taken form to address particular environmental concerns in American communities.[47]

Members of AACE began seeking information on Nyanza and obtained the July 1981 summary report prepared by DEQE during discussions with EPA about priority sites in Massachusetts.[48] Acting on this report, AACE planned a health survey of neighborhood residents early in the spring of 1982. Its members canvassed door to door in the neighborhood across the railroad tracks from Nyanza, completing a total of 47 household questionnaires covering 169 individuals. This survey brought out information on 8 cases of cancer, 3 miscarriages, and 1 birth defect. Also cited in the study were the following responses: 27 of seasonal water in the basement of the homes surveyed; 6 of children or pets coming home from the site with dye or chemical contamination; 7 of severe odor problem from the site; 5 of the mid-1960s evacuation of the neighborhood due to an accident at the Nyanza plant; in addition to assorted observations and minor complaints.[49] The AACE was concerned but uncertain about how to deal with this information. A decision was made to emphasize public education, and the group made a strong effort to obtain additional information on the Nyanza site.

The initial focus of AACE's educational activities was a public forum, held on May 18, 1982. The idea was to have representatives from EPA, DEQE, the Ashland Board of Health, MCL, and the Carr/Connorstone team provide the public with information on assessment of the site, the risks it posed to the people of Ashland, and prospects for the future. Representatives of these groups were contacted and accepted invitations to appear. However, when AACE organizers followed up with a list of twenty-one detailed questions they thought needed to be addressed, EPA, DEQE, and Board of Health spokesmen all canceled their attendance. The well-publicized meeting was held as scheduled, with some 150 interested citizens gathered at the Federated Church. A number of locally elected officials were also on hand to provide some of the information expected from the absent officials.

The meeting did provide for a good flow of general information from AACE members and elected officials and eventually led to the establishment of an informational archive at the Ashland Library. But it also generated in the minds of the more active citizen organizers and the public the image of responsible official avoidance of tough detailed questions. And in the minds of some officials, the experiences of the May 1982 forum conjured up the image of amateurs aggressively pursuing complex questions for which responsible officials had not yet found the answers. Events since 1982 have not completely erased these images from the minds of both sets of participants.

In the meantime, an Ashland woman had written a letter to the Board of Health on April 29, 1982, in which she named five families living near the Nyanza site that had experienced cases of cancer (three fatal). Some of these individuals had played on the site as children. She wondered what could be done to find out if the Nyanza site was responsible. The Board of Health, under the

Source: Dave Granlund, *Middlesex News* (Framingham, Mass.), May 28, 1982.

Figure 6–4

guidance of its professionally trained health agent—a state-licensed sanitarian—handled this request by forwarding the letter to the State Department of Public Health, noting that while the Ashland mortality profiles showed no unusual elevation, "because of the Nyanza site history an epidemiological investigation is warranted."[50] Later, unbeknownst to the Board of Health, AACE conducted its neighborhood health survey. Then came the forum, which the Board of Health declined to attend, since, in the opinion of its health agent, the kind of detailed questions being asked by AACE were premature. Sufficient facts of off-site contamination had not yet been developed by Carr/Connorstone or state agencies, argued the health agent, and appearing in public to say you didn't know the answers was more likely to arouse public anxiety than not appearing at all until the information was in hand.[51] Then came the bombshell (see Figure 6–4).

On May 23, 1982, a dramatic *News* article appeared. An enterprising reporter on the Ashland beat had obtained a copy of the letter reporting the five cancer cases. The story also featured quotes from some members of the five families named in the letter, including this one from the father of one of the cancer victims:

This [the pollution] has been going on for a long time. . . . The people around here have been complaining for the past thirty years, but nobody did anything about it. First, they didn't know anything, and then there were all the politics.[52]

Here was the quintessential public reaction so feared by government officials (who were, of course, the "they" referred to who engaged in all the "politics"). And while it may be convenient for officials to blame the media for provoking such a response, it seems clear that in this case the *News* merely provided an opportunity for the expression of public concern that had been building behind the scenes for some time. The initial story, in any case, was quickly followed by news stories on two local radio stations and an additional nine news articles, editorials, and editorial cartoons over the following two weeks. Many concerned individuals provided the *News* reporter with additional facts, including AACE leaders, who gave a general report on their health survey results.[53] Within four days of the original article, the total number of reported cancer cases in the Nyanza vicinity had grown from the five mentioned in the letter to twenty overall.

Concerned members of the public were not only contacting the *News* reporter to offer information or express their concerns; many were also contacting the Board of Health in the Ashland Town Hall. The record of complaints recorded in the board files is poignant testimony to the state of public anxiety that was dramatized by the media. A summary version of these unsolicited public communications to the Ashland Board of Health follows.

May 24, 1982: WEEI called for reaction to cancer story to be used that evening. . . . [O]ne of the five cancer victims named in the article called, upset her name was revealed, boyfriend didn't know she had cancer. . . . [W]oman called, used to live in the neighborhood, now has cancer, wanted to know if there was a connection, didn't want to give name. . . . Dr. Cutler called from DPH, WEEI had contacted him, surprised to learn names had been used in News, *reported a local X-Ray technician knew of 20 cancer cases locally, hadn't given names and addresses. . . . [W]oman called, husband had leukemia, heard WKOX broadcast, wondered if it was connected to cancer-talk in the newspaper?*

May 25, 1982: mother called, child rides schoolbus which is parked on Megunco Road, might dust particles harm child? . . . [W]oman called, should she plant garden, lives by wetland near site, fears effects on veggies. . . . [W]oman visited office, fears she has mercury poisoning, weaker on one side, extremely tired, liver infection a year ago, enlarged spleen, would like test (health agent suggested she go to her doctor, called her, said she couldn't afford testing).

May 28, 1982: school secretary calls about school buses, told her health agent says it's OK, no carcinogens.

June 1, 1982: woman called to help on cancer survey, daughter died at home, age 18. . . . Camp Winnetaska counselor called, how safe, etc., parents concerned. . . . [W]oman called, News reporter said to contact Board, told her to call DPH. . . . [W]oman called, lived in Nyanza neighborhood all life until recently, cancer diagnosed in April. . . . [W]oman called, lived 30 years near site, mother had mastectomy, remembered playing in "Cough Medicine Brook," ate from garden every year. . . . [W]oman called, she and husband had cancers, volunteered names of two neighbors who died. . . . [W]oman called, husband died of lung cancer, she has cancer now.[54]

This flood of communications overwhelmed the Ashland Board of Health and DPH head John Cutler. The chairman of the Board of Health said, "We don't have the personnel, we don't even have enough phones to handle the kind of calls we've been getting. . . . I don't know how we'd ever be able to manage this kind of thing."[55] Cutler was concerned that the information pouring in was incomplete and not useful unless it adhered to a standard format. He asked the Board of Health to make an effort to systematically gather comparable information for all cancer cases in the neighborhood, matching a checklist of needed information.

The Board of Health, particularly its health agent, seemed determined to cool down the panic reaction. As a result, when it designed a health survey form and released it on June 1, the board deliberately downplayed the connection in people's minds about Nyanza and cancer. The survey's cover letter began as follows:

Dear Resident:

As you know there is an unsubstantiated concern regarding a relationship between the Nyanza site and several reported cancer cases in Ashland. Although we do not feel that there is a connection between the two, if we do not substantiate this positively with facts, the concern will develop into a sensationalized viewpoint resulting in chaos and unnecessary worrying.[56]

Any members of the public who had cancer in their families were invited to call for a form or pick one up at the board office. No effort was made to ensure that all residents within a certain distance of the dumpsite were contacted. Sixty-five forms were distributed. Twenty-two were eventually returned to Cutler in Boston, reporting ten cases of cancer. The Board of Health also forwarded the informal health survey conducted by AACE to the DPH. Cutler told a reporter that much of the information forwarded to him was not usable, since the Board of Health survey did not include the names of the respondents and the AACE survey did not indicate

Granlund—

Source: Dave Granlund, *Middlesex News* (Framingham, Mass.), May 24, 1983.

Figure 6–5

how the sample was drawn or where the households were located in relationship to the site, among other shortcomings. No sound epidemiological study could be conducted without much more and better preliminary information, he stated.[57]

While the DPH was working to assess the cancer situation in Ashland, residents' anxiety increased in August with the release of EPA's Remedial Action Master Plan (RAMP). The plan was sharply critical of the Carr/Connorstone approach, noting significant data gaps, incomplete design for assessment, and conceptual shortcomings in the proposed cleanup solution. The RAMP highlighted the fact that the Carr/Connorstone approach paid little attention to the health and environmental impacts of off-site contamination, the issue of predominant concern to many residents of Ashland.

Coming on the heels of a series of anxiety-provoking revelations, this news reinforced the suspicion that there was a serious health hazard in the town of Ashland—a situation that was skillfully portrayed by the editorial cartoonist of the *News* (see Figure 6–5). The sense of vague menace conjured up by the Nyanza site as symbolized in the Granlund cartoon is an example of a dramatic image that persists in people's minds regardless of subsequent qualifying information or events.[58] Thus, in late summer 1982, people in Ashland knew that they were confronted with a toxic chemical site about which little was yet known, at least with regard to the impact on their health and welfare. In September, the DPH announced that the local health surveys were not technically significant. Since the mortality statistics at hand did not indicate an elevated rate of cancer in Ashland, DPH was unwilling to devote additional resources to determine whether a link existed between the Nyanza site and health problems in the residential neighborhood across the tracks.[59]

Superfund

Into this tinderbox of resentment ("The people around here have been complaining about this for thirty years. . . ")[60] and fear ("I've lost a father, a brother and a cousin so far. . . . Every day I get up and look in the mirror at myself and wonder if I'll be the next one")[61] entered EPA Region I officials working for the Superfund program. For months they had considered taking the lead from the state for assessment and cleanup, after identifying and notifying the seventeen legally responsible individuals and corporate parties in March 1982. Although the formal agreement with the state DEQE was not signed until January 1983, in September 1982 public responsibility for the Nyanza site was taken over by Superfund. From that time, the on-site managers had to deal directly with the individuals, groups, state and local officials, and business interests affected by and concerned with the site. Thus, like the actors in the Pirandello play "Seven Characters in Search of an Author," the EPA Superfund on-site coordinators stepped onto the stage in Ashland without a script. Designated site coordinators and other employees became the individuals with whom the affected Ashland characters began to interact.

Locally affected individuals and groups had different experiences and concerns about the Nyanza site. The generic problem of risk communication is attempting to find effective ways of simultaneously dealing with individuals and groups who are affected by a site in different ways, who are motivated to participate in the risk communication process by alternative considerations, and who tend to resist seeing the problem in any but their own perspective. In the folklore of hazardous waste management professionals, angry and fearful neighbors are probably the most difficult group with which to deal.

But neighbors of hazardous waste sites possess their own folklore about the risks. When news of significant health impacts is raised in their minds, they find it difficult to accept government officials who do not give them reassuring answers. Then there is the folklore of local businessmen, like Camille, who lost money trying to clean up the site ("I wasn't going to wait for any money from the town").[62] Another affected group was the owner operators of the Nyacol plant who did not feel responsible for what previous owners had done. In addition, there were the MCL principals, who had spent approximately $150,000 in engineering consultant fees to clear the way for developing the site for condominiums.

The AACE sought information about who had done what and when on the site, why this was so hard to find out, what the current health hazards were, and when some constructive action might be taken. Local town boosters, including members of the Board of Selectmen, worried about the town's image, fearing a negative business climate in an era of economic decline. Local Board of Health members, in spite of years of efforts to deal with safety and odor problems on Megunco Road, suddenly found themselves on the firing line as the local group responsible for dealing with the health impacts of the site yet powerless to do much about it and reluctant to feed unwarranted public panic. The reporters and editors of the *Middlesex News*, keenly aware of the growing national concern about toxic waste dumps in the aftermath of Love Canal, found that the story brought the issues home to them and their readers. This configuration of personalities, which already occupied the local stage when the Superfund officials made their entrance in the fall of 1982, represents the root of the risk communication problem at Nyanza and other significant hazardous waste sites.

Communication Problems

First, action was taken to restrict direct human contact with the site. Fencing and culverting could be considered a form of implicit risk communication, and the posting of signs such as "Warning: Contaminated Area: Site Contains Hazardous Materials Which May Be Harmful to Public Health—NO TRESSPASSING" was quite explicit along the fence perimeter. It is difficult to evaluate how *successful* the communication of risk was, since trailbike riders and shortcutters vandalized the fence and found ways to get into and traverse the site for fun and convenience. At least those who trespassed on the site were put on notice, in qualitative terms, that they were subjecting themselves to a risk.

The second form of direct contact exposure, via air contaminants,

WARNING
Sudbury River Fish

Fish Contaminated With Mercury
DO NOT EAT

(Spanish)

Pescado Contaminado Con Mercurio
No Se Puede Comer

(Cambodian)

ត្រីពុលជាតិបារត

សូមកុំពិសា

For more information, contact:
Massachusetts Department of Public Health 617-727-0049
Department of Fisheries, Wildlife, and Environmental Law Enforcement 617-366-4470
Department of Environmental Quality Engineering 617-292-5515

Source: Massachusetts Department of Public Health.

Figure 6–6

had been addressed numerous times by Carr/Connorstone, the local Board of Health, state regulators, and the RI/FS of Superfund. All reported either no contaminants or levels well below health standard levels. These findings have been published in fact sheets, formal reports, and the media, as well as communicated repeatedly in public meetings when the question has been posed. How successful this communication has been depends again on how one measures success. When the question is raised at public sessions and the answer is given that measurements indicate no health threat from airborne toxics, the questioners seem to accept it. But the question continues to be asked, so if one measures success in terms of communicating the relevant information to all residents, additional efforts seem to be required. This level of lingering anxiety regarding air exposure is not surprising in the view of the history of the odor problem in the Nyanza neighborhood.

A third route of exposure at Nyanza concerns the catching and eating of mercury-contaminated fish in the Sudbury River and its reservoirs. The problem had been known for years to various state agencies: early measurements of as many as 12.0 ppm (the FDA action level is 1.0 ppm) prompted one official from the Division of Fish and Wildlife to write his counterpart in the Division of Water Pollution Control in January 1972, stating that "these results reflect extremely high concentrations of mercury present and represent the highest that I have seen in Massachusetts to date." He closed his letter with the suggestion: "I believe that these results should also be forwarded to the Metropolitan District Commission and the Massachusetts Department of Public Health for their consideration in relation to perhaps a joint announcement or warning to people that may be fishing in this particular reservoir."[63]

However, although the risks associated with consuming mercury-contaminated fish were noted in various documents, statements to the press, and public meetings, the first formal risk communication in *sixteen years* occurred when the DPH issued a health advisory in March 1986, discussing mercury toxicity, standards for its regulation, and a warning against consuming the fish in certain portions of the Sudbury River and Reservoirs 1 and 2. Pregnant women and children were designated as especially vulnerable to mercury. It was not until the summer of 1986 that temporary warning signs were posted along the banks of the river and reservoirs in well-known public fishing spots. These signs were quickly removed by vandals. Permanently anchored and protected signs, printed in English, Spanish, and Cambodian, were posted in 1987 (see Figure 6–6).

This experience suggests that although informal, episodic communications of risk in documents, public meetings, and the media may reach some part of the general (English-speaking) public, they are not sufficient to put specific, high-risk individuals and population groups on notice with regard to risk. Precise, persistent efforts are necessary and still may yield somewhat uncertain results. These facts also demonstrate a theme mentioned at the outset of this study: that the participation of multiple governmental agencies in addressing a common problem brings a wide range of resources and expertise to bear but prolongs the process.

In this case, it took sixteen years to implement a concrete, specific, risk communication action because the mercury problem, originally discovered by EPA research contractors, was initially under the jurisdiction of the DWPC and later the Hazardous Waste Division of DEQE (once the source of the mercury was determined to be the Nyanza site). But some of the contaminated water was under MDC jurisdiction, which had earmarked the reservoirs as a backup water supply for the Boston-area distribution system. The contaminated fish in the Sudbury River were the concern of DFW, which licensed and promoted sport fishing throughout the Commonwealth. Yet the health effects of eating the DFW fish in the MDC reservoirs was the concern of the Department of Public Health.

Differing test result levels account for part of the delay among the various agencies subsequent to the 1972 letter from DFW to DWPC. By the early 1980s, when Nyanza came under Superfund jurisdiction, continued testing had established repeatedly that the mercury levels in the reservoir fish constituted a true risk to public health. But Superfund's first priority was site containment, and difficult-to-obtain funding was earmarked for that purpose.

Department of Fish and Wildlife officials were reluctant to prohibit sport fishing in the Sudbury River, which, some argued, would not result in the consumption of a large amount of fish (technically, fishing had always been prohibited in the reservoirs). However, DEQE officials argued that small but meaningful numbers of low-income residents of Ashland, Framingham, and other Boston-area towns, including Hispanics and East Asian immigrants, were reported to significantly supplement their weekly food supply with reservoir and river fish. The DPH was concerned about these health impacts but had no budget to post warning signs at the fishing spots. As part of its Superfund matching fund requirement, DEQE paid for the printing and installation of the DPH-produced signs along the affected portions of the river system.

The fourth major area of risk concern and communication was groundwater contaminated by the site. Groundwater moves toward

the Sudbury River and seasonally floods basements of neighborhood residences. This was known early in the site assessment process (as indicated in the 1980 DEQE Preliminary Site Assessment).
Like the other three risks (direct contact, air, and fish), basement
groundwater contamination was mentioned in reports, documents,
public meetings, and the media, and in this informal sense the risk
was communicated to the public in qualitative terms. But AACE
and a number of neighborhood residents persisted in demanding a
more definitive assessment. Officials from the local Board of
Health, DEQE, and EPA had all repeatedly promised, in one form
or another, that adequate health studies would be done to reassure
the residents of the Pleasant Street neighborhood about site impacts, and basement sampling was finally performed in May and
June 1986 by the NUS Corporation under contract to Superfund.
Yet in its technical report to the EPA, NUS indicated a number of
problems with sampling procedures in the field and with laboratory
procedures, which in some cases forced certain assumptions to be
made or estimates to be used instead of hard data.[64]

A number of volatile organics, inorganics, and metals were found
in modest amounts in the basement samples of five homes. In
communicating the risks to the homeowners along with their test
results, EPA merely noted that

> *we have had a toxicologist take a brief look at the data received to*
> *date, and we believe that the trace levels of the several compounds*
> *do not pose an immediate threat to the residents. We are currently*
> *having a toxicologist review all of the data in detail.*[65]

This was the only formal risk communication addressed to the public, and as in the case of the previously discussed risk communications, it was qualitative—in fact, quite vague—in nature. The current review mentioned in this letter to neighborhood residents
consists of passing the data along to the Center for Disease Control's Regional Representative of the Agency for Toxic Substances
and Disease Registry (ATSDR). Under the 1986 SARA amendments, this agency is mandated to conduct health studies at Superfund sites. However, as of July 1987, ATSDR was still designing
criteria and a process for responding to requests for health studies
from citizens.

These facts indicate that, as in the case of mercury in the fish,
considerable inter- and intragovernmental communication and assessment of a risk occurs over a lengthy time period. Communication of risk directly to the public is qualitative in nature, often
vague, and usually episodic and informal. Only occasionally do
formal risk communications take place, with questionable practical

results. In this case, in spite of sampling and laboratory problems, risk managers told people that the risks were no immediate threat. From the perspective of the citizen receivers, these communications have done very little to relieve anxieties concerning the health implications of living near Nyanza. In fact, they may have even increased those anxieties by exposing them to a lengthy, tedious technical process whose outcome is so equivocal.[66]

Having looked at the sources and nature of risk at Nyanza and gained a sense of how those risks have been managed in general terms, we can now consider more analytically the formally documented communication of risk. The bulk of risk assessment communications occurred among governmental officials over lengthy time periods in the form of letters, memos, telephone conversations, and informal discussions in the field and in meetings. This type of communication is not readily accessible to the general public. The assessments of risk posed by the Nyanza site as summarized and formally documented in DEQE's 1980 Preliminary Site Assessment and its 1981 Abstract and in EPA's 1985 Phase I Remedial Investigation/Feasibility Study Report, Addendum, and Record of Decision were intended to inform the public.

Following AACE's public meeting in May 1982, the regular provision of such documents to Ashland residents was established by means of an archive in the public library. DEQE and EPA routinely mail formal reports and other major communications there. The archive is maintained by AACE members and supplemented with some informal communications such as letters, minutes of meetings, and news clippings obtained by AACE in the course of its site-related activities. Thus there does exist a body of documented risk communications.

The following examples, which appear in various documents under such headings as General Conclusions, Health Impacts of the Critical Contaminants, and Health Risk Assessment, are typical of the formally documented communications of risk from Superfund officials to the public:

> *Air does not appear to be affected by the site contamination. . . . The site property does not appear to be the source of low-level organic air contamination detected in the northern industrial area.*
>
> *The site has very little adverse effect upon the health of the neighboring community of Ashland because surface and groundwater are not used for drinking sources. Air does not appear to be significantly affected by the site and thus does not appear to pose any significant adverse health effects on the local population.*
>
> *Inhalation is not considered a major potential route of exposure to aniline in the context of the conditions presently existing.*

Benzene is considered slightly acutely toxic by oral administra-tion. . . . [T]here is assumed to be some potential health risk associ-ated with human skin contact with benzene-contaminated water and soil.

Inhalation of TCE [Trichloroethylene] vapor as a route of ex-posure is not considered a major potential health risk. . . . Inges-tion of water contaminated with TCE constitutes a potential health risk. . . . Skin absorption of TCE does not appear to constitute a potential health risk in the context of the Nyanza Chemical Site setting.

Mercury is the primary critical contaminant associated with this site. It was found in elevated (above background) levels in ambient air, soils, and sediment, groundwater, and surface water. It is widely distributed in the media in and around the Nyanza Chemical Site. . . . [P]ossible chronic inhalation of mercury as a vapor was calculated to be 0.199 ppb/m³. If exposure takes place above this level, there may be some potential for adverse health effects. . . . Since many forms of mercury can be dermally absorbed, exposure to the residuals should be avoided by unprotected individuals. Expo-sure by the general population to low levels of mercury in surface water during recreational activities (swimming) could provide a po-tential exposure pathway.

Lead may enter the body via inhalation of dusts or fumes or by ingestion of lead compounds trapped in the upper respiratory tract or introduced into the mouth on food, fingers, or other objects . . . . [S]ince groundwater is not used for drinking and the municipal water sampled has no lead, the risk of exposure to lead in site-specific circumstances is low or eliminated.

The overall public health impact of exposure to heavy metals ap-pears to be minimal at this point. Although there is a significant groundwater contamination problem, which is probably sludge-related, the absence of an exposed population tends to reduce the risks.[67]

These are prototypical risk communications relating to Su-perfund sites. They are couched in careful, general terms. They apply explicitly to current conditions as established by field re-search data. They incorporate conventional scientific and regula-tory language and standards. When read by governmental regula-tors, most scientists, and industry experts in hazardous materials, they are reassuring; the health risks posed by the Nyanza site appear to be low. But when read by residents of a neighborhood that traditionally considered the site a playground and shortcut as well as a source of chemical odors, they are unnerving. When read by environmental activist members of AACE, the careful qualifica-tions and ready acknowledgments of uncertainties, which are con-

ventional elements of professional and scientific risk discussions,
become evidence of inadequate effort.

Qualitative risk communication generates public skepticism
partly because of the kind of language that is necessarily employed.
When the formal assessments of risk illustrated above were communicated to the public, the comments submitted to EPA focused on
this point. In turn, EPA asked the NUS Corporation to respond to
some of these expressions of public uneasiness. In these examples of
the NUS reply to public comments, we can see the problem of
qualifying language when used in a setting of public fear or mistrust:

> Comment: *"Very little adverse impact should be no adverse im-*
> *pact" or the adverse impact should be explained elsewhere in the*
> *report.*
>
> Reply: *This is meant to be a general statement of the overall*
> *impact of the site on the health of community residents and therefore*
> *generally encompasses all the possible routes of exposure—air, sur-*
> *face water, ground water, and direct contact. Our toxicologists are*
> *reasonably sure that there is no adverse impact through air, surface*
> *water or groundwater exposure; however, due to the various uncer-*
> *tainties with possible direct contact scenarios (such as children play-*
> *ing in it to the point of rubbing it over their extremities, etc., with*
> *possible incidental ingestion) under extreme exposure potential ad-*
> *verse impact could occur.*
>
> Comment: *Is direct contact with sediments transported beyond*
> *the fenced area a hazard or isn't it? "Could" is unacceptable now*
> *that the RI is completed.*
>
> Reply: *As explained above the uncertainties associated with the*
> *potential of direct contact with sediments transported beyond the*
> *site fenced area preclude the use of "is it, or isn't it" type of determi-*
> *nation.*[68]

Thus, the health risks posed by Nyanza may appear to be low, but
under certain circumstances, for certain individuals, they may be
quite high. Governmental risk communicators cannot flatly deny
the validity of this lay approach to the assessment of risk whereby the
analysis is considered incomplete if uncertainties remain. They can
repeat the professionally trained, science-based reasoning behind
their own assessment, they can state probabilities, and they can
argue cost-effectiveness; but they cannot dismiss the possibility that
what *appears* to be a low-risk situation *might* be a high-risk one.
Typically, in such situations a sense of mutual frustration develops.
Government officials resent the public resistance to seeing things
their way; in the minds of officials, this resistance is their risk com-

munication "problem." Citizens resent government officials' insistence on an abstract, rational-analytical approach, which rests on what they perceive as arbitrary assumptions, is hedged with uncertainties, and fails to take into account local residents' experience and history of exposure to a specific site under specific circumstances; in the minds of citizens, the theoretical approach employed by government agencies is their risk communication "problem."

The mutual frustration involved in this typical hazardous waste site risk communications problem manifested itself at Nyanza primarily in disagreement over the remedial action selected by EPA for Phase I. In essence, members of AACE, some neighborhood residents, and the Ashland Board of Health preferred the total removal of contaminants from the site as their first choice of alternatives. But removal from Ashland and disposal elsewhere would be extravagantly expensive, uncertainties existed with regards to available landfill capacity and location, and such an operation would expose Ashland residents to tens of thousands of truckloads of hazardous materials as they moved through town.

The second choice of AACE, the Board of Health, and some residents was to augment the on-site landfill/capping alternative with a lined bottom and a leachate recovery and treatment capacity. But the RI/FS engineering alternative favored by EPA officials included a RCRA-standard cap, an upgradient groundwater diversion, and a plan to de-water and solidify the sludge materials as they were dug up around the site and consolidated on the hillside. Therefore, no bottom liner or leachate recovery system would be needed (since there would be no leachable materials buried), and surface and groundwater would be prevented from reaching the solidified material after burial. EPA's alternative included monitoring to ensure that the structure worked as designed and a fall-back leachate recovery system in case of failure.

In a manner strikingly parallel to the dispute between the Carr/Connorstone team and government regulators—but with the roles reversed—EPA risk managers emphasized the simplicity and cost-effectiveness of their preferred alternative, while citizens and local health officials argued for a more exhaustive, relatively cost-blind approach to the selection of a remedial action for Phase I. While EPA made modifications in its remedial design in response to local and DEQE comments and concerns, it adhered to its basic decision of summer 1985 with regard to Phase I remediation.

Rooted in a divergence of views about the nature and extent of the health risks involved and therefore the nature and extent of remediation needed, this conflict soon provoked local political eruptions. The EPA was determined to proceed with its plan and document it in the required Record of Decision, but the Board of Health was advised by its legal consultant to delay the decision on remedial action rather than accept an alternative they deemed inadequate. This angered the Board of Selectmen, who, already embarrassed by Nyanza's impact on the town's reputation, feared the loss of federal funds if the town did not accept EPA's choice of alternatives.[69] The selectmen basically wanted their constituents to stop rocking the boat so that EPA could eliminate the Nyanza albatross.

The second major issue about which mutual frustrations arose between governmental risk managers and local citizens was the matter of health studies. We have seen the depth of concern and anxiety about cancer that erupted in the spring of 1982 in Ashland. The responses of the local Board of Health and the state Department of Public Health did little to remove this anxiety. Once again, as in the case of the risk communications regarding site remediation, official governmental health assessments were carefully couched in professional and regulatory language, employed conventional scientific standards in epidemiology, and concluded that there were no untoward risks to the neighbors affected by the site. The conclusions of the DPH investigation into the cancer linkage were not communicated in a formal, written document but delivered verbally at a meeting in Ashland hosted by the Board of Health in September 1982 and reported in the *Middlesex News*. In essence, DPH's John Cutler indicated that the cancer mortality statistics for a town the size of Ashland were normal and that therefore no further investment of time and resources was warranted.

Again, to governmental regulators trained in the basics of epidemiology and health statistics, to many scientists, and to science-trained industrial risk management experts, this assessment and communication of the risk would seem reasonable, soundly based, and not at all remarkable: it indicated that there was little or no health risk posed by Nyanza. But for individuals and families with a history of direct exposure to the hazardous materials on the site and a variety of health complaints, whether they could be statistically tied to site exposure or not, this communication of risk was regarded as callous, incomplete, and unconvincing.

Citizens pointed out that there had been no persistent, systematic outreach effort to contact everyone who had lived near the site for a number of years prior to the Nyanza plant closure. Nor had the workers who spent many years there with direct, high-level exposure to many toxics been canvassed. Therefore, the statistical assessment was superficial and misleading. Nor had any account been taken of the personal anxiety and suffering caused by diseases and deaths of nearby residents. Even if these were not *statistically significant*, citizens questioned the detached rationality of those who failed to acknowledge the perceived connection between illness in the community and the Nyanza site. Citizens repeatedly insisted that technical risk communicators must address shared human concerns.

Thus, the Nyanza case illustrates a prototypical risk communication conundrum as a result of the divergent priorities and objectives of the involved parties: science-educated risk managers, trained to be impersonal in their work, make discretionary but logical assumptions, calculate the probabilities and cost-effectiveness of choices, accept uncertainties, render a decision, and move on to the next phase or the next site; citizen-members of affected families, who care deeply, make worst-case but reasonable assumptions, seek certainty and relief regardless of cost, speak out against decisions with which they disagree—and with whose consequences they are forced to live out their lives. Owing to depressed property values for those in proximity to a Superfund site, the possibility of relocating is sharply reduced.

Citizens Advisory Committee

Obstacles to risk communication arise at every Superfund site to some degree, over some issues, at some time in the course of remedial activities. Also, for one reason or another, delay occurs as a matter of course. At Nyanza, the risk communication process was interrupted because of the expiration of the original Superfund legislation and congressional deadlock on passage of a new version. This particular delay, since it was tied to congressional politics rather than bureaucratic immobility, did not result in a deepening of resentment against possible risk managers from Superfund. In fact, by providing a breathing spell, the funding delay allowed local, state, and federal interests to sort out their differences and institute improved risk communication arrangements.

Underlying the disagreements about the selection of remedial measures and the adequacy of health studies was the reality of decision-making power. Only Superfund officials had such power. State officials had some leverage over it, since they had to bear some of the costs and would be responsible for monitoring and maintaining the site after remediation, but local interests had no such bargaining leverage. Town government had a degree of political leverage, but it was limited to pressuring Superfund officials by conventional interest-group methods. And town boards were split over the issue of how far to go in opposition to decisions favored by the EPA, given the fear of losing funding to other, more cooperative Superfund communities. Local citizens participated under an elaborate public participation plan, but the influence of their participation was limited by a lack of time and expertise and a different perspective from EPA risk managers. Citizens could do little more than attend meetings when alternatives were announced and try to take advantage of a short period devoted to public reaction. In the end (at least this was the local experience with remedial selection and health studies), views in opposition to the judgment of the EPA decision makers would not prevail. Thus, dissatisfaction with the ultimate decision-making authority and the degree of sharing that authority underlay technical and procedural disagreements and fueled the resentment and emotions of those who felt powerless.

The DEQE had public participation specialists actively involved at Nyanza. State hazardous waste cleanup policies favored the institutionalization of local interests into community advisory committees, to act as a focus for risk communications. The DEQE seemed to be sensitive to local frustrations and sought ways of reducing structural and procedural obstacles to less controversial risk communications. If this could be accomplished, the basic imbalance in decision-making power might be made more acceptable to the people of Ashland.

At the height of the controversy over EPA's choice of a remedial alternative, a meeting was held on the night of August 20, 1985, at DEQE's suggestion, for the purpose of forming a Citizen's Advisory Committee (CAC). The basic idea of the CAC, according to DEQE officials in attendance, was for the variety of divergent interests on the local level to organize themselves, formulate their own objectives from a communitywide perspective, and act as a focal point for local-state and local-federal communications. The DEQE, through its Community Hazardous Waste Coordinator Program, could provide modest funding for the copying of documents, postage, and telephone expenses. Documents that were sent to the library archive could also be provided directly to the CAC for reliable local distribution and comment. In effect, if the community could organize itself, it could be brought closer to the ongoing decision-making process as it occurred within DEQE and the Superfund program. Local officials, group leaders, and individual citizens of Ashland would have to organize and articulate their diverse interests and participate in a regular fashion in order to make the process of risk communication work to their benefit.

The Ashland participants agreed that the goal of the CAC was *to be the central community organization in the town concerned with the issue of the Nyanza site and its remediation. The group will also be the main contact for state and federal agencies to communicate with to enable local community input into all activities related to the site.[70]*

The group chose to meet monthly and elected a chairman (the chairman of the Board of Health), a vice chairman (the fire chief), a coordinator (Board of Health member), and a secretary (an AACE member). A small but essential step had been taken to improve risk communications at Nyanza.

Eventually, the group's agenda took the following form: (1) announcement and discussion of EPA and DEQE activities at the site; (2) announcement and discussion of local CAC concerns about the site; (3) a call for any other issues of concern to the meeting participants; and (4) agreement on an agenda for the next meeting. This meeting form seemed to function reasonably well in terms of providing a two-way flow of information. That is not to say that *disagreement* did not occur but that *communication* occurred in a regular, orderly fashion.

In the matter of attendance and leadership, local officials other than the coordinator (a Board of Health member) either lost interest over time or came to believe that the town's interests were being pursued satisfactorily without their regular attendance. Leadership of the CAC gravitated to the secretary (an AACE member) due to his diligence, organizational skills, technical knowledge (he is a quality assurance manager at a high-tech firm), patience, and commitment.[71] He was eventually elected chairman. A second member who achieved an important role was the plant manager of Nyacol, who was also the president of the Chamber of Commerce. He was valued by other CAC members for his direct and intimate involvement with the site, his technical expertise, and his own patience and commitment.[72]

After taking a prominent role in initiating the formation of the CAC, DEQE officials were regular attendees, but as part of the supporting cast. It was not EPA policy to form citizen advisory groups at Superfund sites, and at the outset of the CAC process EPA attendance was irregular; it became more regular as the site manager was drawn in and eventually became central to the success of the CAC in the risk communication process. A key impetus to vigorous EPA participation in the Nyanza CAC resulted indirectly from activities of the Community Relations Program in Superfund's Washington headquarters.

Due to inquiries from states, communities, and EPA regional offices, the EPA's Superfund Community Relations Program took advantage of the site activity funding lull to initiate a pilot program aimed at assessing the use of environmental mediators to resolve disputes at Superfund sites. The Nyanza site was chosen to be among the three pilot project sites in the country. Apparently, the Region I selection of Nyanza from among all of its other Superfund sites was due to the timing and intensity of the dispute over the remedial alternative and the ongoing policy dispute with DEQE over the issue of citizen advisory committees. The New England Environmental Mediation Center provided the expertise for the facilitation project, the Community Relations Program provided the funding, and the lull in on-site activities provided the time for EPA, DEQE, and local officials and citizens to discuss and think about how to better resolve conflicts.[73]

The outcome of the pilot project at the Nyanza site was not, properly speaking, a mediated decision, since Superfund officials had sole authority to make decisions concerning the selection of remedial alternatives and (after passage of SARA) health studies. But the project did develop a more uniform set of expectations concerning the possibilities and limitations of local participation in the decision-making process. Consultations of the mediator with EPA, DEQE, and the CAC resulted in a clear set of guidelines for improving the timing, nature, and quality of communication flows through the CAC.[74] While conflicts resulting from risk communications were not eliminated, since the perspectives and interests of participants in the process varied widely, they were no longer caused by a lack of understanding of decision procedures or a lack of structured access to decision information. As a result, the level of frustration and emotion surounding risk communication has dropped considerably since the summer of 1985. It should be noted also that the improved environment for risk communication has been aided considerably by the fact that the EPA on-site representative from 1985 to 1987 was particularly well informed technically as well as unusually patient in dealing with local concerns.[75]

Conclusion

The most significant conclusion to be drawn from the Nyanza case is that governmental risk managers tend to focus intensely on the *site* as a physical assessment and cleanup problem, whereas affected residents focus intensely on their *health*. This discrepancy in focus has major implications for the process of risk communications. Officials preoccupied with technical, fiscal, and administrative procedures can easily come to regard citizens as unsophisticated, emotional, and a serious obstacle to their work. Citizens preoccupied with the impact of years of direct, casual exposure to a highly ranked hazardous waste site, worried about the contaminated groundwater that enters their basements each spring and whether their vegetable gardens draw toxics from the soil, can easily come to regard bureaucratically preoccupied risk managers as uncaring technicians. Some will choose to continue to participate in the communication process, but many more will withdraw from it in anger and frustration.

While the Superfund program is intended to assess and clean up sites, the *method* utilized for this purpose is not an end in itself. The goal of Superfund is the protection of the public health. Whether by means of training, experience, or the selection of especially sensitive individuals to deal directly with Superfund communities, professional risk managers must keep in perspective the fact that the angry, frightened individual who delays a public briefing session is the public that Superfund is intended to serve.

The tendency for official risk managers to focus on the site rather than the population affected by the site is compounded by the sequence and timing of Superfund procedures. In the absence of dramatic, immediate threats to the public, emergency remedial measures may not be taken at a site like Nyanza. Risk management attention is primarily devoted to the issue of *site containment*. While this is orderly, logical, and necessary, the public living in the site vicinity is often more concerned with *off-site impacts*, both past and present. In the case of Nyanza, Superfund personnel concentrated on preparing for and supervising the conduct of the Remedial Investigation and Feasibility Study for site containment (Phase I). It was fortunate that the responsible state agency (DEQE) was attentive enough, aggressive enough, and had sufficient funds to undertake many additional remedial measures such as site fencing, the culverting of Chemical Brook, and the posting of warning signs at fishing spots. The DEQE had the flexibility to respond to other needs that could not be easily addressed by the elaborate, formal Superfund process.

For example, when RI/FS data indicated problems in the area of the Vault, DEQE was able to undertake additional, specialized testing in the area, leading ultimately to a Superfund Emergency Response effort. When two local residents reported seeing, as children, the burial of several hundred barrels on the hill, an event that was not evidenced in the RI/FS data, DEQE initiated a special search in the area indicated. Superfund management has displayed some flexibility in recalling the RI/FS contractor to do basement water sampling, but the basic problem remains that it is easy to become preoccupied with site containment to the detriment of off-site impact management. Residents more concerned with the latter may be impatient when discussing the risks involved in the site containment process. Recognizing this factor and dealing with it effectively are important aspects of improving risk communications.

Finally, the issue of health studies is a significant aspect of risk communication problems at Nyanza and, we suspect, other Superfund sites. *Risk managers* focus on the difficult science of investigating causal linkages between toxics, routes of exposure, effective dosages, and definable health impacts. *Neighbors* focus on health symptoms, which they fear are related to where they live, the air they breathe, and the water in their basements. With what result? A detailed study of the methods and findings of public health investigations at twenty-one hazardous waste disposal sites, including chemicals involved and exposure criteria, has recently been completed by the Universities Associated for Research and Education in Pathology. The summary of the findings of the twenty-one studies stated that "to date, none of the investigations has provided sufficient evidence to support the hypothesis that a causal link exists between exposure to chemicals at a disposal site and latent or delayed adverse health effects in a general populace, although several positive associations have been shown."[76]

Given the conditions at Nyanza, it is extremely unlikely that the demands for scientific evidence and confirmation of the linkage hypothesis could ever be met; even if a comprehensive, systematic effort were to be undertaken, in all likelihood the results would be statistically equivocal. The Board of Health, neighbors, AACE, and CAC have been requesting such a study for at least six years, but it is unlikely that they will get it. Health risk managers, like site risk managers, tend to become preoccupied with their professional roles, namely, to obtain information, identify the obvious problems, and take action for the protection of public health through their canonical systems.

An alternative approach is to deal directly with the health symptoms and concerns of the residents who live or have lived in the vicinity of Nyanza. Under such circumstances it might help to provide local medical practitioners with some specialized workshops in recognizing, diagnosing, and dealing with toxic health impacts. Alternatively, one could bring medical specialists into the community on a regular basis for clinical consultations with people exposed to the chemical hazards as an integral part of a hazardous waste management program. Such public health visitations on a nonspecialized basis (visiting nurses) are a routine part of most community health programs. In sum, on the basis of the experiential response of residents, it makes sense to treat people as patients rather than as objects of study. Until such a shift in the perception of the public health problem at hazardous waste sites occurs, there will continue to be risk communications difficulties.

Endnotes

1. "Chemicals at Nyanza Site: 'They'll Never Get Rid of It,' " *Middlesex News*, May 30, 1982.
2. "Ashland's Nyanza site: Living Near a Chemical Dump," *Middlesex News*, May 30, 1982.
3. "Child's Playground Now a Toxic Nightmare," *Middlesex News*, November 9, 1986.
4. Ashland Advocates for a Clean Environment (AACE), comment on health survey form, May 1982. AACE Archive.
5. Department of Public Health (DPH) to Board of Health, November 26, 1963, Town of Ashland Public Library, Nyanza Archive (hereafter Library Archive).
6. Nyanza, Inc., to Division of Water Pollution Control (DWPC), September 11, 1967. AACE Archive.
7. DWPC, "Memorandum for the Record," November 12, 1969. Library Archive.
8. For correspondence among Nyanza, Inc., the Ashland Board of Public Works, the DWPC, and the Metropolitan District Commission (MDC), see AACE Archive.
9. Nyanza, Inc., to DWPC, December 3, 1971. AACE Archive.
10. DWPC to Nyanza, Inc., October 1, 1970. AACE Archive.
11. DWPC, "Memorandum for the Record," June 14, 1971. AACE Archive.
12. DWPC to Nyanza, Inc., September 6, 1971. AACE Archive.
13. Department of Environmental Quality Engineering (DEQE), "Nyanza Chemical Waste Dump, Megunco Road, Ashland, Massachusetts: Preliminary Site Assessment Report," Woburn, Massachusetts, October 23, 1980. Library Archive.
14. Nyanza, Inc., to DWPC, September 6, 1971; Attorney General, Commonwealth of Massachusetts, to Nyanza, Inc., February 25, 1972; DWPC to Nyanza, Inc., March 28, 1972; DWPC to Nyanza, Inc., August 8, 1972; and Withington, Cross, Park & Groden to DWPC, November 16, 1972. AACE Archive.
15. Ashland Board of Health, note in Nyanza files, November 3, 1972.
16. DPH to Nyanza, Inc., December 7, 1972. Library Archive.
17. Ashland Board of Health, Minutes, meeting of June 3, 1971.
18. JBF Scientific Corporation (JBF), "An Investigation of Mercury Problems in Massachusetts," Burlington, Massachusetts, July 3, 1973. DEQE Files.
19. Camp, Dresser & McKee, Inc. (CDM), "Proposed Pretreatment System for Nyanza, Inc., Ashland, Massachusetts," Boston, Massachusetts, January 1973; and CDM, "Nyanza Inc., Addendum to January 1973 Report," May 5, 1973. DEQE Files.

20. Commonwealth of Massachusetts, Suffolk County Superior Court, "Plaintiff's Memorandum (Equity Number 1977)," Cambridge, Massachusetts, undated. Library Archive.
21. Water Resources Commission to Withington, Cross, Park & Groden, October 8, 1975. DEQE Files.
22. Environmental Protection Agency (EPA) to Withington, Cross, Park & Groden, September 18, 1975. AACE Archive.
23. Robert Lurie, interview, June 6, 1987.
24. EPA, "Overview of Nyanza, Inc.'s Activities," Boston, Massachusetts, February 1985. Library Archive.
25. Connorstone, Inc., to DWPC, October 19, 1979. DEQE Files.
26. DEQE to Edward Camille, March 27, 1980. Library Archive.
27. "Cleanup of 'Top Ten' Continues," *Middlesex News*, November 9, 1981.
28. EPA, "Former Owners/Operators of 'Nyanza' Waste Superfund Site," Boston, Massachusetts, February 1982. AACE Archive.
29. "Planning Board Demands Nyanza Cleanup," *Middlesex News*, June 30, 1982.
30. DEQE, "Nyanza Preliminary Site Assessment."
31. Centers for Disease Control to EPA Region I, August 9, 1983; EPA to Board of Health, September 9, 1983. Library Archive.
32. Connorstone, Inc. to DWPC, October 19, 1979. DEQE Files.
33. Carr Research Laboratory, Inc. (CRL), "On-Site Hydrogeology and Initial Pollutant Studies on the Nyanza Chemical Waste Dump, Ashland, Mass.," Wellesley, Massachusetts, August 14, 1981. DEQE Files. The phasing approach used during the initial stages of investigation at the site were not the same as EPA's later phasing of the project.
34. CRL to DEQE, September 18, 1981. DEQE Files.
35. JBF to DEQE, November 12, 1981. DEQE Files.
36. EPA, "Nyanza—Enforcement Recommendations," Boston, Massachusetts, December 24, 1981. Library Archive.
37. DEQE to Connorstone, Inc., January 28, 1982.
38. CRL, "Discussion Document, DEQE Draft Letter of January 28, 1982: Deep Drilling Program on the Nyanza Chemical Waste Dump, Ashland, Mass.," Wellesley, Massachusetts, February 12, 1982. DEQE Files.
39. CRL to DEQE, May 6, 1982. DEQE Files.
40. CDM, "Nyanza Chemical Site, Ashland, Massachusetts: Remedial Action Master Plan," Boston, Massachusetts, July 1982. Library Archive.
41. "EPA Blasted over Nyanza Site," *Middlesex News*, September 30, 1982.
42. EPA, "Record of Decision," p. 24.
43. There is a substantial literature on risk perception, the communication of risk, and the role of the media in these processes. For a helpful discussion of the main issues and a solid bibliography, see Vincent T. Covello et al., "Communicating Scientific Information About Health and Environmental Risk: Problems and Opportunites from a Social and Behavioral Perspective," in V. Covello, A. Moghissi, and V.R.R. Upporli, eds., *Uncertainties in Risk Assessment and Risk Management* (New York: Plenum Press, 1986).
44. *Chemical Risks: Fears, Facts, and the Media* (Washington, D.C.: The Media Institute, 1985).

45. "Nyanza Dump Subject of Hearing this Month," *Middlesex News*, November 5, 1984.
46. Holly Buckoski, interview, February 25, 1987.
47. Nicholas Freudenberg, "Citizen Action for Environmental Health: Report on a Survey of Community Organizations," *American Journal of Public Health* 74:444–48 (May 1984).
48. Department of Environmental Quality Engineering, "Abstract: Nyanza Chemical Waste Dump, Ashland, Massachusetts," Woburn, Massachusetts, July 1981. Library Archive.
49. Beverly Dort, President, AACE, interview, Ashland, Massachusetts, May 21, 1987. Ms. Dort, after deleting names and addresses, was kind enough to show me the original survey forms with interviewer comments in the margins. AACE Archive.
50. Board of Health to DPH, May 6, 1982. DEQE Files.
51. Michael Sullivan, interviews, May, 29, 1987. Also see "Firm Dodges Public Forum in Ashland," *Middlesex News*, May 19, 1982.
52. "State Probes Ashland Cancer," *Middlesex News*, May 23, 1982.
53. Elizabeth Burt, interview, June, 1, 1987.
54. Board of Health, Ashland, Massachusetts, "Complaints," May 24–June 1, 1982. Board of Health files.
55. "State to Probe Ashland Cancer," *Middlesex News*, May 31, 1982.
56. Board of Health to "Dear Resident," June 1, 1982. Board of Health Files. Also see "Local Officials to Help in Ashland Cancer Probe," *Middlesex News*, June 2, 1984.
57. "State to Probe Ashland Cancer," *Middlesex News*, May 31, 1982.
58. S. Lichtenstein et al., "Judged Frequency of Lethal Events," *Journal of Experimental Psychology: Human Learning and Memory* 4:558–78; and Baruch Fishoff et al., "Lay Foibles and Expert Fables in Judgements About Risk," in T. O'Riordan and R. K. Turner, eds., *Progress in Resource Management and Environmental Planning* (Chichester, England: Wiley, 1981), Volume 3.
59. "Town Told Cancer Rate Normal," *Middlesex News*, September 22, 1982.
60. "State Probes Ashland Cancer," *Middlesex News*, May 23, 1982.
61. "Cancer Plagues Ashland: 17 Cases Discovered Near Waste Dump Site," *Middlesex News*, May 27, 1982.
62. "Chemicals at Nyanza Site: 'They'll Never Get Rid of It,' *Middlesex News*, May 30, 1982.
63. DFW to DWPC, January 3, 1972. AACE Archive.
64. NUS Corporation, "Technical Assistance: Residential Sampling, Nyanza Chemical, Ashland, Massachusetts," Pittsburgh, Pennsylvania, January 26, 1987. AACE Archive.
65. EPA to Beverly Dort, January 29, 1987.
66. "Neighbors Fear Effects of Living Near Nyanza," *Middlesex News*, August 3, 1986.
67. EPA, "Fact Sheet: Nyanza Chemical Superfund Site," Boston, Massachusetts, November 1984 (No. 1,2); NUS, "Phase I RI/FS" (No. 3–7); and NUS, "Addendum to Phase I" (No. 8). AACE Archive.
68. NUS, "Addendum to Phase I," Appendix A.

69. "Ashland Board Irate over Nyanza Opinion," *Middlesex News*, June 21, 1985.
70. Citizens Advisory Committee, minutes of meeting with Board of Health, August 20, 1985. Library File.
71. Richard Brown, interview, June 1, 1987.
72. Thomas Robinson, interview, June 16, 1987.
73. James Arthur, interview, May 18, 1987.
74. New England Environmental Mediation Center, "Evaluation of Pilot Project in Region I," Boston, Massachusetts, 1986. EPA Files.
75. Mary Sanderson, interview, June 1, 1987.
76. Joe W. Grisham, ed., *Health Aspects of the Disposal of Toxic Wastes* (New York: Pergamon Press, 1980), p. 246.

Sources

Interviews

JAMES ARTHUR, New England Environmental Mediation Foundation, May 18, 1987.
RICHARD BROWN, Chairman, Nyanza Citizen's Advisory Committee, June 1, 1987.
HOLLY BUCKOSKI, Ashland Advocates for a Clean Environment (AACE), February 25, 1987.
ELIZABETH BURT, former reporter, *Middlesex News*, June 1, 1987.
RICHARD CAVAGNERO, EPA, former Site Coordinator, June 11, 1987.
RICHARD CLAPP, DPH, June 5, 1987.
SUZANNE CONDON, Department of Public Health (DPH), June 5, 1987.
JAMES CRITSER, former manager, Nyanza, Inc., June 11, 1987.
JOHN CUTLER, DPH, May 27, 1987.
PAT DONAHUE, DEQE, May 16, 1987.
BEVERLY DORT, AACE, May 21, 1987.
HENRY FASSLER, Chairman, Ashland Board of Health, June 4, 1987.
DENNIS GAGNE, EPA, former Site Coordinator, June 11, 1987.
RICHARD GREENE, Ashland Board of Health, May 29, 1987.
MARIE HEATON, former Chairman, Ashland Board of Health, June 4, 1987.
ELAINE KRUGER, DPH, June 5, 1987.
MAUREEN LAVIN, Ashland Board of Health, May 27, 1987.
ROBERT LURIE, President, Nyacol Products, Inc., June 16, 1987.
PETER NICHOLAS, reporter, *Middlesex News*, June 10, 1987.
RICK REYNOLDS, Bureau Chief, *Middlesex News*, June 10, 1987.
THOMAS ROBINSON, Plant Manager, Nyacol Products Inc., June 16, 1987.
MARY SANDERSON, EPA, current Site Coordinator, June 1, 1987.
ARTHUR SHAPIRO, AACE, June 2, 1987.
JOHN P. SULLIVAN, former Health Agent, Ashland, June 15, 1987.
MICHAEL SULLIVAN, former Health Agent, Ashland, May 29, 1987.
SARAH WEINSTEIN, DEQE, May 18, 1987.

Documentary Archives

AACE FILES. Ashland, Massachusetts.
TOWN OF ASHLAND PUBLIC LIBRARY, Nyanza Files. Ashland, Massachusetts.
BOARD OF HEALTH, Nyanza Files. Ashland, Massachusetts.
DEPARTMENT OF ENVIRONMENTAL QUALITY ENGINEERING, Division of Hazardous Waste, Superfund Program, Nyanza Files. Boston, Massachusetts.
ENVIRONMENTAL PROTECTION AGENCY, Nyanza Files. Boston, Massachusetts.

Audiovisual Sources

WCVB-TV, "EPA Releases List of Top Hazardous Waste Sites," evening news, October 23, 1981.
AACE, audiocassette of first public meeting on Nyanza, May 18, 1982; "Nyanza Nite II," videotape of public meeting, January 23, 1987.
MASSACHUSETTS CABLEVISION, Ashland Cable Access, "Nyanza: Rumor and Reality," Out-to-Lunch Production, 1987.

Chronology

1967

August: Massachusetts Division of Water Pollution Control (DWPC) orders Nyanza to implement a Pollution Abatement Program, curbing discharge of wastes into waters of the Commonwealth. Nyanza is ordered to tie into the Metropolitan District Commission (MDC) sewer system by end of 1969.

1969

November: Nyanza is unable to complete its 1969 sewer tie-in. Local sewer authorities and MDC require further improvements in waste effluent limits on pH, mercury, and suspended solids.

1970

May: EPA contractor, JBF Scientific Corporation, begins a study of mercury pollution in Sudbury River and traces it to the Nyanza site.
August: EPA verifies with Nyanza headquarters that approximately 20 pounds per day of mercury are discharged with Nyanza's Ashland wastewater.
October: DWPC orders Nyanza to limit its wastewater discharge of mercury to 0.08 pounds per day.

1971

June: DWPC inspectors find 35' deep sludge pits on the hill adjacent to Nyanza plant.
July: Nyanza completes its tie-in to MDC sewer. Monitoring reveals that pH and other limits are regularly exceeded.
August–September: DWPC orders Nyanza to cease discharge of its wastes into waters from all sources.

1972

March: DWPC contracts with JBF to conduct a site assessment and recommend cleanup activities.

May: MDC fish sampling of the Sudbury River indicates high levels of mercury contamination of fish in reservoirs which are backups for the Boston water supply.

December: The Ashland Board of Health discovers that Nyanza is dumping barrels of spent phenols on the hill. The Department of Public Health (DPH) issues a citation.

1973

July: A JBF report is submitted to DWPC. It assesses the extent of site contamination, groundwater contamination, and surface runoff contamination from disposal of mercury sludges on site.

October: DWPC orders Nyanza to draw up plans to cease all surface and groundwater pollution.

1974

February: Camp, Dresser & McKee (CDM), engineering consultant to Nyanza, submits a remedial plan to the DWPC.

September: A consent decree gives Nyanza one year to carry out the CDM remedial plan.

1975

June: The Department of Environmental Quality Engineering (DEQE) is created, absorbing the DWPC.

September: Nyanza granted a one-year extension for compliance.

1976

September: Nyanza sells two parcels of its property to a local industrial developer to gain financing for remedial activities.

November: Nyanza fails to comply with consent decree. The Attorney General's Office asks Suffolk Superior Court for a contempt citation.

1977

January/February: Nyanza is found in contempt and ordered to sell more property to finance a $50,000 cleanup escrow account to be held by the attorney general.

July: Nyanza sells several buildings and lots to Nyacol, Inc., a separate corporation manufacturing colloidal silica.

November: Nyanza closes its Ashland dye operations and ceases to do business on the site.

1978

December: Nyanza sells the remainder of its property to a local industrialist. The new owner hires a local engineer to propose a final cleanup plan.

1979

September: Using the $50,000 escrow money, engineering plans are submitted for a cleanup.

December: Massachusetts passes a Hazardous Waste Management Act.

1980

January: DEQE finds the proposed cleanup plans inadequate and performs its own preliminary site assessment.

November: The Nyanza site is included on the DEQE's list of confirmed hazardous waste sites.

1981

February: Nyacol and a local developer sell the hill property to the MCL Corporation, who plan to build condominiums on the site.

August: MCL consulting engineers submit an upgraded site assessment and remediation plan.

October: The Nyanza site is placed on the EPA Superfund National Priority List.

1982

January: DEQE finds MCL's plan for the site cleanup inadequate.

March: EPA hires CDM to prepare a Remedial Action Master Plan (RAMP).

May: Ashland Advocates for a Clean Environment (AACE), a citizen's group, holds a public forum on the Nyanza problem. Board of Health, DEQE, and EPA representatives decline to attend. Local elected officials and 150 citizens discuss Nyanza and pose questions.

June: Board of Health and DPH undertake a study of cancer cases reported to be linked to wastes from Nyanza.

August: CDM presents RAMP. The estimated cost of the remedial investigation and feasibility study (RI/FS) is $400,000.

September: DPH tells the Ashland Board of Health that, for a town of its size, the cancer rate is statistically normal.

November: DEQE fences the hill site to help control direct contact. The hill site has been used by children as a playground and shortcut to Ashland High School, 2,000 feet south of the hill.

1983

January: EPA and DEQE sign a Superfund contract, giving EPA lead responsibility as Superfund administrator.

May: EPA and RI/FS contractor NUS Corporation of Pittsburgh meet to discuss plans with Ashland officials, AACE, and citizens.

December: EPA holds a public meeting for comment on the RI/FS scope of work plans.

1984

September: Mercury-contaminated soil is discovered in a neighbor's yard.

November: EPA holds public meeting on draft RI/FS. Local residents call for total excavation of on-site contaminants and disposal off site. EPA also asks for off-site impact studies, citing contaminated soils and groundwater in basements.

1985

April: A public meeting is held for comment on the final RI/FS study by NUS. AACE and local citizens fear on-site solutions and the lack of attention to off-site impacts.

July: EPA selects the remedial action to clean up sludges on site: bury them under capped disposal pits at top of hill, divert groundwater from site.

August: Under DEQE initiative, a Citizen's Advisory Committee (CAC) is formed in Ashland. Local boards are divided about the remedial action selected by EPA. EPA suspends Superfund monies pending reauthorization of Superfund legislation.

November: EPA signs a Record of Decision regarding the selection of their preferred remedial action.

1986

April: NUS conducts on-site assessment studies and samples groundwater in neighborhood basements.

June: DEQE samples the old Nyanza waste sump and discovers high levels of nitrobenzene and contaminated sludges in concrete "vault."

December: Superfund Amendments and Reauthorization Act (SARA) is passed and signed by President Reagan.

1987

January: EPA reports to the Nyanza neighbors that contaminants found in their basements pose no immediate threat to their health.

February: EPA begins preparing final plans to conform to the Corps of Engineers specifications to carry out on-site cleanup, dispose of wastes on the hill, and cap the buried materials.

August: Emergency Response Team begins work to clean up the concrete "vault" by using an inflatable building and a high-temperature portable incinerator to destroy nitrobenzene and decontaminate the soils. Final plans are announced to be ready for bid by the fall. Construction work on Phase I, on-site containment, is expected to begin early in 1988.

Acronyms

AACE	Ashland Advocates for a Clean Environment
ATSDR	Agency for Toxic Substances and Disease Registry
CAC	Citizen's Advisory Committee
CDM	Camp, Dresser, & McKee (engineering consultants)
DEQE	Department of Environmental Quality Engineering
DFW	Division of Fish and Wildlife
DPH	Department of Public Health
DWPC	Division of Water Pollution Control

EPA	Environmental Protection Agency
JBF	JBF Scientific Corporation (EPA contractor)
MCL	MCL Corporation (development firm)
MDC	Metropolitan District Commission
MGL	Massachusetts General Laws
NPL	National Priority List
ppb/m^3	parts per billion per cubic meter
ppm	parts per million
RAMP	Remedial Action Master Plan
RCRA	Resource Conservation and Recovery Act
RI/FS	Remedial Investigation/Feasability Study
SARA	Superfund Amendments and Reauthorization Act (1986)

Selected Bibliography

ANDELMAN, J. B., AND D. W. UNDERHILL. *Health Effects from Hazardous Waste Sites*. Chelsea, Mich.: Lewis Publishers, 1987.

COVELLO, V., A. MOGHISSI, L. LAVE, AND V. UPPULURI, eds. *Uncertainty in Risk Assessment, Risk Management and Decisionmaking*. New York: Plenum Press, 1986.

GIRSHAM, J. W., ed. *Health Aspects of the Disposal of Waste Chemicals*. New York: Pergamon Press, 1986.

GREENBERG, M., AND R. ANDERSON. *Hazardous Waste Sites: The Credibility Gap*. New Brunswick: Center for Urban Policy Research, Rutgers University, 1984.

LESTER, J., AND C. DAVIS, eds. *Hazardous Waste Politics and Policy*. Westport, Conn.: Greenwood Press, 1987.

LEVINE, A. G. *Love Canal: Science, Politics, and People*. Lexington, Mass.: D.C. Heath, 1982.

LIPSET, S., AND W. SCHNEIDER. *The Confidence Gap: Business, Labor and Government*. New York: The Free Press, 1983.

MEDIA INSTITUTE. *Chemical Risks: Fears, Facts, and the Media*. Washington, D.C.: Media Institute, 1985.

NATIONAL RESEARCH COUNCIL. *Risk Assessment in the Federal Government: Managing the Process*. Washington, D.C.: National Academy Press, 1983.

ROSENBAUM, W. *The Politics of Public Participation in Hazardous Waste Management*. Durham, N.C.: Duke University Press, 1983.

U.S. ENVIRONMENTAL PROTECTION AGENCY. *National Priorities List Fact Book*. Washington, D.C.: Office of Emergency and Remedial Response, HW-7.3, June 1986.

U.S. CONGRESS. *Superfund Strategy*. Washington, D.C.: Office of Technology Assessment, OTA-ITE-252, April 1985.

Chapter 7

CONCLUSION: BRIDGING THE TECHNICAL AND CULTURAL PERSPECTIVES ON RISK

The five cases presented in our study represent diverse types of environmental risks and vary from local events (the Nyanza Superfund site) to national events (the EDB episode). While each case is best understood in its own social, political, and historical context, there are some general conclusions that can be drawn. Some of our findings are consistent with the conventional understanding of risk communication problems, while other conclusions stress the importance of developing new ways to frame these problems. The principal findings that apply to the five cases are summarized in Figure 7–1.

One important finding in the five cases is that the communication of risk that occurs during an environmental controversy takes many forms. Risk messages emanate from different sources through formal and informal channels. This is not a novel discovery in itself, as most discussions of risk communication recognize the importance of multiple audiences and acknowledge that the public is a highly diverse aggregation of individuals. However, beyond the diversity of the audience, multiple generators of risk information—including nonofficial sources—play a key role in the overall risk communication scenario. Moreover, our analysis suggests that risk communications in their social context resemble

Portions of this chapter were adapted from A. Plough and S. Krimsky, "The Emergence of Risk Communication Studies: Social and Political Context," *Science, Technology & Human Values* 12 (3&4): 4–10 (Summer/Fall 1987).

Figure 7–1 General Conclusions from the Risk Communication Cases

1. For an Environmental Controversy Risk Communications Are Like Tangled Webs
2. Experts Rarely Dominate Risk Communications of a Highly Politicized Event
3. The Community Context Plays a Vital Role in the Public's Response to Risk Communication
4. The Media Places the Issue of a Risk Event in Bold Relief; simplification; dramatization and polarization are common, whereas inaccuracy and explicit bias are rare.
5. There are Two Logics of Risk: The Technical Versus The Cultural

tangled webs, in contrast to a parallel series of sender/receiver interactions.

There are many different types of risk communicators; some base their legitimacy in their role as official representatives of agencies with regulatory responsibility; others derive legitimacy as representatives of affected parties or potential bearers of risk.

The metaphor of the tangled web was chosen to highlight the following characteristics of a risk communication event. First, one cannot anticipate which of several possible sources of communication will dominate a risk controversy. Official risk communicators are one of many potentially influential voices. Second, risk communication messages become entangled and may result in unpredictable social outcomes. Moreover, since there is a constant interplay among messages, communications from any source may change over time. Third, risk messages come in many forms (from literal to symbolic forms; from print to television) reflecting the diverse modes of communication.

Our studies demonstrate that under particular circumstances it is quite difficult for any single communicator to establish the boundaries of risk communication for an issue. For example, in the EDB case, EPA had a clear message that was presented by a number of able official communicators: "Calm down! There is no health crisis in the short term. The agency is resolving the long-run risk through a responsible, science-based process." Simultaneously, other communicators representing industry, state public health departments, Congress, and environmental advocacy groups also communicated clear messages, many of which ran counter to that of EPA. When there is a mix of conflicting messages, and an emotional issue like a carcinogen in the food chain, a communication process may result that is nearly impossible for any one communicator to control.

In the ice minus case, the negative symbolism of "mutant potatoes" loomed like an ominous cloud in the minds of the Tulelake potato farmers who had prior experiences with the economic consequences of crop stigmatization.

We found that technical information often does not play a dominant role in a risk communication controversy. As much as technical experts would like science to drive the risk communication process, this is unlikely to be the case when there is uncertainty in the risk assessment and important social and economic contexts that shape the risk debate.

The ASARCO/Tacoma case provides a good example. Although the workshops run by EPA presented the technical issues comprehensively, the economic threat of plant closure, disputes between local experts, and the high uncertainty in the risk assessment models played a large role in determining the meaning of the risk to local participants. Radon provides a different example in which the technical data suggest significant risk. In this case scientific information alone seems not to play a determining role in an individual's decision to test or to act on the findings of a high level of radon in the home. In the ice minus case, the science gave no reason for concern over a small field test. Yet the risk arguments were carried by other (metascientific) considerations, such as the limitations of science in predicting the outcome of releasing genetically engineered organisms or the ethical norm of not tampering with nature.

From the technical or agency perspective, the public's interest in risk communication may appear ambiguous or even paradoxical at times. On one hand people want to know the "bottom line" on risk. "What are the risks to me and my family?" Those affected seek a simplification and personalization of the technical risk assessment. On the other hand, people are skeptical of the confident, pat answer. They seek to know the limits of confidence for a risk estimate—the degree of consensus and divergence among experts. There is no easy answer to this dilemma. It mirrors the situation of a physician who must tell a patient the prognosis of a dread disease. Sometimes it is explained in terms of optimistic simplicity. At other times the appropriate explanation requires a realism consisting of complexity and ambiguity. Knowing the audience and its needs may be the key.

How official risk communicators should best deal with scientific uncertainty is an important issue but is not clearly resolved by our analysis. By making scientific uncertainty explicit, communicators reinforce anxiety and reduce the public's confidence in science. On the other hand, if scientific uncertainty is presented as an unavoidable outcome of risk assessment, it can generate confidence in the honesty of the process. It may also build trust and diffuse the efforts of antagonists who play on the weak links in the technical basis of any risk assessment.

The case studies suggest the importance of visible and independent scientific validation of the technical decisions supporting official risk communications. For example, in the ice minus case, explicit references to the external scientific review panel in EPA risk communications significantly helped to reduce polarization among scientists and also improved trust within the lay community. In the EDB case, the perception of bias in EPA's risk assessment process weakened the legitimacy of its official risk messages. There was no external validation of this process.

Local contexts, including specific economic and historical factors, largely shape risk communication associated with a potentially hazardous event. Our analysis suggests that citizen responses to a risk event are strongly influenced by prior experiences with either the particular hazard or a related event. Critical events may set the terms for a debate well before the formal risk communication activities have begun. In the EDB case, the death of a worker from acute exposure to the pesticide and the ensuing public debate on occupational hazards created a first impression of the risk for Congress and the media. This influenced the public's reception of EPA's risk communications later in the year.

Environmental risks in a community can evoke concerns about equity, the moral responsibility of government, and participatory democracy. The Nyanza case suggests the importance of these factors in local Superfund-related risk communications where a community group formed to voice these concerns. In the ice minus case, Monterey County officials were angered at a genetics firm and EPA for not giving the community adequate notice about the field test. Informed consent of citizens in the county became a key issue.

The availability or presentation of clear information to a community is not always sufficient to elicit the public's interest in a risk event. People often require time to process information. Sometimes the public does not respond to a risk communication until after it is provocated by some social catalyst or dramatic event. Before one labels the public as disinterested about a risk issue, communicators should consider whether the proper catalyst has been presented. Different participants follow different time trajectories in their consideration of a risk. This will vary from community to community. For example, the community outrage over pesticide use in Hawaii was triggered by the discovery of EDB in well water in 1983, but was not fully expressed until two years later, in 1985. The Radon, Ice Minus, and Nyanza cases demonstrate the interaction of critical factors, the emergence of catalytic events, and the timing of responses to risk communication.

In the five cases studied, we have found that the media tend to dramatize but do not reconstruct a risk communication. They highlight the existing uncertainties, dissonances, and conflicts. The media are a great equalizer of perspective on risk. Expert and lay commentaries are both presented as legitimate risk communication. This was very explicit in the ice minus case, as the message of danger in the print media was carried almost exclusively by nonexperts. The media also play an important role in fixing images (Nyanza cartoons; EDB-contaminated cereals removed from the shelves; the ASARCO arsenic-emitting smokestack; the genetics firm that violated EPA rules; an indicted EPA official) that determine how risk information will be processed.

One of the most significant findings we draw from the studies is that there are two competing models for the interpretation of risk information—one technical and one cultural. Differences between expert and lay perceptions of risk are based on alternative definitions of the importance of technical and cultural approaches to risk assessment. Our analysis suggests that effective risk communication does not treat the cultural model as an error to be corrected. Indeed, the successful risk communicator must act as a translator between the two models. This conclusion implies that the clarity of risk information, selecting the proper channel, or ensuring adequate public education does not fully address the divergence of technical and popular attitudes toward risk. At the root of this divergence is the distinction between technical and cultural rationality of risk, which is not emphasized in current risk perception research or the risk communication approaches of official agencies largely based on that research. This distinction plays such a central role in understanding risk communication that it deserves a more extensive discussion.

Risk communication evolved out of a need of risk managers to gain public acceptance for policies grounded in risk assessment methodologies. The prevailing view of many experts and risk managers is that local communities and the general public react to limited, false, or inadequate information. These lay groups exercise a personal or democratic prerogative in response to "bad" information that is inconsistent with the more fully informed conclusions of risk assessment experts on whom policymakers depend for developing rational responses to complex problems. Frequently, the exercise of local democracy and personal choice is at odds with the rationality of technical experts. Quantitative risk analysis, rather than narrowing differences, may actually exacerbate antagonisms between the technosphere (the culture of experts) and the demosphere (popular culture). Casting the issues in a technical language reduces the possibility of a dialogue between the public and elites.

Recent studies in the risk perception literature reinforce the conception that rationality and democracy are antagonistic to one another. There are many areas where public perceptions about hazards are inconsistent with so-called objective information. For example, people are said to exhibit too little concern about some hazards (smoking, auto accidents, geological radon, and exposure to sunlight) and too much concern about others (nuclear power, toxic wastes, pesticides, and genetic engineering).

Recently, the EPA convened experts from its principal divisions to determine which events, technologies, and situations represented the greatest environmental risks.[1] The results were not surprising. The experts' inventory of environmental priorities did not correlate positively with the agency's regulatory priorities. The EPA was allocating a large share of its resources for reducing adverse environmental effects in areas its own experts did not consider to warrant the most attention.

The discrepancy between what experts deem most important and what the public demands of its government raises difficult policy questions. Two things deserving respect, namely, scientific rationality and democracy (the rights of local communities to express their will on issues of health and safety), are in conflict. How does one proceed? The options that policymakers usually consider include (1) circumventing the public by avoiding disclosure, by distraction, by preemption, or by citing social contract doctrine according to which agencies represent the public through their elected officials and can decide for the people; (2) appealing to some exemplary and independent authoritative body that will apply the rational decision framework and secure public confidence; or (3) communicating the risks and educating the public into thinking about the problem the way the experts do (e.g., public perception must be brought into conformity with scientific rationality).

Of the three options, only the last is directed specifically at reducing the opposition between the demosphere and the technosphere from the experts' perspective. The emphasis is on a restricted notion of informed democratic practice. Within this context, risk communication is the responsibility of elites and falls into the general rubric of "public understanding of science." The success of risk communication is measured by the degree to which popular attitudes reflect the technical rationality of risk and the extent that popular behavior conforms to technocratic values. A lack of convergence between public perceptions and "objective risk" is attributed to a failure of risk communication.

The rapid growth in research on risk perception began to cast doubt about the public education model. Popular conceptions of risk resisted the conclusions of elites despite clear presentation of the "facts." Studies appeared that purported to explain the discrepancies between expert and lay perceptions of risk. Variables that are intended as proxies for cultural determinants were introduced to account for the differences. Two events with the same risk (probability of mortality) evoke different risk perceptions in experimental studies. One event is *perceived as* voluntary while the other is *perceived as* involuntary. Lay people do not compare events strictly in terms of actuarial risks.

Psychologists began codifying these and other factors that appeared to explain the discrepancies between technical risk assessment and public perceptions. This has resulted in a labeling of risk events according to the restricted logic of cognitive science. The conventional risk communication approach was modified to accommodate adjustment parameters (voluntary vs. involuntary; familiarity vs. unfamiliarity; natural vs. man-made). However, instead of building a culture-based theory of risk perception, psychometricians isolated the cultural factors and treated them as another variable in an experimentally derived technical framework. Every risk event possesses objective hazard estimates and certain qualities that are afforded an ontological status. Thus, a risk event that is voluntary would not be compared on pure rational grounds to one that is involuntary. This system preserves the dichotomy between expert judgment and lay perception of risk. It merely categorizes "irrationalities" and does not explore the cultural underpinnings of risk perception. Moreover, cultural noise affecting the popular response to risk is rationalized. This form of the analysis treats the cultural inputs into risk perception as deviant but comprehensible. Risk communication is still viewed as information transfer from experts to lay people.

A cultural approach that seriously considers popular behavior and symbolic dimensions distinguishes two forms of rationality applied to risk: technical and experiential. Both make contributions to the problem of constructing and analyzing a risk event, but neither is sufficient. Deviance is not the appropriate metaphor to understand differences between the demosphere and the technosphere. Once these distinct modes of rationality are understood, the problem of risk communication is transformed; the problem becomes one of mutual understanding and mutual learning. This cultural model is based on the notion that expert and popular approaches to a risk event can each be logical and coherent on its own terms but may exhibit differences in how the problem is articulated, in the factors relevant to the analysis, and in who the experts are.

Technical rationality rests on explicitly defined sets of principles and scientific norms. These include hypotheticodeductive methods, a common language for measurement, and quantification and comparison across risk events. In its more advanced forms technical rationality encompasses a mature theory with predictive power. The emphasis is on objective (nonpersonal) inputs rather than subjective (experiential) information. Perceived responses to risk are important only in understanding the extent to which ordinary people's ideas deviate from the truth. Logical consistency is an imperative. Two events that have an identical risk profile are treated the same—they are interchangeable.

One of the common mistakes in attempting to codify the public attitudes about risk is to measure people's responses to hypothetical questions. Cultural rationality can only be understood when people's cognitive behavior is observed as they are threatened by an actual risk event. It is only then that the full range of factors come into play that create a complete picture of a public response. To understand cultural rationality, one must address anthropological and phenomenological issues as well as behavioral ones. From the perspective of technical rationality, risk can be studied independently of context. Mary Douglas provides some insight on this point: "The question of acceptable standards of risk is part of the question of acceptable standards of morality and decency, and there is no way of talking seriously about the first while evading the task of analyzing the cultural system in which the second take their form."[2]

Lay people bring many more factors into a risk event than do scientists. For technical experts, the event is denuded of elements that are irrelevant to the analytical model. Figure 7–2 illustrates some of the differences. Many events that are deemed to have very low or insignificant risk by experts are viewed as serious problems by the laity. Burial of low-level radioactive wastes and releasing a natural organism minus a few genes into the environment are among such cases. Where there has been discussion of rationality, it has focused on the scientific grounds of a decision. And yet there are clear instances of reasonable decision making at the community level that are inconsistent with expert opinion. Once it is accepted that two inconsistent decisions can be rational and consistent *on independent criteria*, it is possible to reach beyond the deviant model of risk communication.

Cultural reason does not deny the role of technical reason; it

Figure 7–2   Factors Relevant to the Technical and Cultural Rationality of Risk

Technical Rationality	Cultural Rationality
Trust in scientific methods, explanations, evidence	Trust in political culture and democratic process
Appeal to authority and expertise	Appeal to folk wisdom, peer groups and traditions
Boundaries of analysis are narrow and reductionist	Boundaries of analysis are broad and include the liberal use of analogy and historical precedent
Risks are depersonalized with emphasis on statistical probability	Risks are personalized with the emphasis on the family and the community
Appeal to consistency and universality	Focus on particularity with less concern for consistency
Where there is controversy, in science, resolution follows status	Popular culture does not follow the prestige principle
Those impacts that cannot be specified are irrelevant	Unanticipated or unarticulated risks are relevant

SOURCE: A. Plough and S. Krimsky, "The Emergence of Risk Communication Studies: Social and Political Context," *Science, Technology & Human Values* 12(3&4):4–10 (Summer/Fall 1987).

simply extends it. The former branches out, while the latter branches in. Cultural rationality does not separate the context from the content of risk analysis. Technical rationality operates as if it can act independently of popular culture in constructing the risk analysis, whereas cultural rationality seeks technical knowledge but incorporates it within a broader decision framework.

If these forms of rationality are unalterably antagonistic, technical reason and popular response to risk may be truly incommensurable. Both forms of rationality must be capable of responding to a process of mutual learning and adjustment. If the technosphere begins to appreciate and respect the logic of local culture toward risk events and if local culture has access to a demystified science, points of intersection will be possible.

This study examined only five cases of risk communication and cannot attempt to represent all of the issues that will confront risk communicators. Our cases were selected because these were important sentinel issues for a single regulatory agency. In this respect they are not typical of the universe of risk communication. There are far more simple and uneventful examples of risk communication in EPA's experience. However, the detailed analysis of the relationship between the history, process, and outcome of a risk communication problem, and the highlighting of the sources of sometimes unexpected controversy in the cases, illustrates the complexity of risk communication. It trivializes this complexity to view the differences between the technical and cultural rationality of risk as a manifestation of deviance of the latter with respect to the former.

Endnotes

1. U.S. Environmental Protection Agency, *Unfinished Business: A Comparative Assessment of Environmental Problems* (Washington, D.C.: EPA, February 1987).
2. Quoted in James S. Short, "The Social Fabric at Risk: Toward the Social Transformation of Risk Analysis," *American Sociological Review* 49:720 (December 1984).

BIBLIOGRAPHY

ALE, BEN J. "Technology Acceptance: A Contribution to a Discussion." *Risk Analysis* 6(3):267–268 (September 1986).

ALLEN, FREDERICK W. "Towards a Holistic Appreciation of Risk: The Challenge for Communicators and Policymakers." *Science, Technology, & Human Values* 12(3&4):138–143 (Summer/Fall 1987).

ALTMORE, MICHAEL. "The Social Construction of a Scientific Controversy: Comments on Press Coverage of the Recombinant DNA Debate." *Science, Technology & Human Values* 7(41):24–31 (Fall 1982).

BAIRD, BRIAN N. "Tolerance for Environmental Health Risks: The Influence of Knowledge, Benefits, Voluntariness, and Environmental Attitudes." *Risk Analysis* 6(4):425–435 (December 1986).

BARAM, MICHAEL. "Implementation and Evaluation of Regulations." In H. Otway, and M. Peltu, eds., *Regulating Industrial Risks*. London: Butterworths, 1985.

BARAM, MICHAEL. "Chemical Industry Accidents, Liability, and Community Right to Know." *American Journal of Public Health* 76(5):568–572 (May 1986).

———. "Risk Communication and the Law for Chronic Health and Environmental Hazards." *The Environmental Professional* 8(2):165–178 (1986).

———. "Risk Communication: Moving from Theory to Law to Practice." Paper presented at the annual conference of the Society for Risk Analysis, Boston, Massachusetts, November 11, 1986.

BARTELS, DITTA, AND MARCEL BLANC. "The Political Agenda in Cancer Virus Risk Assessment." Manuscript.

BAYER, RONALD. "Notifying Workers at Risk: The Politics of the Right-to-Know." *American Journal of Public Health* 76(11):1352–56 (November 1986).

BERRY, J., K. PORTNEY, M. BABLITCH, AND R. MAHONEY. "Public Involvement in Administration: The Structural Determinants of Effective Citizen Participation." *Journal of Voluntary Action Research* 13(2):7–23 (April 1984).

BINGHAM, GAIL. *Resolving Environmental Disputes: A Decade of Experience*. Washington, D.C.: The Conservation Foundation, 1986.

BLYSKAL, JEFF, AND MARIE BLYSKAL. "Making the Best of Bad News: How

Corporations in Crisis Use the Press." *Washington Journalism Review* 7:51–55 (December 1985).

BROWN, PHIL. "Popular Epidemiology: Community Response to Toxic Waste-Induced Disease in Woburn, Massachusetts." *Science, Technology, & Human Values* 12(3&4):78–85 (Summer/Fall 1987).

BUCKLE, L., AND S. THOMAS-BUCKLE. "Placing Environmental Mediation in Context: Lessons from 'Failed' Mediations." *Environmental Impact Assessment Review* 6(1):55–70 (March 1986).

BURGER JR., EDWARD J. *Health Risks: The Challenge of Informing the Public*. Washington, D.C.: Media Institute, 1984.

BURKETT, WARREN. "Science and Environmental Reporting: On a Roll." *The Quill* 74(2):6–7 (February 1986).

CHRISTOFFEL, TOM. "Grassroots Environmentalism under Legal Attack: Dandelions, Pesticides, and a Neighbor's Right-to-Know." *American Journal of Public Health* 75(5):565–7 (May 1985).

COMBS, BARBARA, AND PAUL SLOVIC. "Newspaper Coverage of Causes of Deaths." *Journalism Quarterly* 56(4):837–43 (Winter 1979).

CONGRESSIONAL RESEARCH SERVICE, LIBRARY OF CONGRESS. *Risk: Assessment, Acceptability and Management*. Proceedings of a Seminar for the Subcommittee on Science, Research & Technology, November 1981. Washington, D.C.: Government Printing Office, 1981.

CONN, W.D., AND N. R. FEIMER. "Communicating with the Public on Environmental Risk: Integrating Research and Policy." *The Environmental Professional* 7(1):39–47 (1985).

COUCH, S., AND J. KROLL-SMITH. "The Chronic Technical Disaster: Toward a Social Scientific Perspective." *Social Science Quarterly* 66(3):564–575 (September 1985).

COUNCIL ON ENVIRONMENTAL QUALITY. *Public Opinion on Environmental Issues: Results of a National Public Opinion Survey*. Washington, D.C.: Council on Environmental Quality, 1980.

———. "Special Report: Risk Assessment and Risk Management." In *Environmental Quality: 15th Annual Report of the Council on Environmental Quality*. Washington, D.C.: Council on Environmental Quality, 1984.

COVELLO, V. "The Perception of Technological Risks: A Literature Review." *Technological Forecasting and Social Change* 23(4):285–297 (August 1983).

COVELLO, V., J. MENKES, AND J. NEHNEVAJSA. "Risk Analysis, Philosophy and the Social and Behavioral Sciences: Reflections on the Scope of Risk Analysis Research." *Risk Analysis* 2(2):53–58 (June 1982).

COVELLO, V., W. G. FRAMM, J. RODRICKS, AND R. TARDIFF, eds. *The Analysis of Actual vs. Perceived Risk*. New York: Plenum Press, 1983.

COVELLO, V., AND J. MUMPOWER. "Risk Analysis and Risk Management: An Historical Perspective." *Risk Analysis* 5(2):103–120 (June 1985).

COVELLO, V., D. VON WINTERFELDT, AND P. SLOVIC. "Communicating Scientific Information About Health & Environmental Risks: Problems and Opportunities from a Social and Behavioral Perspective." In V. Covello, L. Lave, A. Moghissi, and V. Uppuluri, eds., *Uncertainty in Risk Assessment, Risk Management, and Decisionmaking*. New York: Plenum Press, 1986.

COVELLO, V., L. LAVE, A. MOGHISSI, AND V. UPPULURI, eds. *Uncertainty in Risk Assessment, Risk Management, and Decisionmaking*. New York: Plenum Press, 1986.

COVELLO, V., J. MENKES, AND J. MUMPOWER, eds. *Risk Evaluation and Management*. New York: Plenum Press, 1986.

CREIGHTON, J. L. *Public Involvement Manual: Involving the Public in Water & Power Resource Decisions*. Washington D.C.: U.S. Department of the Interior, Water and Power Resources Service, 1980.

CUMMING, R. B. "Risk and the Social Sciences." *Risk Analysis* 2(2):47–48 (June 1982).

DAVIES, CLARENCE, VINCENT T. COVELLO, AND FREDERICK W. ALLEN, eds. *Risk Communication*. Proceedings of the National Conference on Risk Communication held in Washington, D.C., January 29–31, 1986. Washington, D.C.: The Conservation Foundation, 1987.

DAVIES, RICHARD. "The Effectiveness of the Sizewell B Public Inquiry in Facilitating Communication about the Risks of Nuclear Power." *Science, Technology, and Human Values* 12(3&4):102–10 (Summer/Fall 1987).

DAVIS, CHARLES. "Public Involvement in Hazardous Waste Siting Decisions." Paper presented at the 1986 Annual Meeting of American Political Science Association, Washington, D.C., August 1986.

DERBY, S. L., AND R. L. KEENEY. "Risk Analysis: Understanding 'How Safe Is Safe Enough'?" *Risk Analysis* 1(3):217–224 (September 1981).

DIERKES, M., S. EDWARDS, AND R. COPPOCK, eds. *Technological Risk: Its Perception and Handling in the European Community*. Cambridge, Massachusetts: Oelgeschlager, Gunn & Hain, Inc., 1980.

DOUGLAS, MARY, AND AARON WILDAVSKY. "How Can We Know the Risks We Face? Why Risk Selection Is a Social Process." *Risk Analysis* 2(2):49–51 (June 1982).

DOUGLAS, MARY. *Risk Acceptability According to the Social Sciences*. New York: Russell Sage, 1985.

DUTTON, DIANA B. "Medical Risks, Disclosure, and Liability: Slouching Toward Informed Consent." *Science, Technology, & Human Values* 12(3&4):48–59 (Summer/Fall 1987).

EARLE, T. C. "Risk Communication: A Marketing Approach." Paper presented at the National Science Foundation/EPA Workshop on Risk Perception and Risk Communication, Long Beach, California, December 1984.

EARLE, T. C., AND G. CVETKOVICH. *Risk Judgement and the Communication of Hazard Information: Toward a New Look in the Study of Risk Perception*. BHARC (400/83/017). Seattle, Washington: Battelle Human Affairs Research Centers, 1983.

EDWARDS, WARD, AND D. VON WINTERFELDT. "Public Values in Risk Communication." Paper presented at the National Science Foundation/EPA Risk Perception and Risk Communication Workshop in Long Beach, California, December 1984.

EISER, J. R., ed. *Social Psychology and Behavioral Medicine*. New York: Wiley & Sons, 1982.

EVANS, J., C. PETITO, AND D. GRAVALLESE. "Cleaning Up the Gilson Road Hazardous Waste Site: A Case Study." Unpublished paper, John F. Kennedy School of Government, Harvard University, February 1986.

FESSENDEN-RADEN, JUNE, CAROLE BISOGNI, AND KEITH PORTER. "Enhancing Risk Management by Focusing on the Local Level: An Integrated Approach." Unpublished paper, Cornell University, 1986.

FESSENDEN-RADEN, JUNE, JANET M. FITCHEN, AND JENIFER S. HEATH. "Providing Risk Information in Communities: Factors Influencing What Is Heard and Accepted." *Science, Technology, and Human Values* 12(3&4):94–101 (Summer/ Fall 1987).

FIELDS, BERNARD, MALCOLM A. MARTIN, AND DAPHNE KAMELY, eds. *Genetically Altered Viruses and the Environment*. Cold Spring Harbor, New York: Cold Spring Harbor Laboratory, 1985.

FISCHHOFF, B. "For Those Condemned to Study the Past: Heuristics and Biases in Hindsight." In D. Kahneman, A. Tversky, and P. Slovic, eds. *Judgement Under Uncertainty: Heuristics and Biases*. New York: Cambridge University Press, 1982.

———. "Judged Lethality: How Much People Seem to Know Depends Upon How They're Asked." *Risk Analysis* 3(4):229–236 (December 1983).

———. "Perceived vs. Actual Risks: When Experts and Laymen Disagree." *The Journalist* 8–11 (Spring 1984).

———. "Setting Standards: A Systematic Approach to Managing Public Health & Safety Risks." *Management Science* 30(7):823–43 (July 1984).

———. "Managing Risk Perceptions." *Issues in Science and Technology* 2(1):83– 96 (Fall 1985).

———. "Protocols for Environmental Report: What to Ask the Experts." *The Journalist* 11–15 (Winter 1985).

———. "Treating the Public with Risk Communications: A Public Health Perspective." *Science, Technology, & Human Values* 12(3&4):13–19 (Summer/Fall 1987).

FISCHHOFF, B., P. SLOVIC, S. LICHTENSTEIN, S. READ, AND B. COMBS. "How Safe Is Safe Enough? A Psychometric Study of Attitudes Towards Technological Risks and Benefits." *Policy Sciences* 9(2):127–152 (April 1978).

FISCHHOFF, B., C. HOHENEMSER, R. KASPERSON, AND R. KATES. "Handling Hazards," *Environment* 20(7):16–20, 32–37 (September 1978).

FISCHHOFF, B., S. LICHTENSTEIN, P. SLOVIC, S. DAILEY, AND R. L. KEENEY. *Acceptable Risk*. New York: Cambridge University Press, 1981.

FISCHHOFF, B., P. SLOVIC, AND S. LICHTENSTEIN. "Lay Foibles and Expert Fables in Judgments About Risk." *The American Statistican* 36(3):240–255 (August 1982).

FISCHHOFF, B., P. SLOVIC, AND S. LICHTENSTEIN. "The Public vs. The Experts: Perceived vs. Actual Disagreements About Risks of Nuclear Power." In V. Covello et al., eds. *The Analysis of Actual vs. Perceived Risks*. New York: Plenum Press, 1983.

FISCHHOFF, B., S. WATSON, AND C. HOPE. "Defining Risk." *Policy Sciences* 17(2):123–139 (October 1984).

FISCHHOFF, B., O. SVENSON, AND P. SLOVIC. "Active Responses to Environmental Hazards: Perceptions and Decision Making." In D. Stokols and I. Altman, eds. *Handbook of Environmental Psychology*, Vol. 2. New York: Wiley & Sons, 1986.

FITCHEN, JANET. "Cultural Aspects of Environmental Problems: Individualism and Chemical Contamination of Groundwater." *Science, Technology, & Human Values* 12(2):1–12 (Spring 1987).

FITZGERALD, M., W. LYONS, AND P. FREEMAN. "Hazardous Waste Disposal in a Federal System: Public Attitudes and Intergovernmental Responsibility." Paper presented at the annual meeting of the Southern Political Science Association, Atlanta, Georgia, November 1986.

FREIMUTH, V. S., AND J. P. VAN NEVEL. "Reaching the Public: The Asbestos Awareness Campaign." *Journal of Communications* 31(2):155–67 (Spring 1981).

FREIMUTH, V. S., R. GREENBERG, J. DeWITT, AND R. ROMANO. "Covering Cancer: Newspapers and the Public Interest." *Journal of Communications* 24(1):62–73 (Winter 1984).

FREIMUTH, VICKI S., TIMOTHY EDGAR, AND SHARON L. HAMMOND. "College Students' Awareness and Interpretation of the AIDS Risk." *Science, Technology, & Human Values* 12(3&4):37–40 (Summer/Fall 1987).

FREUDENBERG, NICHOLAS. "Citizen Action for Environmental Health: Report on a Survey of Community Organizations." *Journal of Community Health* 10(1):3–9 (Spring 1985).

FRIEDMAN, SHARON. "Blueprint for Breakdown: Three Mile Island and Media Before the Accident." *Journal of Communications* 31(2):116–28 (Spring 1981).

GIBBS, LOIS. *Love Canal: My Story*. Albany: State University of New York Press, 1982.

GREENBERG, MICHAEL, AND RICHARD ANDERSON. *Hazardous Waste Sites: The Credibility Gap*. New Brunswick, N.J.: Center for Urban Policy Research, Rutgers University, 1984.

GRIFFITHS, RICHARD F., ed. *Dealing with Risk: The Planning, Management, and Acceptability of Technological Risk*. New York: Wiley & Sons, 1982.

GRIMA, A. P., P. TIMMERMAN, C. FOWLE, AND P. BYER. *Risk Management and EIA: Research Needs and Opportunities*. Hull, Quebec: Canadian Environmental Assessment Research Council, 1986.

GROSS, SIDNEY. "Chemical Accidents Need Not Be Media Disasters." *Chemical Business* 7(3):22–26 (March 1985).

HARRIS, J. "Toxic Waste Uproar: A Community History." *Journal of Public Health Policy* 4(2):181–201 (1983).

HARRISON, E. BRUCE. "Cancer from Cake Mix? The Public Demands Personal Answers." *Public Relations Journal* 41(1):32–3 (January 1985).

———. "When Worrying Works." *Public Relations Journal*, 41(11):18–37 (November 1985).

HEATH, J., AND J. FESSENDEN-RADEN. "The Relationship Between Risk Management Intervenors and the Community." Paper presented at the Society for Risk Analysis meeting, Washington, D.C., October 1985. In Lester Lave, ed., *Enhancing Risk Management*. New York: Plenum Press, 1987.

HILGARTNER, STEPHEN, AND DOROTHY NELKIN. "Communication Controversies over Dietary Risks." *Science, Technology, & Human Values* 12(3&4):41–47 (Summer/Fall 1987).

HOHENEMSER, C., AND JEANNE KASPERSON, eds. *Risk in the Technological Society*. Boulder, Colorado: Westview Press, 1982.

HOHENEMSER, C., R. W. KATES, AND P. SLOVIC. "The Nature of Technological Hazards." *Science* 220:378–84 (April 22, 1983).

HUGHES, ROBERT G., AND STEVEN P. SCHWARTZ. "Communities and Toxic Wastes: Further Evidence." *Journal of Public Health Policy* 4(4):514–18 (December 1983).

HUMPHREYS, PATRICK. "Value Structures Underlying Risk Assessments." In H. Kunreuther, ed. *Risk, A Seminar Series*. Laxenburg, Austria: International Institute for Applied Systems Analysis, 1981.

JASANOFF, SHEILA. "EPA's Regulation of Daminozide: Unscrambling the Messages on Risk." *Science, Technology, & Human Values* 12(3&4):116–124 (Summer/Fall 1987).

JOHNSON, BARRY L. "Health Risk Communication at the Agency for Toxic Substances and Disease Registry." [editorial] *Risk Analysis* 7(4):409–12 (December 1987).

JOHNSON, F. R. *Does Risk Information Reduce Welfare? Evidence from the EDB Food Contamination Scare*. Paper #870130, Economics Department, U. S. Naval Academy, Annapolis, Maryland.

————. *Bad News & Perceived Risks: Homeowner Responses to Indoor Radon*. Paper #860605, Economics Department, U.S. Naval Academy, Annapolis, Maryland.

————. "Radon Risk Information and Voluntary Protection: Evidence from a Natural Experiment." *Risk Analysis* 7(1):97–107 (March 1987).

————. "Economic Costs of Misinforming About Risk: The EDB Scare and The Media." *Risk Analysis* (forthcoming).

KAHNEMAN, D., AND A. TVERSKY. "Intuitive Prediction: Biases & Corrective Procedures." In D. Kahneman, A. Tversky, and P. Slovic, eds., *Judgement Under Uncertainty: Heuristics and Biases*. New York: Cambridge University Press, 1982.

KALIKOW, BARNETT. "Environmental Risk: Power to the People." *Technology Review* 87(7):55–61 (October 1984).

KASPERSON, R. E., AND J. X. KASPERSON. "Determining the Acceptability of Risk." In J. Rogers, and D. Bates, eds., *Risk: A Symposium on the Assessment and Perception of Risk to Human Health*. Ottawa: Royal Society of Canada, 1983.

KASPERSON, R. E., AND K.D. PIJAWKA. "Societal Responses to Hazards and Major Hazardous Events: Comparing Natural and Technological Hazards." *Public Administration Review*, Special Issue 45:7–19 (January 1985).

KASPERSON, R. E. "Six Propositions on Public Participation and Their Relevance for Risk Communication." *Risk Analysis* 6(3):275–281 (September 1986).

KASPERSON, R. E., AND I. PALMLUND. "Evaluating Risk Communication." Unpublished paper, CENTED, Clark University, January 1987.

KATES, ROBERT, AND JEANNE X. KASPERSON. "Comparative Risk Analysis of Technological Hazards: A Review," *Proceedings of the National Academy of Sciences, U.S.A.* 80(11):7027–38 (November 1983).

KATES, ROBERT. "Success, Strain & Surprise." *Issues in Science and Technology* 2(1):46–58 (Fall 1985).

KATES, R. W., C. HOHENEMSER, AND J. X. KASPERSON, eds. *Perilous Progress: Managing the Hazards of Technology*. Boulder, Colorado: Westview Press, 1985.

KEENEY, R. L., AND D. VON WINTERFELDT. "Improving Risk Communication." *Risk Analysis* 6(4):417–424 (December 1986).

KLAIDMAN, S. *Health Risk Reporting*. Roundtable Workshop on the Media and the Reporting of Risks to Health, Workshop Summary Report. Washington D.C.: Institute for Health Policy Analysis, Georgetown University Medical Center, 1985.

KLOTZ, M. L., NEIL D. WEINSTEIN, AND PETER M. SANDMAN. "Promoting Remedial Response to the Risk of Radon: Are Information Campaigns Enough?" Manuscript.

KRAFT, M., AND R. KRAUT. "The Impact of Citizen Participation on Hazardous Waste Policy Implementation: The Case of Clermont County, Ohio." *Policy Studies Journal* 14(1):52–61 (September 1985).

KUNREUTHER, H. "Decision Making for Low Probability Events: A Conceptual Framework." In H. Kunreuther, ed., *Risk: A Seminar Series*. Laxenburg, Austria: International Institute for Applied Systems Analysis, 1981.

KUNREUTHER, H., P. KLEINDORFER, AND R. YAKSICK. "Compensation as a Tool for Improving Risk Communication." Paper presented at the National Science Foundation/EPA workshop on Risk Perception and Risk Communication, Long Beach, California, December 1984.

KUNREUTHER, H., AND P. SLOVIC. "Decision Making in Hazard and Resource Management." In R. W. Kates, and I. Burton, eds., *Geography, Resources and Environment: Vol. II, Themes from the Work of Gilbert F. White*. Chicago: University of Chicago Press, 1986.

KWEIT, R. AND M. G. KWEIT. "Bureaucratic Decision-Making: Impediments to Citizen Participation." *Polity* 12(4):647–66 (Summer 1980).

LAVE, LESTER B. "Regulating Risks." *Risk Analysis* 4(2):79–80 (June 1984).

LESTER, JAMES, AND A. BOWMAN. "Subnational Hazardous Waste Policy Implementation: A Test of the Sabatier-Mazmanian Model." Paper presented at the annual Applied Geography Conference, U.S. Military Academy, West Point, New York, October 1986.

LINDELL, M., AND T. EARLE. "How Close Is Close Enough: Public Perception of the Risks of Industrial Facilities." *Risk Analysis* 3(4):245–253 (December 1983).

LITAI, D., D. LANNING, AND N. RASMUSSEN. "The Public Perception of Risk." In V. Covello, et al., eds., *The Analysis of Actual vs. Perceived Risks*. New York: Plenum Press, 1983.

LYNN, F., AND M. BRUCKS. "Public Programs for Risk Communication." Paper presented at the National Science Foundation/EPA workshop on Risk Perception and Risk Communication, Long Beach, California, December 1984.

MACCOBY, N., AND D. SOLOMON. "Heart Disease Prevention: Community Studies." In R. Rice and W. Paisley, eds., *Public Communication Campaigns*. Beverly Hills: Sage Publications, 1981.

MACGREGOR, D. "Understanding Public Reaction to Risk Analysis." Paper presented at the National Science Foundation/EPA workshop on Risk Perception and Risk Communication, Long Beach, California, December 1984.

MACGREGOR, D., AND P. SLOVIC. "Perceived Acceptability of Risk Analysis as a Decision-Making Approach." *Risk Analysis* 6(2):245–256 (June 1986).

MACLEAN, DOUGLAS. "Risk and Consent: Philosophical Issues for Centralized Decisions." *Risk Analysis* 2(2):59–67 (June 1982).

MAYER, F., AND H. LEE, eds. *Environmental Risk Management: Research Needs and Opportunities*. Report E-86-10 Discussion Paper Series. Cambridge: Kennedy School of Government, Harvard University, October 1986.

MAZUR, ALLAN. "Disputes Between Experts." *Minerva* 11(2):243–262 (April 1973).

———. "Media Coverage & Public Opinion on Scientific Controversies." *Journal of Communication* 31(2):106–115 (Spring 1981).

———. *The Dynamics of Technical Controversy*. Washington D.C.: Communications Press, 1981.

———. "The Journalists & Technology: Reporting about Love Canal and Three Mile Island." *Minerva* 22(1):45–66 (Spring 1984).

———. "Putting Radon on the Public's Risk Agenda." *Science, Technology, & Human Values* 12(3&4):86–93 (Summer/Fall 1987).

MCGUIRE, W. "The Communication-Persuasion Model & Health-Risk Labeling." In M. Morris, M. Mazis, and I. Barolsky, eds., *Product Labeling & Health Risks*. Cold Spring Harbor, New York: Cold Spring Harbor Laboratory, 1980.

MEDIA INSTITUTE. *Chemical Risks: Fears, Facts and the Media*. Washington D.C.: Media Institute, 1985.

MEYER, M., AND K. SOLOMON. "Risk Management in Local Communities." *Policy Sciences* 16(3):245–66 (February 1983).

MORGAN, M. GRANGER. "Risk Research: When Should We Say 'Enough'?" [editorial] *Science* 232:917 (May 23, 1986).

MORRIS, M., M. MAZIS, AND I. BAROFSKY, eds. *Product Labeling and Health Risks*. Cold Spring Harbor, New York: Cold Spring Harbor Laboratory, 1980.

NEEDLEMAN, CAROLYN. "Ritualism in Communicating Risk Information." *Science, Technology, & Human Values* 12(3&4): 20–25 (Summer/Fall 1987).

NELKIN, DOROTHY. *Selling Science*. San Francisco: W.H. Freeman, 1987.

O'BRIEN, D., AND D. MARCHAND. *The Politics of Technology Assessment*. Lexington, Massachusetts: Lexington Books, 1982.

O'BRIEN, R., M. CLARKE, AND S. KAMIENIECKI. "Open and Closed Systems of Decision Making: The Case of Toxic Waste Management." *Public Administration Review* 44(4):334–340 (July/August 1984).

O'HARE, MICHAEL. "Bargaining and Negotiation in Hazardous Material Management." In P. R. Kleindorfer, and H. Kunreuther, eds., *Insuring and Managing Hazardous Risks: From Seveso to Bhopal and Beyond*. Berlin: Springer-Verlag, 1987.

O'RIORDAN, T. "Risk-Perception Studies and Policy Priorities." *Risk Analysis* 2(2):95–100 (June 1982).

O'RIORDAN, T. "Approaches to Regulation." In H. Otway, and M. Peltu, eds., *Regulating Industrial Risks: Public, Experts, and Media*. London: Butterworths, 1986.

———. "Experts, Risk Communication, and Democracy." *Risk Analysis* 7(2):125–31 (June 1987).

OTWAY, H., AND D. VON WINTERFELDT. "Beyond Acceptable Risk: On the Social Acceptability of Technologies." *Policy Sciences* 14(3):247–256 (June 1982).

OTWAY, H., AND K. THOMAS. "Reflections on Risk Perception and Policy." *Risk Analysis* 2(2):69–82 (June 1982).

OTWAY, H., AND M. PELTU, eds. *Regulating Industrial Risks: Public, Experts and Media*. London: Butterworths, 1986.

OZONOFF, DAVID, AND LESLIE BODEN. "Truth and Consequences: Health Agency Responses to Environmental Health Problems." *Science, Technology, & Human Values* 12(3&4):70–77 (Summer/Fall 1987).

PERROW, C. *Normal Accidents: Living with High Risk Technologies.* New York: Basic Books, 1984.

PLOUGH, ALONZO, AND SHELDON KRIMSKY. "The Emergence of Risk Communication Studies: Social and Political Context." *Science, Technology, & Human Values* 12(3&4):4–10 (Summer/Fall 1987).

PORTNEY, K. "The Perception of Health Risk and Opposition to Hazardous Waste Treatment Facility Siting: Implications for Hazardous Waste Management and Policy from Survey Research." *Papers and Proceedings of the Applied Geography Conference* 9:114–23 (1986).

PORTNEY, K. "The Role of Economic Factors in Lay Perceptions of Risk: A Look at Citizen Attitudes with Special Reference to Hazardous Waste Treatment Facility Siting." In Charles E. Davis, and James P. Lester, eds., *Dimensions of Hazardous Waste Politics and Policy.* New Haven: Greenwood Press, forthcoming.

PRESS, FRANK. "Comments on Risk Communication." Paper presented at the Conference on Risk Communication, Washington, D.C., 1986.

PROTHROW-STITH, DEBORAH, HOWARD SPIVAK, AND ALICE J. HAUSMAN. "THE Violence Prevention Project: A Public Health Approach." *Science, Technology, & Human Values* 12(3&4):67–69 (Summer/Fall 1987).

RAYNER, S., AND R. CANTOR. "How Fair Is Safe Enough? The Cultural Approach to Societal Technology Choice." *Risk Analysis* 7(1):3–9 (March 1987).

REGENS, J., T. DIETZ, AND R. RYCROFT. "Risk Assessment in the Policy-Making Process: Environmental Health & Safety Protection." *Public Administration Review* 43(2):137–145 (March/April 1983).

RENN, ORTWIN. "Risk Assessment: Scope & Limitations." In H. Otway and M. Peltu, eds., *Regulating Industrial Risks: Public, Experts and Media.* London: Butterworths, 1986.

ROBINS, J., P. LANDRIGAN, F. ROBINS, AND L. FINE. "Decision-Making Under Uncertainty in the Setting of Environmental Health Regulations." *Journal of Public Health Policy* 6(3):322–328 (September 1985).

RODRICKS, J. V. "Risk Assessment at Hazardous Waste Disposal Sites." *Hazardous Waste* 1(3):333–362 (Fall 1984).

ROSENBAUM, WALTER A. "The Politics of Public Participation in Hazardous Waste Management." In James P. Lester and Ann Bowman, eds., *The Politics of Hazardous Waste Management.* Durham, N.C.: Duke University Press, 1983.

ROTHMAN, STANLEY, AND S. ROBERT LICHTER. "The Nuclear Energy Debate: Scientists, the Media and the Public." *Public Opinion* 5(4):47–52 (August/September 1982).

RUBIN, BERNARD, ed. *When Information Counts: Grading the Media.* Lexington, Mass.: Lexington Books, 1985.

RUBIN, D., AND LANDERS, S. "National Exposure & Local Cover-up: A Case Study." *Columbia Journalism Review* 8(2):17–22 (Summer 1969).

RUBIN, DAVID, AND VAL HENDY. "Swine Influenza and the News Media." *Annals of Internal Medicine* 87(6):769–774 (December 1977).

RUCKELSHAUS, W. D. "Science, Risk, and Public Policy." *Science* 221:1026–1028 (September 9, 1983).

———. "Risk in a Free Society." *Risk Analysis* 4(3):157–162 (1984).

———. "Risk, Science & Democracy." *Issues in Science & Technology* 1(3):19–38 (Spring 1985).

RUSHEFSKY, M. "Reducing Risk Conflict by Regulatory Negotiation: A Preliminary Evaluation." Paper presented at the annual Meeting of the American Political Science Association, Washington, D.C., August 1986.

RYCROFT, ROBERT W., JAMES L. REGENS, AND THOMAS DIETZ. "Acquiring and Utilizing Scientific and Technical Information to Identify Environmental Risks." *Science, Technology, & Human Values* 12(3&4):125–130 (Summer/Fall 1987).

SAGAN, L. "Beyond Risk Assessment." *Risk Analysis* 7(1):1–2 (March 1987).

SAGOFF, MARK. "Values and Preferences." *Ethics* 96(2):301–316 (January 1986).

SANDMAN, P., AND M. PADEN. "At Three Mile Island." *Columbia Journalism Review* 18(2):43–58 (July/August 1979).

SANDMAN, P. "Environmental Emergencies: Are Journalists Prepared?" *SIPIscope* 13(4):1–11 (September/October 1985).

SANDMAN, P., AND J. VALENTI. "Scared Stiff—Or Scared Into Action." *Bulletin of the Atomic Scientists* 42(1):12–16 (January 1986).

SANDMAN, P., D. SACHSMAN, M. GREENBERG, M. JURKAT, A. GOTSCH, AND M. GOCHFELD. "Environmental Risk Reporting in New Jersey Newspapers." Environmental Risk Reporting Project, Industry/University Cooperative Center for Research in Hazardous and Toxic Substances, New Jersey Institute of Technology, January 1986.

SANDMAN, P. "Getting to Maybe: Some Communications Aspects of Hazardous Waste Facility Siting." *Seton Hall Legislative Journal* 9(2):437–65 (Spring 1986).

SANDMAN, P. *Explaining Environmental Risk*. Washington, D.C.: U.S. Environmental Protection Agency, Office of Toxic Substances, 1986.

SCHULTE, P., AND K. RINGEN. "Notification of Workers at High Risk: An Emerging Public Health Problem." *American Journal of Public Health* 74(5):485–491 (May 1984).

SCHWARTZ, S., P. WHITE, AND R. HUGHES. "Environmental Threats, Communities and Hysteria." *Journal of Public Health Policy* 6(1):58–77 (March 1985).

SCIENTISTS' INSTITUTE FOR PUBLIC INFORMATION. "Covering Toxics." *SIPIscope* 13(4) (November/December 1983).

SHARLIN, H. "EDB: A Case Study in Communicating Risk." *Risk Analysis* 6(1)61–8 (March 1986).

SHORT, J. S. "The Social Fabric at Risk: Toward the Social Transformation of Risk Analysis." *American Sociological Review* 49:711–25 (December 1984).

SHRADER-FRECHETTE, K. S. *Risk Analysis and Scientific Method: Methodological and Ethical Problems with Evaluating Societal Hazards*. Boston: D. Reidel Publishing Co., 1985.

SLEEPER, D. "Do the Media Too Often Miss the Message?" *Conservation Foundation Newsletter* (January 1979):1–8.

SLOVIC, P. "Perception of Risk." *Science* 236:280–85 (April 17, 1987).

SLOVIC, P., B. FISCHHOFF, AND S. LICHTENSTEIN. "Weighing the Risks." *Environment* 21(3):14–20,36–9 (April 1979).

———. "Risky Assumptions." *Psychology Today* 14(1):44–48 (June 1980).

————. "Informing the Public About the Risks from Ionizing Radiation." *Health Physics* 41(4):589–598 (October 1981).

————. "Why Study Risk Perception?" *Risk Analysis* 2(2):83–93 (June 1982).

SMITH, T. "Let Them Eat Cake Mix." *Vital Speeches of the Day* 52(20):622–625 (August 1, 1986).

STARR, C. "Social Benefit versus Technological Risk." *Science* 165:1232–38 (September 19, 1969).

————. "Risk Management, Assessment and Acceptability." *Risk Analysis* 5(2):97–102 (June 1985).

STARR, C., AND C. WHIPPLE. "Risks of Risk Decisions." *Science* 208:1114–1119 (June 6, 1980).

STEPHENS, M., AND N. EDISON. "News Media Coverage of Issues During the Accident at Three Mile Island." *Journalism Quarterly* 259(2):199–204 (Summer 1982).

STOTO, M. "Disagreement Among Experts in Risk Communication." Unpublished paper, Kennedy School of Government, Harvard University, Cambridge, Massachusetts, October 1985.

SUTTON, S. R. "Fear-arousing Communications: A Critical Examination of Theory & Research." In J. Eiser, ed., *Social Psychology & Behavioral Medicine*. Chichester: Wiley & Sons, 1982.

SWAN, J. "Uncovering Love Canal." *Columbia Journalism Review* 17(5):46–51 (January/Febuary 1979).

THOMAS, LEE. "The Challenge of Community Involvement." *EPA Journal* 11(10):2–11 (December 1985).

————. "Risk Communication: Why We Must Talk About Risk." *Environment* 28(40):4–5 (March 1986).

TIEMANN, A. "Risk, Technology, and Society." *Risk Analysis* 7(1):11–13 (1987).

TVERSKY, A., AND D. KAHNEMAN. "The Framing of Decisions and the Psychology of Choice." *Science* 211:235–271 (January 30, 1981).

TWENTIETH CENTURY FUND. *Science in the Streets*. New York: Priority Press, 1984.

U.S. CONGRESS, COMMITTEE ON INTERIOR AND INSULAR AFFAIRS, SUBCOMMITTEE ON ENERGY AND THE ENVIRONMENT. *Accident at the Three Mile Island Nuclear Power Plant*. Washington, D.C.: Government Printing Office, 1979.

U.S. ENVIRONMENTAL PROTECTION AGENCY, OFFICE OF EMERGENCY AND REMEDIAL RESPONSE. *Community Relations in Superfund: A Handbook*. Washington, D.C.: U.S. Environmental Protection Agency, Office of Solid Waste and Emergency Response, 1983.

————. *Risk Assessment and Management: Framework for Decisionmaking*. Washington D.C.: Environmental Protection Agency, 1984.

————. *Unfinished Business: A Comparative Assessment of Environmental Problems*. Washington, D.C.: U.S. Environmental Protection Agency, Office of Policy Analysis, Office of Policy, Planning, and Evaluation, 1987.

————. Office of Information Resources Management and Office of Toxic Substances. *Risk Assessment Management, Communication: A Guide to Selected Sources*. Washington, D.C.: Government Printing Office, 1987.

VALENTINE, JEANETTE. "Communicating Social Risk: The Challenge of Preventing Teenage Pregnancy." Manuscript.

VERTINSKY, I., AND P. VERTINSKY. "Communicating Environmental Health Risk Assessment and Other Risk Information: Analysis of Strategies." In H. Kunreuther, ed., *Risk: A Seminar Series*. Laxenburg, Austria: International Institute for Applied Systems Analysis, 1982.

VON WINTERFELDT, D., AND R. KEENEY. "Improving Risk Communication." Paper presented at the National Science Foundation/EPA workshop on Risk Perception and Risk Communication, Long Beach, California, December 1984.

VON WINTERFELDT, D., P. SLOVIC, AND V. COVELLO. "Risk Perception, Risk Acceptance, and Risk Communication: Assessment of the State of Knowledge." Final Report, National Science Foundation, Division of Policy Research and Analysis, May 1, 1986.

WARD, BUD. "Communicating on Environmental Risk." *The Environmental Forum* 4(9):7–11 (January 1986).

WARREN, B., D. KOTELCHUCK, AND J. CARAVANOS. "Community Health Survey Near a Contaminated Sanitary Landfill." Paper presented at the American Public Health Association, November 1985.

WEINBERG, A. "Science and its Limits: The Regulator's Dilemma." *Issues in Science & Technology* 2(1):59–72 (Fall 1985).

WEINSTEIN, N. "Why It Won't Happen to Me: Perceptions of Risk Factors and Susceptibility." *Health Psychology* 3(5):431–457 (1984).

WEINSTEIN, N., AND P. SANDMAN. "Recommendations for a Radon Risk Communication Program." Office of Science and Research, New Jersey Department of Environmental Protection, November 1985.

WERTZ, DOROTHY C., AND JOHN C. FLETCHER. "Communicating Genetic Risks." *Science, Technology, & Human Values* 12(3&4):60–66 (Summer/Fall 1987).

WHELAN, E. M. "Toxic Error." *The Quill* 73(11):10–18 (December 1985).

WYNNE, B. "A Case Study: Hazardous Waste in the European Community." In H. Otway and M. Peltu, eds., *Regulating Industrial Risks: Public, Experts and Media*. London: Butterworths, 1986.

YANKELOVICH, D. "Bringing the Public to the Table." Speech presented at the National Conference of the Domestic Policy Association, Ann Arbor, Michigan, February 1983.

ZIMMERMAN, RAE. "The Management of Risk." In V. Covello et al., eds., *Risk Evaluation & Management*. New York: Plenum Press, 1986.

ZIMMERMAN, RAE. "A Process Framework for Risk Communication." *Science, Technology, & Human Values* 12(3&4):131–137 (Summer/Fall 1987).

INDEX

Advanced Genetic Sciences (AGS), 8, 76, 78, 104
  frostban proposal of, 79
  ice minus history of, 80–85
  public relations of, 104–106
  unauthorized testing of, 94
AFL/CIO, in EDB case study, 18–19
Agency for Toxic Substances and Disease Registry (ATSDR), 277
Agitation, and risk communication, 14
Agriculture, in Hawaii, 50, 58–59
Agriculture, Dept. of, Hawaiian, 52
Agriculture, U.S. Dept. of (USDA)
  in EDB case study, 14, 23–24
  in ice minus case study, 85
Air contaminants, in Nyanza case study, 273, 275 (*See also* Toxic emissions)
Air Quality Standards Division, of EPA, 217
Ajax, Robert, 202, 203–204, 217
Alexander, Martin, 88
Allen, Alfred M., 212–213
Altman, Beth, 41
American College of Nuclear Physicians, in radon case study, 143
American Grain Products Processing Institute (AGPPI), 43
American Mining Congress, 143
American Smelting and Refining Companies (ASARCO), Tacoma, Wash., 180
  arsenic emissions from, 180, 186
  arsenic processing operations of, 219
  closure of, 200, 217–219
  control technology at, 197
  measurement of arsenic output from, 215

as national issue, 200
  pollution controls at, 189
American Smelting and Refining Companies (ASARCO)/Tacoma case study, 9
  chronology for, 226–236
  historical context for, 183–186
  legal and regulatory structure for, 186–190
  potential control options for, 192
  risk assessment in
    alternatives for, 197–199
    problem of, 180–192
    uncertainties in health effects, 193–197
  risk communication in, 199–202
    dynamics of workshop participation, 206–209
    EPA evaluations, 210–211
    goals of public workshops, 203–204
    press evaluations, 214–216
    public evaluations, 212–214
    public hearings, 216–217
    reactions to smelter closing, 217–219
    social context of public workshops, 204–205
  technical information in, 300
Americans Protecting the Environment (APE), 205
Ames, Bruce, 34–35, 47
Anderson, Djuanna, 83, 99, 101, 106
Anderson, Elizabeth, 202, 203–204
Anderson, Jack, 19
Animal and Plant Health Inspection Service (APHIS), of USDA, 23
Arlington, Mass., 158

321